# Lecture Notes in Mechanical Engineering

**Lecture Notes in Mechanical Engineering (LNME)** publishes the latest developments in Mechanical Engineering—quickly, informally and with high quality. Original research or contributions reported in proceedings and post-proceedings represents the core of LNME. Volumes published in LNME embrace all aspects, subfields and new challenges of mechanical engineering.

To submit a proposal or request further information, please contact the Springer Editor of your location:

**Europe, USA, Africa:** Leontina Di Cecco at Leontina.dicecco@springer.com
**China:** Ella Zhang at ella.zhang@cn.springernature.com
**India, Rest of Asia, Australia, New Zealand:** Swati Meherishi at swati.meherishi@springer.com

Topics in the series include:

- Engineering Design
- Machinery and Machine Elements
- Mechanical Structures and Stress Analysis
- Automotive Engineering
- Engine Technology
- Aerospace Technology and Astronautics
- Nanotechnology and Microengineering
- Control, Robotics, Mechatronics
- MEMS
- Theoretical and Applied Mechanics
- Dynamical Systems, Control
- Fluid Mechanics
- Engineering Thermodynamics, Heat and Mass Transfer
- Manufacturing Engineering and Smart Manufacturing
- Precision Engineering, Instrumentation, Measurement
- Materials Engineering
- Tribology and Surface Technology

**Indexed by SCOPUS, EI Compendex, and INSPEC.**

All books published in the series are evaluated by Web of Science for the Conference Proceedings Citation Index (CPCI).

To submit a proposal for a monograph, please check our Springer Tracts in Mechanical Engineering at https://link.springer.com/bookseries/11693.

Adam Hamrol · Marta Grabowska ·
Paulina Rewers · Ivan Pavlenko ·
Grigore Marian Pop
Editors

# Advances in Manufacturing V

## Volume 3 - Quality Engineering: Factory of the Future

Springer

*Editors*
Adam Hamrol
Faculty of Mechanical Engineering
Poznań University of Technology
Poznań, Poland

Marta Grabowska
Faculty of Mechanical Engineering
Poznań University of Technology
Poznań, Poland

Paulina Rewers
Faculty of Mechanical Engineering
Poznań University of Technology
Poznań, Poland

Ivan Pavlenko
Sumy State University
Sumy, Ukraine

Grigore Marian Pop
Technical University of Cluj-Napoca
Cluj-Napoca, Romania

ISSN 2195-4356 ISSN 2195-4364 (electronic)
Lecture Notes in Mechanical Engineering
ISBN 978-3-032-21700-4 ISBN 978-3-032-21701-1 (eBook)
https://doi.org/10.1007/978-3-032-21701-1



This Springer imprint is published by the registered company Springer Nature Switzerland AG
The registered company address is: Gewerbestrasse 11, 6330 Cham, Switzerland

# Preface

This volume of Lecture Notes in Mechanical Engineering gathers accepted papers presented at the 9th International Scientific-Technical Conference MANUFACTURING 2026, held in Poznan, Poland, on May 19–21, 2026. The conference was hosted by the Faculty of Mechanical Engineering, Poznan University of Technology, Poland.

The primary objective of the MANUFACTURING 2026 was to showcase cutting-edge advancements across the broad spectrum of mechanical engineering, providing a dynamic forum for the exchange of innovative ideas and expert opinions. Focused on the theme of the Factory of the Future, the conference serves as a vital platform for dialogue on industrial evolution and science-industry synergy. Its global reach continues to expand, as evidenced by the consistently increasing participation from international research centers. Furthermore, the event fosters the integration of academia with the economic sector, encouraging collaboration with industry leaders and management professionals interested in implementing research results into business practice. The scope encompassed the following areas:

- Mechanical Engineering.
- Production Engineering.
- Quality Engineering.
- Measurement and Control Systems.
- Biomedical Engineering.

The global significance of the conference is reflected in the diversity of the submissions. The Organizing Committee received 219 contributions from 37 countries. To ensure the highest academic standards, all papers underwent a rigorous double-blind peer review process. Each manuscript was evaluated by at least three independent international experts, with a strict limit of five assignments per reviewer to guarantee depth and quality of feedback. Following this comprehensive appraisal, 120 papers were accepted for publication (an acceptance rate of 55%), representing the work of 347 authors from 25 countries. A total of 109 articles were published in the proceedings Advances in Manufacturing V, while 11 selected papers appeared in the journal Management and Production Engineering Review, issued under the auspices of the Committee on Production Engineering of the Polish Academy of Sciences and the Polish Society for Production Management.

The review process focused on originality of paper, significance of presented results, comprehensibility and quality of the presentation, relevance for this conference and overall impression. Furthermore, reviewers meticulously assessed the timeliness of the research, the adequacy of the literature review, methodological rigor, and the clarity of conclusions, as well as the quality of English. Feedback provided to authors included specific recommendations for improving methodology, data presentation, and editorial quality. Acceptance required at least two positive reviews and a full resolution of all reviewers' concerns. This international effort involved experts from 30 countries across Europe, North America, South Asia, and North Africa. Additionally, all manuscripts were checked using anti-plagiarism software, which allowed the similarity index to be less than 15%.

The book Advances in Manufacturing V is organized into five volumes that correspond to the main conference disciplines mentioned above.

The Advances in Manufacturing V — Volume 3 — Quality Engineering: Factory of the Future presents current research and technology trends in the topics in quality engineering, with a special focus on issues relating to Industry 4.0 and 5.0. Chapters were carefully selected to highlight important advancements in the field of sustainable development, in the context of product and business planning, ergonomics and the principles, best practices of the circular economy. Written by scientists and practitioners around the world, this book is intended to promote quality engineering, covering human and environmental aspects of industrial production, and risk-based management methods. This book gathers 23 chapters prepared by 68 authors from 8 countries.

We wish to express our sincere gratitude to the members of the Scientific and Program Committees for their dedication and expertise during the review process. We also extend our thanks to the authors, sponsors, and all contributors whose commitment was instrumental in making MANUFACTURING 2026 a resounding success.

May 2026

Adam Hamrol
Marta Grabowska
Paulina Rewers
Ivan Pavlenko
Grigore Marian Pop

# Organization

## Steering Committee

### General Chair

| | |
|---|---|
| Adam Hamrol | Poznan University of Technology, Poland |

### Chairs

| | |
|---|---|
| Bartosz Gapiński | Poznan University of Technology, Poland |
| Magdalena Hryb | Poznan University of Technology, Poland |
| Justyna Trojanowska | Poznan University of Technology, Poland |

## Conference Founders

| | |
|---|---|
| Zenobia Weiss (Deceased) | Poznan University of Technology, Poland |
| Adam Hamrol | Poznan University of Technology, Poland |

## Scientific Committee

| | |
|---|---|
| Katarzyna Antosz | Rzeszow University of Technology, Poland |
| Christopher Brown | Worcester Polytechnic Institute, USA |
| Anna Burduk | Wrocław University of Science and Technology, Poland |
| Suthep Butdee | Rajamangala University of Technology Krungthep, Thailand |
| Marcin Butlewski | Poznan University of Technology, Poland |
| Olaf Ciszak | Poznan University of Technology, Poland |
| Reggie Davidrajuh | University of Stavanger, Norway |
| Ewa Dostatni | Poznan University of Technology, Poland |
| Jan Duda | Cracow University of Technology, Poland |
| Mosè Gallo | University of Naples Federico II, Italy |
| Bartosz Gapiński | Poznan University of Technology, Poland |
| Józef Gawlik | Cracow University of Technology, Poland |
| Adam Gąska | Cracow University of Technology, Poland |
| Arkadiusz Gola | Lublin University of Technology, Poland |

| | |
|---|---|
| Anna Timofiejczuk | Silesian University of Technology, Poland |
| Justyna Trojanowska | Poznan University of Technology, Poland |
| Wiesław Urban | Bialystok University of Technology, Poland |
| Leonilde Varela | University of Minho, Portugal |
| Nikola Vitković | University of Niš, Serbia |
| Michał Wieczorowski | Poznan University of Technology, Poland |
| Ewa Więcek-Janka | Poznan University of Technology, Poland |
| Szymon Wojciechowski | Poznan University of Technology, Poland |
| Ralf Woll | BTU Cottbus-Senftenberg, Germany |
| Naiming Xie | Nanjing University of Aeronautics and Astronautics, China |
| Jiehan Zhou | Shandong University of Science and Technology, China |

## Program Committee

| | |
|---|---|
| Adam Gąska | Cracow University of Technology, Poland |
| Adam Hamrol | Poznan University of Technology, Poland |
| Agnieszka Kaczmarek-Pawelska | University of Zielona Góra, Poland |
| Agnieszka Stachowiak | Poznan University of Technology, Poland |
| Alejandro Pereira | University of Vigo, Portugal |
| Aleksandar Miltenovic | University of Nis, Serbia |
| Alice Chlupová | Czech Academy of Sciences, Czech Republic |
| Alzbeta Sapietova | University of Žilina, Slovakia |
| Ancuta Pacurar | Technical University of Cluj-Napoca, Romania |
| André Guimarães | University of Beira Interior, Portugal |
| Andrzej Jardzioch | West Pomeranian University of Technology in Szczecin, Poland |
| Angelika Hofer | Free University of Bozen-Bolzano, Italy |
| Anna Borucka | Military University of Technology Warsaw, Poland |
| Anna Burduk | Wroclaw University of Science and Technology, Poland |
| Anna Karwasz | Poznan University of Technology, Poland |
| Anna M. Deptuła | Opole University of Technology, Poland |
| Anna Saniuk | University of Zielona Góra, Poland |
| Arkadiusz Gola | Lublin University of Technology, Poland |
| Bartosz Gapinski | Poznan University of Technology, Poland |
| Bogdan Dybała | Wroclaw University of Science and Technology, Poland |
| Christopher Brown | Worcester Polytechnic Institute, USA |

| | |
|---|---|
| Cristina Stefana Miron-Borzan | Technical University of Cluj-Napoca, Romania |
| Dagmara Łapczyńska | Wroclaw University of Science and Technology, Poland |
| Damian Grajewski | Poznan University of Technology, Poland |
| Damian Krenczyk | Silesian University of Technology, Poland |
| Dan-Sorin Comsa | Technical University of Cluj-Napoca, Romania |
| Daniel Medyński | Witelon Collegium State University, Poland |
| Dariusz Mikołajewski | Kazimierz Wielki University, Poland |
| Djahida Belayadi | National School of Cybersecurity, Algeria |
| Dražan Kozak | University of Slavonski Brod, Croatia |
| Edward Kozłowski | Lublin University of Technology, Poland |
| Emilia Campean | Technical University of Cluj-Napoca, Romania |
| Emilia Sabau | Technical University of Cluj-Napoca, Romania |
| Eryk Szwarc | Koszalin University of Technology, Poland |
| Ewa Dostatni | Poznan University of Technology, Poland |
| Ewa Więcek-Janka | Poznan University of Technology, Poland |
| Gabriel Stolárik | Technical University of Košice, Slovakia |
| Grzegorz Pająk | University of Zielona Góra, Poland |
| Hanna Łosyk | University of Zielona Góra, Poland |
| Hubert Wojciechowski | Poznan University of Technology, Poland |
| Izabela Kudelska | Poznan University of Technology, Poland |
| Izabela Rojek | Kazimierz Wielki University, Poland |
| Jacek Diakun | Poznan University of Technology, Poland |
| Jakub Kaščák | Technical University of Košice, Slovakia |
| Jakub Kopowski | Kazimierz Wielki University, Poland |
| Jerzy Jozwik | Lublin University of Technology, Poland |
| Joanna Cyganiuk | University of Zielona Góra, Poland |
| Joanna Helman | Wroclaw University of Science and Technology, Poland |
| Joanna Wilczarska | Bydgoszcz University of Science and Technology, Poland |
| Jozef Husár | Technical University of Košice, Slovakia |
| Jozef Török | Technical University of Košice, Slovakia |
| Józef Kuczmaszewski | Lublin University of Technology, Poland |
| Justyna Patalas-Maliszewska | University of Zielona Góra, Poland |
| Kamil Krot | Wroclaw University of Science and Technology, Poland |
| Kamil Leksycki | University of Zielona Góra, Poland |
| Kamil Musiał | Wroclaw University of Science and Technology, Poland |
| Karol Dąbrowski | University of Zielona Góra, Poland |
| Katarzyna Antosz | Rzeszow University of Technology, Poland |

| | |
|---|---|
| Katarzyna Cheba | West Pomeranian University of Technology in Szczecin, Poland |
| Katarzyna Kazimierska-Drobny | Kazimierz Wielki University, Poland |
| Katarzyna Skrzypek | University of Zielona Góra, Poland |
| Khrystyna Berladir | Sumy State University, Ukraine |
| Konrad Kobiela | Cracow University of Technology, Poland |
| Krzysztof Kalinowski | Silesian University of Technology, Poland |
| Krzysztof Nowacki | Silesian University of Technology, Poland |
| Krzysztof Żywicki | Poznan University of Technology, Poland |
| Leonilde Varela | University of Minho, Portugal |
| Lucia Knapčíková | Technical University of Košice, Slovakia |
| Łukasz Rymaniak | Poznan University of Technology, Poland |
| Magdalena Hryb | Poznan University of Technology, Poland |
| Magdalena Rybaczewska-Błażejowska | Kielce University of Technology, Poland |
| Magdalena Żukowska | Poznan University of Technology, Poland |
| Małgorzata Śliwa | University of Zielona Góra, Poland |
| Marcin Suszynski | Poznan University of Technology, Poland |
| Marek Kočiško | Technical University of Košice, Slovakia |
| Marek Macko | Kazimierz Wielki University, Poland |
| Maria Rosienkiewicz | Wroclaw University of Science and Technology, Poland |
| Marian Bartoszuk | Opole University of Technology, Poland |
| Marius Sorin Pustan | Technical University of Cluj-Napoca, Romania |
| Marks Gorohovs | Riga Technical University, Latvia |
| Marta Grabowska | Poznan University of Technology, Poland |
| Marta Harničárová | Slovak University of Agriculture in Nitra, Slovakia |
| Michał Jakubowicz | Poznan University of Technology, Poland |
| Michał Rogalewicz | Poznan University of Technology, Poland |
| Mihai Dragomir | Technical University of Cluj-Napoca, Romania |
| Milan Edl | University of West Bohemia, Czech Republic |
| Mircea Fulea | Technical University of Cluj-Napoca, Romania |
| Mukund Harugade | Padmabhooshan Vasantraodada Patil Institute of Technology, India |
| Natalya Shramenko | Lviv Polytechnic National University, Ukraine |
| Nataša Náprstková | University of Jan Evangelista in Ústí nad Labem, Czech Republic |
| Nguyen Hong Vi | Vietnamese German University, Vietnam |
| Nikola Vitković | University of Nis, Serbia |
| Olaf Ciszak | Poznan University of Technology, Poland |
| Olha Kalman | Technical University of Košice, Slovakia |

| | |
|---|---|
| Paulina Rewers | Poznan University of Technology, Poland |
| Paweł Buń | Poznan University of Technology, Poland |
| Paweł Szczygieł | Kielce University of Technology, Poland |
| Paweł Zmarzły | Kielce University of Technology, Poland |
| Pedro Torres | Polytechnic Institute of Castelo Branco, Portugal |
| Piotr Gąska | AGH University of Krakow, Poland |
| Piotr Kotlarz | Kazimierz Wielki University, Poland |
| Piotr Trojanowski | West Pomeranian University of Technology in Szczecin, Poland |
| Piotr Ziembicki | University of Zielona Góra, Poland |
| Przemysław Zawadzki | Poznan University of Technology, Poland |
| Radosław Wichniarek | Poznan University of Technology, Poland |
| Răzvan Păcurar | Technical University of Cluj-Napoca, Romania |
| Roman Kielec | University of Zielona Góra, Poland |
| Sandra Grabowska | Silesian University of Technology, Poland |
| Sebastian Saniuk | University of Zielona Góra, Poland |
| Sebastian Skoczypiec | Cracow University of Technology, Poland |
| Sergej Hloch | Technical University of Košice, Slovakia |
| Sławomir Kłos | University of Zielona Góra, Poland |
| Stanislaw Legutko | Poznan University of Technology, Poland |
| Stella Hrehova | Technical University of Košice, Slovakia |
| Sven Maricic | Juraj Dobrila University of Pula, Croatia |
| Sylvia Kusmierczak | University of Jan Evangelista in Ústí nad Labem, Czech Republic |
| Szymon Wojciechowski | Poznan University of Technology, Poland |
| Thanh Tran | RMIT University, Vietnam |
| Tomasz Bednarek | Kazimierz Wielki University, Poland |
| Vijaya Kumar Manupati | NITIE Mumbai, India |
| Vitalii Ivanov | Sumy State University, Ukraine |
| Wiesław Kuczko | Poznan University of Technology, Poland |
| Wiktor Szot | Kielce University of Technology, Poland |
| Wojciech Babirecki | University of Zielona Góra, Poland |
| Wojciech Macek | Gdańsk University of Technology, Poland |
| Yuliia Denysenko | Sumy State University, Ukraine |
| Zbigniew Wiśniewski | Lodz University of Technology, Poland |
| Zuzana Mitaľová | Technical University of Košice, Slovakia |

## Special Sessions and Special Session Organizing Committee

### Efficiency-Driven Manufacturing: Tools and Trends in the Era of Industry 4.0

| | |
|---|---|
| Justyna Trojanowska | Poznan University of Technology, Poland |
| Leonilde Varela | University of Minho, Portugal |
| Vijaya Kumar Manupati | NITIE Mumbai, India |
| Thanh Tran | RMIT University, Vietnam |
| Jozef Husár | Technical University of Košice, Slovakia |

### Intersecting Quality Management and Mechanical Design for Manufacturing Sustainability

| | |
|---|---|
| Florin Popişter | Technical University of Cluj-Napoca, Romania |
| Diana Cristina Dragomir | Technical University of Cluj-Napoca, Romania |
| Corina Julieta Birleanu | Technical University of Cluj-Napoca, Romania |
| Mircea Claudiu Fulea | Technical University of Cluj-Napoca, Romania |

### Human Factors in Manufacturing

| | |
|---|---|
| Marcin Butlewski | Poznan University of Technology, Poland |
| Beata Mrugalska | Poznan University of Technology, Poland |
| Wiktoria Czernecka | Poznan University of Technology, Poland |
| Fryderyk Wachowiak | Poznan University of Technology, Poland |
| Krzysztof Hankiewicz | Poznan University of Technology, Poland |

### Circularity in Advanced Production Systems: Materials, Methods, and Digital Transformation

| | |
|---|---|
| Marek Kočisko | Technical University of Košice, Slovakia |
| Lucia Knapčikova | Technical University of Košice, Slovakia |
| Jakub Kaščak | Technical University of Košice, Slovakia |
| Jozef Török | Technical University of Košice, Slovakia |

## Smart Manufacturing and Sustainability: Interdisciplinary Approaches in the Era of Industry 4.0/5.0

| | |
|---|---|
| Radosław Maruda | University of Zielona Góra, Poland |
| Sebastian Saniuk | University of Zielona Góra, Poland |
| Katarzyna Skrzypek | University of Zielona Góra, Poland |
| Daria Moskwa - Bęczkowska | Kielce University of Technology, Poland |
| Karol Dąbrowski | University of Zielona Góra, Poland |

## From Additive Manufacturing to Healthcare Innovation through Digital Transformation in Biomedical Engineering

| | |
|---|---|
| Filip Górski | Poznan University of Technology, Poland |
| Răzvan Păcurar | Technical University of Cluj-Napoca, Romania |
| Nikola Vitkovic | University of Niš, Serbia |
| Michał Rychlik | Poznan University of Technology, Poland |

## Intelligent Adaptive Systems in Integrated Production Management Processes

| | |
|---|---|
| Mariusz Piechowski | Merito WSB University in Poznan, Poland |
| Małgorzata Jasiulewicz-Kaczmarek | Poznan University of Technology, Poland |
| Katarzyna Antosz | Rzeszow University of Technology, Poland |
| Ryszard Wyczółkowski | Silesian University of Technology, Poland |
| Arkadiusz Gola | Lublin University of Technology, Poland |

## Intelligent Solutions for Industry of the Future

| | |
|---|---|
| Izabela Rojek | Kazimierz Wielki University in Bydgoszcz, Poland |
| Ewa Dostatni | Poznan University of Technology, Poland |
| Dariusz Mikołajewski | Kazimierz Wielki University in Bydgoszcz, Poland |
| Nataša Náprstková | University of Jan Evangelista in Ústí nad Labem, Czech Republic |

## Grey Systems Modeling for Engineering Solutions and Business Optimization

| | |
|---|---|
| Magdalena Hryb | Poznan University of Technology, Poland |
| Agnieszka Kujawińska | Poznan University of Technology, Poland |
| Naiming Xie | Nanjing University of Aeronautics and Astronautics, China |
| Liangyan Tao | Nanjing University of Aeronautics and Astronautics, China |
| Rafał Mierzwiak | Poznan University of Technology, Poland |

## Intelligent Methods Supporting Manufacturing Systems Efficiency

| | |
|---|---|
| Anna Burduk | Wroclaw University of Science and Technology, Poland |
| Damian Krenczyk | Silesian University of Technology, Poland |
| Dagmara Łapczyńska | Wroclaw University of Science and Technology, Poland |
| Kamil Krot | Wroclaw University of Science and Technology, Poland |

# Contents

## Industry 4.0 & 5.0

# Quality Management Methods

# Mapping Innovation Capability: A Diagnostic Approach for Assessing and Enhancing Organizational Capabilities

Nenad Vladić, Damjan Maletič, and Matjaž Maletič(✉)

Faculty of Organizational Sciences, University of Maribor, Kranj, Slovenia
matjaz.maletic@um.si

**Abstract.** This study operationalizes the Systems Framework for Sustainable Innovation Capability (SFSIC) as a maturity-based diagnostic tool for assessing sustainable leadership, organizational enabling mechanisms, and innovation capability. The research is based on a quantitative single-case study conducted in a high-technology manufacturing company, using a structured questionnaire derived from the SFSIC framework. The results show that all three dimensions reached a high level of maturity (Level 4 – Integrated), indicating a systematic alignment between leadership practices, organizational routines, and innovation-related activities. This demonstrates the applicability of the SFSIC as a diagnostic framework for evaluating systemic alignment within organizations. The proposed maturity-based diagnostic provides organizations with a structured approach to assessing innovation maturity and identifying development priorities that support adaptability and long-term value creation.

**Keywords:** Sustainable Leadership · Innovation capability · Maturity assessment

## 1 Introduction

Innovation and sustainability are widely recognized as key drivers of competitiveness and long-term organizational success in increasingly complex and rapidly changing environments. Unpredictable technological change, global interdependence, and rising societal expectations require organizations to innovate while simultaneously acting in socially and environmentally responsible ways. Innovation supports growth and competitiveness, while sustainability ensures that progress creates economic, social, and environmental value—making their integration essential for future-oriented organizational development [1]. Leadership is central to this integration, as it shapes how organizations balance competing priorities and align sustainability goals with innovation efforts. Empirical research shows that leadership stimulates innovation [2] and strengthens organizational resilience [3]. It also fosters psychological safety and experimentation, which support creativity and innovative behaviour [4, 5].

Recent studies conceptualize innovation capability not as a static set of factors, but as a dynamic system shaped by leadership, organizational mechanisms, and emergent


A. Hamrol et al. (Eds.): MANUFACTURING 2026, LNME, pp. 3–17, 2026.
https://doi.org/10.1007/978-3-032-21701-1_1

outcomes [6, 7]. From this perspective, leadership provides strategic direction and values, while organizational mechanisms embed them into routines that enable innovation capability to evolve across different levels of the organization [5, 8]. However, much of this literature remains conceptual, offering limited guidance on how such systemic relationships can be empirically assessed and compared across organizations.

Despite these advances, gaps remain in empirical operationalization of the relationship between sustainability-oriented leadership and innovation capability. Studies often examine these constructs in isolation and approach sustainable leadership through normative or value-based lenses [8, 9]. As a result, organizations lack structured diagnostic tools to assess how sustainability-oriented leadership and innovation capability align and mature over time. To address this gap, the Systems Framework for Sustainable Innovation Capability (SFSIC) conceptualizes innovation capability as a dynamic, feedback-driven system composed of three interdependent dimensions [10]:

(1) Sustainable Leadership, which provides strategic direction and values;
(2) Organizational Enabling Mechanisms, which embed leadership principles into organizational culture, structure, and learning practices; and
(3) Innovation Capability, which reflects tangible outcomes such as product, process, managerial, and individual innovation.

While the SFSIC model offers strong conceptual and qualitative foundations, its diagnostic potential has not yet been empirically tested. This study addresses that gap by operationalizing the SFSIC as a maturity-based diagnostic framework, enabling organizations to assess the development of their innovation capability. The maturity perspective assumes that innovation evolves through identifiable stages—from ad hoc and fragmented activities to fully integrated, learning-oriented systems [6, 11]. By mapping maturity levels across the three SFSIC dimensions, organizations can identify strengths, detect gaps, and plan targeted improvements.

The empirical study was conducted in a high-tech manufacturing company well suited to examining the integration of sustainability-oriented leadership and innovation capability. Using a structured survey instrument derived from the three SFSIC dimensions, the analysis assesses the maturity of sustainable leadership, organizational enabling mechanisms, and innovation capability. The study demonstrates how leadership and enabling mechanisms translate into measurable innovation capability within a systemic framework, while also providing a diagnostic tool for identifying capability development priorities. This study pursues three interrelated objectives:

(1) to operationalize the SFSIC as a maturity-based diagnostic framework;
(2) to empirically apply the proposed maturity model in a high-technology manufacturing organization to assess the development of sustainable leadership, organizational enabling mechanisms, and innovation capability; and
(3) to derive diagnostic implications by illustrating how maturity-based assessment can support targeted capability development in practice.

By doing so, the study contributes a structured maturity-based diagnostic approach that bridges conceptual systems theory and the empirical assessment of innovation capability in manufacturing contexts. The remainder of the paper is organized as follows. Section 2 outlines the theoretical background, including the foundations of the SFSIC and the concept of the maturity model. Section 3 describes the research methodology. Section 4 presents the empirical findings, followed by the discussion in Sect. 5 and the concluding remarks in Sect. 6.

## 2 Theoretical Background

### 2.1 Systems Framework for Sustainable Innovation Capability (SFSIC)

The Systems Framework for Sustainable Innovation Capability (SFSIC) provides an integrative model that explains how sustainability-oriented leadership, organizational mechanisms, and innovation capability interact within a dynamic system. Developed through in-depth qualitative research in a high tech manufacturing context [10], the framework conceptualizes innovation capability as an emergent property of three interdependent and mutually reinforcing dimensions. As such, it offers a unified systems-based perspective on the co-evolution of leadership, organizational mechanisms, and innovation capability.

**Table 1.** Systems Framework for Sustainable Innovation Capability (SFSIC) integrating leadership, enabling mechanisms, and innovation capability.

| **Systems Framework for Sustainable Innovation Capability (SFSIC)** | | |
|---|---|---|
| **Sustainable leadership** | **Organizational Enabling Mechanisms** | **Innovation Capability** |
| • Futures-Oriented Strategic Intent<br>• Distributed Ownership and Governance<br>• Shared and Systemic Sensemaking<br>• Ethical and Relational Leadership Practices<br>• Collective Stewardship Orientation<br>• Continuous Feedback Culture<br>• Inclusive, values-based decision-making<br>• Long-Term Vision for Intergenerational Equity | • Agile, Adaptive and Resilient Systems<br>• Collaborative Knowledge Infrastructure<br>• Integration of Digital and Social Systems<br>• Wellbeing-oriented HR systems<br>• Learning as Organizational Capability<br>• Open Innovation Interfaces<br>• Transparent Digital Processes<br>• Psychological Safety and Feedback Loops | • Strategic Innovation Alignment<br>• Empowered Innovation Culture<br>• Organizational Learning for Innovation<br>• Structured and Agile Ideation<br>• External Integration and Open Innovation<br>• Responsiveness and Agility<br>• Sustainability-Oriented Innovation<br>• Employee Wellbeing and Engagement as Innovation Drivers |
| »» **Learning and Leadership Reinforcement: Continuous loops of reflection, adaptation, and capability renewal.** «« | | |

The first dimension, Sustainable Leadership, captures the principles and practices through which leaders articulate long-term direction, embed core values, and create the conditions for responsible innovation. It encompasses behaviours such as participative decision-making, ethical judgment, and systems-oriented sensemaking, which collectively foster an organizational environment that supports experimentation and learning [8, 12, 13].

The second dimension, Organizational Enabling Mechanisms, comprises the structural, procedural, and cultural elements that translate leadership intent into organizational practices. These mechanisms—including knowledge-sharing routines, learning processes, flexible structures, and developmental HR practices—mediate the relationship between leadership principles and the development of innovation capability [9, 22–24]. In doing so, they ensure that sustainability-oriented values are consistently embedded within everyday routines and decision-making processes.

The third dimension, Innovation Capability, refers to the observable manifestations of an organization's capacity to innovate, as reflected in product, process, managerial, and individual innovativeness. Individual innovativeness refers to employees' willingness to generate and implement new ideas and to experiment with alternative approaches [14]. Together, these outcomes signal the organization's ability to generate and implement novel ideas that enhance adaptability and long-term value creation [6].

The three dimensions interact through continuous learning loops and internal feedback processes. Leadership shapes enabling mechanisms, which in turn influence innovation capability, while innovation outcomes themselves reinforce or challenge existing leadership practices and routines. This reciprocal interaction reflects the logic of dynamic capabilities [7], emphasizing ongoing adaptation and renewal rather than linear cause–effect relationships. Accordingly, the SFSIC conceptualizes leadership, organizational mechanisms, and innovation capability as elements of a co-evolving system. By integrating these elements into a unified framework, the SFSIC bridges the gap between normative theories of sustainable leadership and their practical operationalization in organizational contexts, and provides a theoretical foundation for the maturity-based diagnostic developed in the following section.

## 2.2 From Conceptual Framework to Maturity Model

While the SFSIC framework provides a systemic conceptual understanding of how sustainable leadership enables innovation capability, its practical diagnostic potential requires further operationalization. In response, this study extends the SFSIC framework into a maturity-based model that allows organizations to assess their current level of development across the three dimensions. The maturity approach assumes that organizational innovation capability evolves through identifiable stages of integration, systematization, and organizational learning [6, 11].

Each maturity level reflects the degree to which sustainability-oriented leadership, enabling mechanisms, and innovation practices are embedded within organizational structures and routines. Rather than treating innovation capability as a fixed attribute, the maturity perspective views it as a developmental process driven by cumulative learning and systemic alignment. Drawing on prior maturity models in quality and innovation management [15, 16], as well as more recent empirical studies applying maturity-based approaches to innovation capability assessment [17, 18], the SFSIC Maturity Model defines five progressive levels of development. Lower levels are characterized by fragmented and largely reactive practices, whereas higher levels reflect integrated routines, structured feedback mechanisms, and continuous improvement. Table 2 summarizes these levels and their key characteristics.

**Table 2.** SFSIC maturity model – levels of organizational innovation capability.

| Dimension | Level 1: Initial | Level 2: Emerging | Level 3: Established | Level 4: Integrated | Level 5: Transformative |
|---|---|---|---|---|---|
| **Sustainable Leadership** | Leadership is reactive, fragmented, and short-term oriented; sustainability not considered. | Initial awareness of sustainability; occasional participative or ethical practices. | Leadership supports shared values and collaboration, but practices are not yet systemically aligned. | Sustainability principles embedded in strategy, culture and communication; consistent leadership behaviours. | Leadership embodies long-term systems thinking; sustainability becomes part of organizational identity and purpose. |
| **Organizational Enabling Mechanisms** | Learning and innovation processes are inconsistent, informal, dependent on individuals. | Basic mechanisms (training, communication, knowledge sharing) exist but lack coordination. | Structured and documented mechanisms exist across functions; routines support learning and innovation. | Mechanisms are aligned with leadership intent; culture supports experimentation, dialogue, and continuous learning. | Organization operates as an adaptive learning system; integrated processes, digital tools, external collaboration, and regenerative practices. |
| **Innovation Capability** | Innovation occurs sporadically and depends on individual initiative. | Early product or process innovations appear, but organizational support is limited. | Innovation established across product, process, managerial, and individual levels. | Innovation is strategically managed; supported by enabling systems and cross-functional routines. | Innovation embedded in culture and daily work; continuous renewal, agility, and sustainability-driven innovation. |

Note: Descriptions summarise maturity levels derived from the diagnostic instrument applied in this study.

The maturity model translates the three dimensions into measurable stages of organizational development based on diagnostic indicators derived from the survey instrument. By adopting a maturity-based perspective, the SFSIC framework evolves from a conceptual model into a practical management tool that enables organizations to identify strengths, address gaps, and guide targeted interventions for sustainable innovation. This operationalization bridges theory and practice by offering a standardized yet adaptable structure for evaluating innovation capability in manufacturing contexts. From an applied and strategic perspective, the maturity-based operationalization of the SFSIC framework is conceptually consistent with the principles of innovation management systems outlined in ISO 56002:2019 [19], particularly with respect to structured capability development, leadership involvement, and continuous improvement.

# 3 Methodology

## 3.1 Research Design

We applied a quantitative single-case study design to operationalize and empirically validate the Systems Framework for Sustainable Innovation Capability (SFSIC) as a maturity-based diagnostic model. While the SFSIC was originally developed through qualitative inquiry and conceptual reasoning [10], the present study advances the framework by introducing a structured measurement instrument that captures organizational maturity across its three key dimensions: Sustainable Leadership, Organizational Enabling Mechanisms, and Innovation Capability. The combination of conceptual development and quantitative validation follows established recommendations for building robust theory in management and innovation research [20, 21].

A case study strategy was selected because it enables the examination of complex and interconnected organizational systems in real contexts [21, 22]. The company operates in the high-technology manufacturing sector and is internationally recognized for its innovation culture and advanced technological capabilities. This organizational context provides a suitable setting for testing the diagnostic potential of the SFSIC model in a dynamic and knowledge-intensive environment.

While the single-case design and sample size ($N = 68$) enable an in-depth and context-sensitive examination of the proposed diagnostic framework, they limit the statistical generalizability of the findings. The results should therefore be interpreted as analytically generalizable insights derived from a specific organizational context, in line with established case study research principles [21].

The quantitative component consists of a structured survey instrument derived directly from the three theoretical dimensions of the SFSIC. Each dimension is operationalized through multiple items designed to capture behavioural, structural, and process-related characteristics. All items were rated on a five-point Likert scale ranging from 1 ("strongly disagree") to 5 ("strongly agree"), enabling the calculation of maturity scores and comparisons across dimensions. This approach reflects standard procedures for transforming conceptual constructs into empirical indicators [23, 24].

The research design aligns closely with the logic of the SFSIC by ensuring direct correspondence between the survey dimensions and the three conceptual pillars of the model. By combining contextual depth with quantitative measurement, the study bridges the gap between conceptual theory and empirical operationalization, thereby advancing the application of maturity modelling to the study of innovation capability and sustainable leadership [6, 11].

## 3.2 Case Company Description

To preserve confidentiality, the analyzed organization is referred to as Company Zeta. It is a medium-sized high-technology firm with approximately 250 employees, specializing in the development of advanced measurement and data acquisition systems used across industries such as automotive, aerospace, and energy. Over the past two decades, the company has built a strong global reputation for technological excellence and product innovation, earning international design and innovation awards.

Headquartered in Central Europe, Company Zeta operates within an extensive global network of partners. It represents a relevant case for examining how sustainable leadership and enabling mechanisms foster innovation capability within a high-tech, knowledge-intensive context.

### 3.3 Data Collection and Instrument

Quantitative data were collected using a structured questionnaire designed to measure the three core dimensions of the Systems Framework for Sustainable Innovation Capability (SFSIC): (1) Sustainable Leadership, (2) Organizational Enabling Mechanisms, and (3) Innovation Capability. The instrument was developed by adapting items from established scales in leadership, innovation, and sustainability research, thereby ensuring conceptual alignment with the SFSIC framework and content validity.

All items were rated on a five-point Likert scale ranging from 1 ("strongly disagree") to 5 ("strongly agree"). The survey was administered electronically across different departments and hierarchical levels. Participation was voluntary and anonymous, and respondents were informed about the research purpose and confidentiality in accordance with established ethical standards [25].

The final instrument comprised 70 items. A total of 68 valid responses were collected, representing approximately 35% of the target population. The dataset contained no missing values, as incomplete questionnaires were excluded prior to analysis. For analysis, individual item scores were aggregated to calculate mean values for each of the three SFSIC dimensions. These mean scores were then mapped onto predefined maturity levels (see Sect. 3.3), producing an overall maturity profile that highlights strengths and areas requiring further development. Table 3 summarizes the main constructs, subdimensions, and key reference sources used in designing the measurement instrument.

All three dimensions of the Systems Framework for Sustainable Innovation Capability (SFSIC) were operationalized in accordance with the conceptual structure presented in Table 1. To align the empirical instrument with the theoretical framework, related items were combined into broader analytical categories (e.g., collaboration, learning, digitalization, and innovation types), thereby ensuring conceptual consistency and parsimony.

### 3.4 Measurement and Maturity Mapping

To translate the survey results into a structured diagnostic overview, a maturity mapping approach was applied. This approach enables the transformation of mean Likert scores into categorical maturity levels, thereby illustrating the degree of development within each of the three SFSIC dimensions: Sustainable Leadership, Organizational Enabling Mechanisms, and Innovation Capability. For each respondent, item scores were aggregated at the construct level to calculate mean values for each dimension. These mean values were then assigned to one of five predefined maturity levels based on threshold

**Table 3.** Constructs and reference sources used for instrument design.

| Construct | Subdimension | Reference sources |
|---|---|---|
| **Sustainable Leadership** | – | McCann and Holt [13] |
| **Organizational Enabling Mechanisms** | Teamwork & Communication | Saunila [6]. |
| | Learning & Knowledge Sharing | Liao et al. (2007); Lin (2007); Saunila [6]; Calantone et al. (2002). |
| | HR Development & Wellbeing | Saunila [1]; Avery and Bergsteiner [7]. |
| | External Collaboration & Regeneration | Saunila [6]; Liao et al. (2007); Tidd & Bessant (2009). |
| | Digitalization & Individual Empowerment | Nair & Bhattacharyya (2022); Tidd and Bessant [11]; Lin (2007); Saunila [6]. |
| **Innovation Capability** | Product innovativeness | Liao et al. (2007); Lin (2007); Calantone et al. (2002). |
| | Process innovativeness | Liao et al. (2007). |
| | Managerial innovativeness | Liao et al. (2007). |
| | Individual innovativeness | Hurt et al. [14]. |

intervals aligned with established maturity modelling practices [6, 15, 16], which are typically used to support diagnostic interpretation rather than statistical classification. The levels represent a progression from fragmented, largely reactive practices to fully integrated and learning-oriented systems. By applying this mapping approach, the SFSIC dimensions are translated into observable indicators of organizational maturity, enabling a diagnostic assessment of development gaps and areas of relative strength.

For each dimension, a maturity index (MI) was calculated by averaging item scores within that dimension:

$$MI_d = \frac{\sum_{i=1}^{n_d} x_i}{n_d} \tag{1}$$

where $MI_d$ denotes the maturity index for dimension $d$, $x_i$ is the score for item $i$, and $n_d$ is the number of items in that dimension. The resulting indices form the basis of the organization's maturity profile, enabling comparative interpretation of development across the three SFSIC dimensions. Items removed during the reliability and validity assessment were excluded from all subsequent maturity index calculations to ensure that maturity scores were based solely on statistically robust indicators.

Table 4 summarizes the five maturity levels and the criteria used to classify each dimension. This mapping procedure operationalizes the SFSIC as a quantitative diagnostic instrument by linking survey-based evidence to maturity evaluations. By translating

**Table 4.** Maturity level thresholds.

| Maturity level | Mean score range | Description |
|---|---|---|
| **Level 1 – Initial** | 1.00–1.80 | Practices are largely ad hoc, inconsistent, and dependent on individual effort rather than organizational systems. |
| **Level 2 – Emerging** | 1.81–2.60 | Basic awareness exists, but processes remain informal and are not yet embedded in organizational routines. |
| **Level 3 – Defined** | 2.61–3.40 | Core mechanisms and leadership practices are defined and partially implemented across the organization. |
| **Level 4 – Integrated** | 3.41–4.20 | Practices are systematically implemented, integrated across departments, and supported by consistent leadership commitment. |
| **Level 5 – Transformative** | 4.21–5.00 | Continuous improvement and learning mechanisms are fully embedded; innovation and sustainability become self-reinforcing. |

survey scores into maturity levels, the model enables organizations to track development over time and identify targeted improvement priorities aligned with sustainable leadership and innovation capability [6, 8, 11].

## 4 Results and Discussion

### 4.1 Descriptive Statistics and Reliability

This section presents the quantitative results of the study. Table 5 reports descriptive statistics, internal consistency (Cronbach's α), and construct validity for the measured constructs and subdimensions, while Table 6 presents the resulting maturity profile across the three SFSIC dimensions.

Specifically, Table 5 summarizes descriptive statistics (mean and standard deviation) together with key indicators of internal consistency and construct validity for each construct and its subdimensions. All Cronbach's alpha values exceed the recommended threshold of .70, while CITC values range from .40 to .88, indicating satisfactory internal reliability and adequate item–scale relationships across all dimensions. Factor loadings ranged from .50 to .91, indicating that the items adequately represent their intended constructs. During the reliability and validity assessment, three items were removed to enhance the robustness and internal consistency of the measurement scales. Mean values ranged from 3.8 to 4.2, indicating generally high levels across the three SFSIC dimensions. Exploratory Factor Analysis (EFA) was applied in line with the exploratory, developmental, and diagnostic orientation of the study and the use of a single-case research design. Future studies should validate the structure using Confirmatory Factor Analysis (CFA) in larger and more diverse samples.

**Table 5.** Descriptive statistics, reliability, and factor structure of the constructs.

| Construct | Subdimension | M | SD | α (Cronbach's) | CITC range | Factor loading range |
|---|---|---|---|---|---|---|
| **Sustainable Leadership** | – | 4.1 | .75 | .94 | .47–.88 | .74–.91 |
| **Organizational Enabling Mechanisms** | Teamwork & Communication | 3.8 | .86 | .94 | .70–.86 | .77–.91 |
| | Learning & Knowledge Sharing | 3.8 | .70 | .84 | .49–.67 | .63–.81 |
| | HR Development & Wellbeing | 3.9 | .85 | .86 | .54–.80 | .67–.87 |
| | External Collaboration & Regeneration | 4.0 | .54 | .74 | .40–.67 | .57–.84 |
| | Digitalization & Individual Empowerment | 4.0 | .64 | .81 | .47–.67 | .61–.80 |
| **Innovation Capability** | Product innovation | 3.9 | .53 | .76 | .41–.65 | .59–.81 |
| | Process innovation | 4.1 | .54 | .78 | .52–.63 | .71–.81 |
| | Managerial innovation | 3.9 | .68 | .86 | .48–.73 | .60–.82 |
| | Individual innovation | 4.2 | .59 | .81 | .35–.78 | .50–.86 |

## 4.2 Maturity Assessment Results

The diagnostic analysis based on the SFSIC framework provided a maturity profile of Company Zeta across three key dimensions—Sustainable Leadership, Organizational Enabling Mechanisms, and Innovation Capability. The results revealed differentiated levels of development within this profile (Table 6).

**Table 6.** Mean scores and corresponding maturity levels.

| Construct | Subdimension | Mean Score | Maturity level |
|---|---|---|---|
| **Sustainable Leadership** | – | 4.1 | Level 4 – Integrated |
| **Organizational Enabling Mechanisms** | Teamwork & Communication | 3.8 | Level 4 – Integrated |

(*continued*)

**Table 6.** (*continued*)

| Construct | Subdimension | Mean Score | Maturity level |
|---|---|---|---|
| | Learning & Knowledge Sharing | 3.8 | Level 4 – Integrated |
| | HR Development & Wellbeing | 3.9 | Level 4 – Integrated |
| | External Collaboration & Regeneration | 4.0 | Level 4 – Integrated |
| | Digitalization & Individual Empowerment | 4.0 | Level 4 – Integrated |
| **Innovation Capability** | Product innovation | 3.9 | Level 4 – Integrated |
| | Process innovation | 4.1 | Level 4 – Integrated |
| | Managerial innovation | 3.9 | Level 4 – Integrated |
| | Individual innovation | 4.2 | Level 4 – Integrated |

The table presents the maturity assessment of Company Zeta according to the SFSIC framework. All constructs and subdimensions were rated at Level 4 (Integrated), with mean scores ranging from 3.8 to 4.2.

## 5 Discussion

### 5.1 Interpretation of Maturity Profiles

The classification of all constructs at Level 4 (Integrated) indicates that practices are systematically implemented, integrated across departments, and supported by sustained leadership commitment. Although all three dimensions reached Level 4 (Integrated), closer inspection of the maturity profiles reveals meaningful variation, with sustainable leadership receiving consistently high evaluations, indicative of stable leadership practices that support organizational functioning. The greatest variation occurred within the organizational enabling mechanisms dimension, which is expected given its broader scope and larger number of subdimensions. Within this dimension, collaboration- and learning-related subdimensions scored lower ($M = 3.8$) than others ($M = 4.0$–$4.2$), indicating stronger development of established operational routines than of knowledge-sharing and learning practices. Innovation capability followed a similar maturity pattern, reinforcing the systemic interconnectedness conceptualized in the SFSIC framework. Notably, the highest mean score was observed for individual innovation capability, reflecting employees' creative behaviours and personal engagement in generating new ideas.

Although the assessed maturity levels were generally high, lower scores within the organizational enabling mechanisms related to collaboration and learning subcategories indicate a specific developmental gap. Given their central role in idea exchange, shared understanding, and knowledge development, the comparatively lower maturity of these mechanisms suggests that participatory leadership behaviours, idea circulation, constructive disagreement, and continuous skill development are not yet as systematically embedded as established organizational routines. The results therefore imply that, while structural routines are well established, everyday mechanisms for sharing, refining, and transforming ideas into collective learning remain less consistently developed. This pattern aligns with prior studies that conceptualize innovation capability as unevenly developed across structural and relational dimensions, even at advanced stages of maturity [25].

These interpretations should be considered in light of methodological limitations, as the analysis is based on a single-case study with a relatively small sample (N = 68). Accordingly, the findings should be interpreted as analytically generalizable insights derived from a specific organizational context. Future research should apply the proposed maturity-based diagnostic across multiple organizations and larger samples, including confirmatory statistical validation, to further strengthen the robustness and generalizability of the model.

## 5.2 Theoretical Implications

This study makes three main theoretical contributions to the literature on sustainable leadership and innovation capability. First, it provides empirical support for the SFSIC model by demonstrating that sustainable leadership, organizational enabling mechanisms, and innovation capability operate as interdependent elements of a single system. The strong alignment between leadership maturity and the development of organizational mechanisms reinforces earlier theoretical work on the systemic nature of leadership influence [8] and is consistent with more recent research showing that leadership shapes organizational structures, learning processes, and coordination routines in an integrated manner [3, 5].

While earlier research has typically treated sustainable leadership as a primarily normative or values-oriented concept [9], this study advances sustainable leadership theory by demonstrating how its influence becomes embedded in concrete organizational practices. By assessing maturity across three interconnected dimensions, this study shows how sustainability-oriented leadership manifests as a measurable, system-level capability within organizations.

Second, the findings contribute to the understanding of innovation capability as a dynamic and continuously evolving construct. Consistent with dynamic capabilities theory [7] and prior work emphasizing behavioural and structural components of innovation capability [22], the observed variation in maturity levels indicates that innovation capability develops through processes of alignment, learning, and system-wide integration.

Third, the study advances theory by providing a maturity-based operationalization of the SFSIC, specifying measurable levels and thresholds that enable comparative assessment of systemic alignment across leadership, enabling mechanisms, and innovation outcomes. Beyond these substantive theoretical contributions, the study also contributes methodologically by demonstrating how a systems-based framework can be translated into a structured diagnostic instrument suitable for empirical assessment.

### 5.3 Practical Implications

The maturity profile indicates that leadership alone is insufficient, as its effects materialize only when supported by well-developed enabling mechanisms. The lower maturity observed in collaboration and learning indicates a specific gap that limits the extent to which leadership intentions translate into everyday work practices. Strengthening these mechanisms—particularly cross-functional communication and reflective learning—would help close this gap and reinforce organization-wide innovation capability. In this sense, the maturity assessment helps identify where alignment between sustainable leadership, organizational enabling mechanisms, and innovation capability is strong, as well as where gaps remain. Addressing these gaps—particularly those related to collaboration and learning mechanisms—would generate system-wide improvements in innovation capability.

The SFSIC maturity model also provides a structured tool for monitoring organizational progress over time. By converting survey responses into maturity levels, organizations can observe how changes in sustainable leadership, organizational enabling mechanisms, or innovation capability influence the overall maturity profile and alignment among the three components. This supports evidence-based decision-making, reduces reliance on fragmented improvement initiatives, and enables internal benchmarking to facilitate the diffusion of effective practices across the organization.

## 6 Conclusion

This study demonstrates that the SFSIC framework can be effectively operationalized as a maturity-based diagnostic tool for assessing organizational innovation capability. The proposed approach translates a complex, systems-oriented perspective on innovation capability into a practical assessment method, enabling organizations to identify their developmental position as well as areas requiring further improvement. The maturity model offers a structured way to capture how key organizational conditions for innovation develop over time and provides a foundation for more evidence-based capability-building efforts. In doing so, the study further reinforces the view of innovation capability as a dynamic, maturity-driven phenomenon that evolves through alignment, learning, and system-wide integration. Future research could extend this work by applying the model across multiple organizational contexts or by using longitudinal research designs to examine how maturity evolves as organizations implement change initiatives. Overall, the SFSIC maturity-based diagnostic extends the innovation management literature by offering a structured, manufacturing-relevant approach to assessing systemic alignment across leadership, enabling mechanisms, and innovation outcomes.

## References

1. Adams, R., Jeanrenaud, S., Bessant, J., Denyer, D., Overy, P.: Sustainability-oriented innovation: A systematic review. Int. J. Manag. Rev. **18**, 180–205 (2016). https://doi.org/10.1111/ijmr.12068

2. Hughes, D.J., Lee, A., Tian, A.W., Newman, A., Legood, A.: Leadership, creativity, and innovation: A critical review and practical recommendations. Leadersh. Q. **29**, 549–569 (2018). https://doi.org/10.1016/j.leaqua.2018.03.001
3. Hillmann, J.: Disciplines of organizational resilience: Contributions, critiques, and future research avenues. Rev. Manag. Sci. **15**, 879–936 (2021). https://doi.org/10.1007/s11846-020-00384-2
4. Frazier, M.L., Fainshmidt, S., Klinger, R.L., Pezeshkan, A., Vracheva, V.: Psychological safety: A meta-analytic review and extension. Pers. Psychol. **70**, 113–165 (2017). https://doi.org/10.1111/peps.12183
5. Sajjad, A., Eweje, G., Tappin, D.: Sustainability leadership: An integrative review and conceptual synthesis. Bus. Strateg. Environ. **33**, 123–144 (2024). https://doi.org/10.1002/bse.3631
6. Saunila, M.: Innovation capability in achieving higher performance: Perspectives of management and employees. Tech. Anal. Strat. Manag. **29**, 903–916 (2017). https://doi.org/10.1080/09537325.2016.1259469
7. Teece, D.J.: Business models and dynamic capabilities. Long Range Plan. **51**, 40–49 (2018). https://doi.org/10.1016/j.lrp.2017.06.007
8. Avery, G.C., Bergsteiner, H.: Sustainable Leadership: Honeybee and Locust Approaches. Routledge, Abingdon, UK (2011)
9. Tideman, S.G., Arts, M.C., Zandee, D.P.: Sustainable leadership: Towards a workable definition. J. Corp. Citizsh. **2013**, 17–33 (2013). https://doi.org/10.9774/GLEAF.4700.2013.ma.00004
10. Vladić, N., Maletič, D., Maletič, M.: A Systems perspective on sustainable leadership and innovation capability: Building organizational resilience in a high-tech company. Systems. **13**, 1075 (2025). https://doi.org/10.3390/systems13121075
11. Tidd, J., Bessant, J.: Managing Innovation: Integrating Technological, Market and Organizational Change. John Wiley & Sons, Chichester, UK (2009)
12. Liao, Y.: Sustainable leadership: A literature review and prospects for future research. Front. Psychol. **13**, 1045570 (2022). https://doi.org/10.3389/fpsyg.2022.1045570
13. McCann, J.T., Holt, R.A.: Defining sustainable leadership. Int. J. Sustain. Strateg. Manag. **2**, 204–210 (2010). https://doi.org/10.1504/IJSSM.2010.032561
14. Hurt, H.T., Joseph, K., Cook, C.D.: Scales for the measurement of innovativeness. Human Comm Res. **4**, 58–65 (1977). https://doi.org/10.1111/j.1468-2958.1977.tb00597.x
15. Fraser, P., Moultrie, J., Gregory, M.: The use of maturity models/grids as a tool in assessing product development capability. In: IEEE International Engineering Management Conference, pp. 244–249. IEEE, Cambridge, UK (2002). https://doi.org/10.1109/IEMC.2002.1038431
16. CMMI Institute: CMMI Development V2.0: Capability Maturity Model Integration. CMMI Institute, Pittsburgh, PA (2018)
17. Inkow, M.: Methodological aspects of measuring the innovation maturity of enterprises-proposal of the author's own innovation maturity model. Eur. Res. Stud. J. **XXVII**, 499–510 (2024). https://doi.org/10.35808/ersj/3444
18. Cubo, C., Oliveira, R., Fernandes, A.C., Sampaio, P., Carvalho, M.S., Afonso, P.: An innovative maturity model to assess supply chain quality management. IJQRM. **40**, 103–123 (2023). https://doi.org/10.1108/IJQRM-06-2021-0184
19. ISO: International Organization for Standardization (ISO): ISO 56002:2019 Innovation management — Innovation management system — Guidance, https://www.iso.org/standard/68221.html, (2019).
20. Edmondson, A.C., Mcmanus, S.E.: Methodological fit in management field research. Acad. Manag. Rev. **32**, 1246–1264 (2007). https://doi.org/10.5465/amr.2007.26586086

21. Yin, R.K.: Case Study Research and Applications: Design and Methods. SAGE Publications, Thousand Oaks, CA, USA (2018)
22. Baxter, P., Jack, S.: Qualitative case study methodology: Study design and implementation for novice researchers. Qual. Rep. **13**, 544–559 (2008). https://doi.org/10.46743/2160-3715/2008.1573
23. Bryman, A.: Social Research Methods. Oxford University Press, Oxford (2016)
24. Hair Jr., J.F., Hult, G.T.M., Ringle, C.M., Sarstedt, M.: A Primer on Partial Least Squares Structural Equation Modeling (PLS-SEM). SAGE Publications, Thousand Oaks, CA, USA (2021)
25. Saunila, M.: Innovation capability in SMEs: A systematic review of the literature. J. Innov. Knowl. **5**, 260–265 (2020). https://doi.org/10.1016/j.jik.2019.11.002

# An Approach to Quality Control of Heat-Treated Parts Towards a Smart Manufacturing

Justyna Patalas-Maliszewska[1(✉)], Sławomir Kłos[1], Michał Bazel[2], and Hanna Łosyk[1]

[1] University of Zielona Góra, Licealna 9, 65-417 Zielona Góra, Poland
J.Patalas-Maliszewska@iim.uz.zgora.pl

[2] Seco Warwick SA, Sobieskiego 8, 66-200 Świebodzin, Poland

**Abstract.** Smart manufacturing can be treated as an enterprise that adopts Industry 4.0/Industry 5.0 (I4.0/I5.0) technologies with the objective of enhancing process efficiency. A key challenge in this context is the automation of product quality control through the integration of selected I4.0/I5.0 technologies. This paper provides a comprehensive overview of the smart manufacturing concept and reviews the currently available methods and tools that support quality control of heat-treated components. Next, a novel approach to quality control for heat-treated parts is proposed. Specifically, using the example of the heat treatment process, a catalog of potential quality defects was developed, and a procedure for their automatic detection was defined prior to their occurrence, utilizing artificial intelligence (AI) in accordance with the Smart Manufacturing concept. Next, the experimental studies involving the heat treatment of 84 rollers manufactured from 20MnCr5 and 18CrNiMo7–6 are provided. Subsequently, the challenges associated with the implementation of this approach were identified.

**Keywords:** Smart Manufacturing · Quality control · Low pressure carburizing · Defects after heat treatment

## 1 Introduction

The manufacturing market plays a key role in global economies, as evidenced by the employment of 14% of the total global workforce [1]. This market is estimated to reach $732.2 billion by 2027 [2]. The development of Industry 4.0/5.0 (I4.0/I5.0) technologies and the transformation towards smart manufacturing is changing existing production systems and creating new opportunities and challenges for enterprises. The advanced level of production automation in the I4.0/I5.0 era allows for higher efficiency of production systems, increased productivity and the ability to personalize products through the collaboration of machines [3]. In addition, organizations must adapt to changing legal regulations, as well as market requirements, standards, and customer expectations. Organizations therefore strive to satisfy customers by providing them with products and services that meet or exceed their expectations [4]. Next, organizations must adapt to changing legal regulations, as well as market requirements, standards, and customer expectations. Organizations therefore strive to satisfy customers by providing them with products and services that meet or exceed their expectations [5].


A. Hamrol et al. (Eds.): MANUFACTURING 2026, LNME, pp. 18–29, 2026.
https://doi.org/10.1007/978-3-032-21701-1_2

Heat and thermochemical treatment of manufactured parts often occurs at the very end of the production chain and often involves large production batches exceeding thousands parts. Therefore, a significant portion of the total production costs has already been incurred for the batch that reaches the heat treatment chamber. This process is extremely important, as the risk of defects at this stage generates significant losses for the company. Furthermore, a single heat treatment process in an industrial furnace chamber can process semi-finished products from many different production orders, so the risk of conducting a defective process can expose the company to high penalties resulting from delayed order fulfillment. Finally, a key point is that the heat treatment process cannot always be repeated, and defective parts must be suitable only for disposal.

Despite the growing importance of quality control in manufacturing processes, there is still little research available to support the quality control process in heat treatment processes. The paper [6] used conventional nondestructive testing (NDT) techniques to verify the accuracy and efficiency of heat treatment. The results showed that in the case of the analyzed methods, the ultrasonic method showed the highest properties in verifying the correctness of chemical heat treatment, while the acoustic method was indicated as a method not suitable for adaptation in real conditions. In the paper [7], an Machine Learning-based decision model was developed and then implemented in real production conditions in the automotive industry for the process of thermal tempering of glass. The developed model (based on the Ridge Regression algorithm) demonstrated high prediction performance, which allowed for cost savings by reducing the number of destructive tests.

Therefore, there is still a research gap in the context of defining a novel approach to quality control in the heat treatment process according to the concept of smart manufacturing. The main objective of this study is to develop a new concept for quality control of heat-treated parts, with a focus on applying I4.0/I5.0 technologies, specifically AI. The following research questions (RQ) are formulated thus:

**RQ1:** What are key parameters to quality control before and after the technological heat treatment process?

**RQ2:** What type of methods and techniques, according to the concept of Industry 5.0 are the most effective for developing an automated approach to quality control in heat treatment process?

**RQ3:** What are the future challenges and direction for an improvement the quality control in heat treatment process?

The main contributions of this study in the context of increasing the level of automation in the quality control process are as follows:

- Distinguishing the key parameters to quality control before and after the technological heat treatment process based on the case study.
- Providing the real use case in the laboratory conditions in order to develop the procedure for automated quality control process.
- Determining further research directions in the field of examining the condition of details before heat treatment and the proper selection of heat treatment parameters in order to reduce defective products.

## 2 Quality Control in Heat Treatment Process

In order to illustrate the heat treatment process, the course of this process is described in detail using the example of a real manufacturing company.

During the heat treatment process, each previously manufactured part changes a number of its mechanical properties. Under the influence of the specified technological parameters in the heat or thermochemical treatment process, conscious and planned changes occur at the atomic structure level. A new or modified crystallographic network are created. A new phases and structure occurred. Heat-treated parts exhibit improved mechanical parameters and can better fulfill their role independently or as a component of a larger mechanism. Heat treatment is therefore an essential process in the production chain of many components such as shafts, gears, etc.

One of the most important types of thermochemical treatment is carburizing. This is a process in which atomic carbon is introduced into the base material, most often steel. The supplied carbon creates a suitably thick carburized layer, which later becomes a hardened layer during the hardening process. Key technological parameters in the carburizing process include: carburizing and austenitizing temperatures, exposure time at both temperatures, the atmosphere from which the atomic carbon is produced, and the time of adding this atmosphere. An integral part of carburizing is also the hardening operation, which involves the rapid removal of heat from the workpieces arranged in the charge, the aim of which is to create appropriate non-equilibrium structures, most often high-carbon martensite with a specific amount of retained austenite in the carburized layer, and bainite structures or a mixture of bainite and pearlite in the core of each component. The entire thermal operation is usually completed by low tempering.

Among carburizing technologies, the two most popular are: atmospheric and low-pressure carburizing (vacuum carburizing). The goal of these technologies is to create a carburized surface layer of the component, ultimately resulting in a hardened layer. However, atmospheric carburizing and low-pressure carburizing differ in the way of realization so they are implemented in different industrial furnaces.

In the case of low-pressure carburizing (LPC), this is a non-equilibrium process due to the very low pressure prevailing almost throughout the process, oscillating at a level of several hundred pascals (3–8 mbar), while in atmospheric carburizing, atmospheric pressure prevails throughout. The atmosphere from which free atomic carbon is generated in low-pressure carburizing is based on acetylene, for practical reasons also with admixtures of gases such as ethylene, hydrogen, or nitrogen. Figure 1 shows the course of the low-pressure carburizing process through graphs of changes in pressure temperature, pressure and acetylene supply to the vacuum furnace chamber.

The typical temperature range within which the carburizing process (Fig. 1a-1c) can be performed is specified in the material data sheets for individual steel grades, and often also in internal regulations and standards. The most common temperature range for carburizing is 920–980 °C, which is fully possible for devices (furnaces) offering low-pressure carburizing technology, while the upper temperature limit for furnaces performing atmospheric carburizing is 950–960 °C. In the case of both new, higher-alloy carburizing steels and low-pressure carburizing furnaces, temperatures above 980 °C, or even above 1050 °C, are often required, so to perform carburizing, it is necessary to choose a device with vacuum technology.

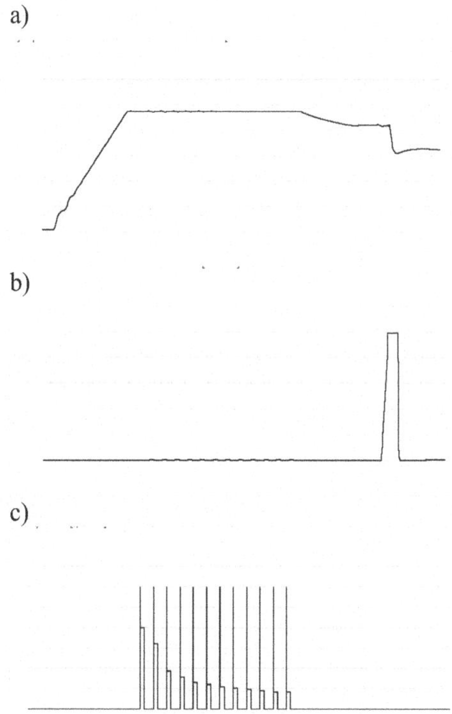


**Fig. 1** a) Temperature and b) pressure changes in the vacuum furnace chamber, c) acetylene feeding for low pressure carburizing.

Qualitative analysis of heat-treated parts includes examining possible geometric changes that may result from stresses in the processed parts. Deformations in parts can be caused by uneven temperature distributions (thermal stresses) occurring during rapid heating or cooling of the part. Large differences in the surface and core temperatures of the part result in differences in volumetric expansion or contraction, ultimately leading to stresses. Structural transformations (stresses during phase transformations, e.g., the transformation of austenite to martensite during hardening) are associated with changes in the specific volume of the steel. The non-simultaneous occurrence of these transformations in different parts of the part's cross-section causes internal stresses.

Post-heat treatment quality inspection focuses primarily on analyzing the structural and mechanical properties and changes in the geometry of the processed parts (deformation analysis). One of the most commonly used methods for assessing the quality of heat-treated parts is hardness testing because this parameter is directly related to the microstructure and strength of the steel after processing. Static methods: Brinell, Rockwell, or Vickers are used to measure the hardness of surface and core (all) and to determine the effective depth of the hardened layer in the cross-section of the part (Vickers). Metallographic testing is used to assess the material's microstructure, which determines its properties. Non-destructive visual testing and surface inspection are also performed to detect any cracks, scale, or significant deformations. Heat treatment (especially hardening) can lead to deformation and dimensional changes. Geometry testing involves checking key dimensions and tolerances using measuring instruments (calipers, micrometers, measuring machines) to assess straightness/flatness, shape and positional deviations, and cylindricity. Chemical composition analysis includes verifying elemental composition for compliance with requirements (e.g., after carburization).

So, based on knowledge and experience gained from the implementation of the carburizing process, a set of parameters (Sec. 3.1) and defects (Sect. 3.2) has been identified that require qualitative assessment.

# 3 An Approach to Quality Control of Heat-Treated Parts

## 3.1 Parameters and Quality Defects to Quality Control in Heat Treatment Process

Answering to the QR1, in the table the list of parameters that should be controlled along with their accuracy is formulated.

**Table 1** List of parameters.

| Measured parameter | Accuracy | Description |
|---|---|---|
| Mass | 0,1 g | Measuring the mass before and after the carburizing process can provide valuable information on the effectiveness of steel saturation with carbon, the depth of the layer, and the uniformity of carburization of details in the working space of the furnace. |
| Deformation: diameter | 0,05 mm | Measuring deformation/strain is important for further technological processes and the operational properties of parts. Frequent monitoring of deformation can also contribute to process optimization through better batch arrangement, selecting appropriate process temperatures, and selecting the appropriate cooling media and their parameters. Measurements should be taken before and after the process. Additionally, diameter and length measurements can be used to verify the parts' compliance with the technical drawing. |
| Deformation: length | 0,5 mm | |
| Deformation: runout | 0,05 mm | |
| Temperature | 1 °C | The measurement should be performed with the deformation measurement in mind, which should be performed at precisely defined component and ambient temperatures. The measurement should only be performed after the heat treatment process. |
| Parts counting | 1 pcs | Counting of details to check their compliance with the order |

Next, the quality defects of the products before and after carburizing are defined. Most defective parts can be identified before the heat treatment process begins using image analysis. A common defect that can be quickly identified using image analysis is mechanical damage to parts intended for heat treatment (Fig. 2).

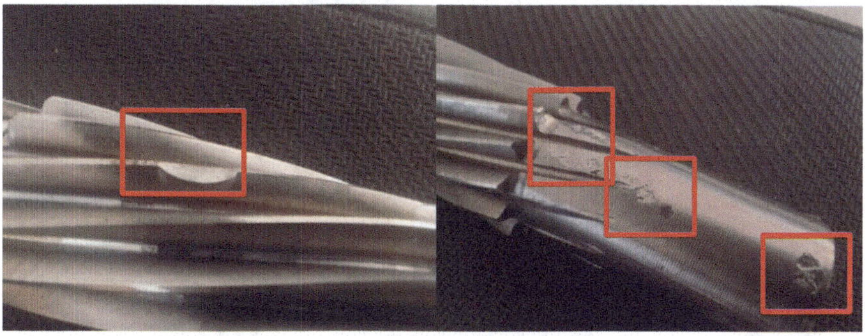

**Fig. 2** Examples of mechanical damage to parts.

Another very significant problem is corrosion, which occurs as a result of improper storage of parts. Figure 3 shows examples of corroded parts before heat treatment. As shown in the photos, corrosion can be a major problem if it occurs in the gear slots as it can be difficult to visually identify.

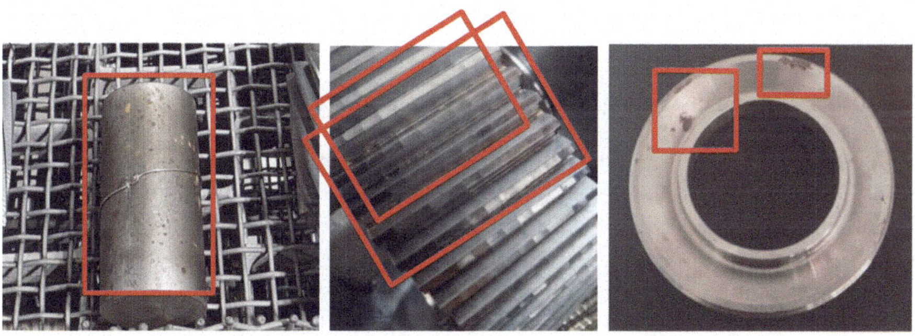

**Fig. 3** Examples of corrosion of details identified before heat treatment.

Contaminated or corroded parts that undergo heat treatment can be permanently damaged. Figure 4 shows examples of corroded and heat-treated parts contaminated with coolant (oil residue).

In addition to defects resulting from contamination or material quality, defects resulting from improperly performed heat treatment or damage to the installation (e.g., a leaky vacuum furnace) may occur after heat treatment. Figure 5 shows examples of defects resulting from improper heat treatment.

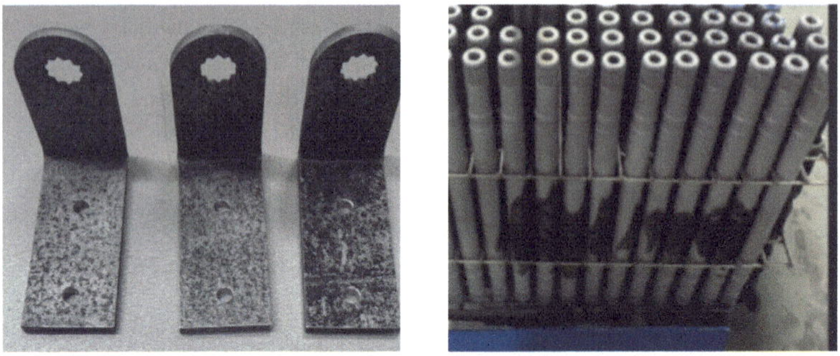

**Fig. 4** Example of parts corroded and contaminated with coolant after heat treatment.

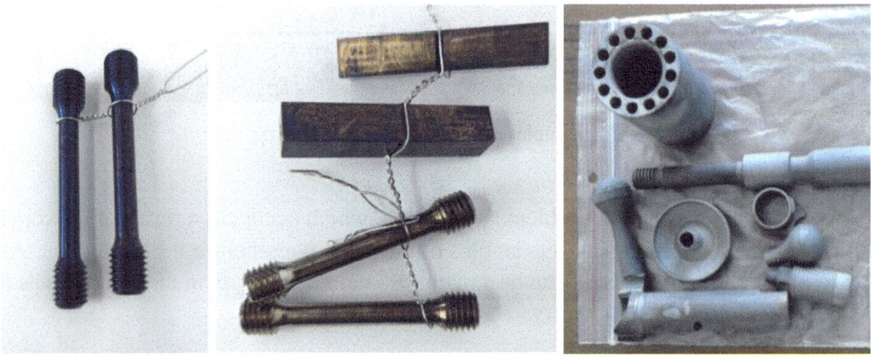

**Fig. 5** Examples of defective details after improper heat treatment (so-called detail coloring).

## 3.2 Procedure for assessing the quality of heat-treated parts

Searching for an answer to the QR2 and based on the assumptions of the Smart Manufacturing concept and the research findings, as well as knowledge and experience in the carburizing process (Sect. 3.1), a procedure for assessing the quality of heat-treated parts was developed.

I4.0/I5.0 technologies have the potential to eliminate waste in the production process, by eliminating everything that does not add value [8] and the implementation of quality forecasting processes, which is one of the three current trends in production management (along with predictive maintenance and fault diagnostics) [9]. The term Quality 4.0 (Q4.0) appears in the literature, which is a combination of digital technology with complex techniques and intelligent procedures that allow customers to provide customers with products of the highest quality [10]. Although Q4.0 integrates traditional quality control methods with I.4.0/I5.0 technologies such as the Internet of Things (IoT), big data analytics and artificial intelligence (AI), human input in the inspection process remains essential [5]. Machine learning (ML) and deep learning (DL) play an important role in Q4.0 [5]. Building an ML-based quality control model should be based on four elements: process and product quality description, quality forecasting, quality classification, and optimization of process parameters [11].

Therefore, the procedure for assessing the preparation of parts for heat treatment is based on the AI technique, especially on the method focus on objects recognition, and includes the following steps.

**Stage 1:** Building a catalogue of quality defects prior and after heat treatment process.

**Stage 2:** Acquiring data from products prepared for processing by taking photos using a collaborative robot.

**Stage 3:** Classification of samples according to the defects identified in the defect catalog prior to heat treatment.

**Stage 4:** Applying of the developed AI-based automatic defect recognition model to eliminate defects in the carburizing process before they occur.

**Stage 4:** Providing heat treatment process.

**Stage 5:** Acquiring data from products after processing by taking photos using a collaborative robot.

**Stage 6:** Automatic visual inspection using the prepared procedure and dimensional inspection of parts during furnace charge preparation (loading parts into containers using a robot with image analysis, weight, and dimensions of individual parts).

### 3.3 Experimental Study

According to the procedure, in the experimental studies, the 84 rollers (Fig. 6) from two types of steel: 20MnCr5 and 18CrNiMo7–6 was produced in the research unit at the Institute of Mechanical Engineering of University of Zielona Góra, Poland.

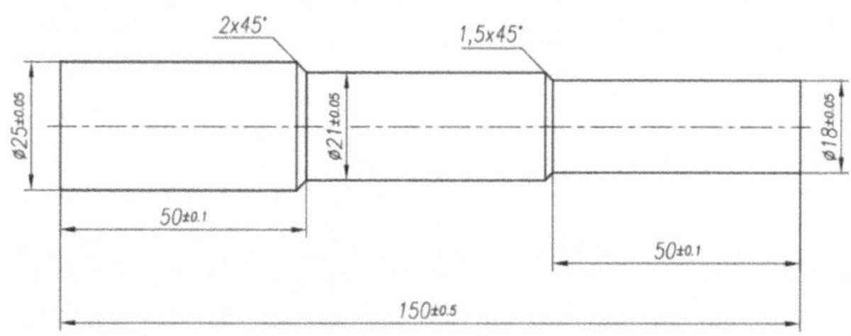


**Fig. 6** Technical drawing of the object of research.

According to stage 1 the catalogue of quality defects prior and after heat treatment process is developed (Table 2).

The first example of the quality defects is presented on the Fig. 7.

If parts are not properly prepared before heat treatment, their surface may be visually unacceptable after the process. Furthermore, processing unwashed parts can result in the furnace becoming dirty.

We are currently developing a database of images based on the defect catalog (Table 2) to facilitate further research on the development of an appropriate AI-based model.

**Table 2** The catalogue of quality defects.

| Quality defects prior heat treatment process | Quality defects after heat treatment process | Description |
|---|---|---|
| Corrosion of details | Black deposits, pitting on the surface of details | Iron oxides form an insulating layer on the surface of the part, preventing the adsorption and diffusion of carbon atoms into the surface layer. As a result of corrosion, the carburization layer on the part may be uneven or absent. |
| None (furnace chamber leak) | Color / discoloration of details | A vacuum furnace chamber leak can cause ambient air to be drawn into the furnace chamber, introducing oxygen and moisture into the furnace atmosphere. This leak results in reduced hardness and wear resistance on the surface of the workpieces. Thin, colored oxide layers form on the steel surface due to insufficient protective atmosphere (vacuum). Leaks can also result in a flaking layer of iron oxides on the workpiece surface due to the oxidizing atmosphere. This can also result in a dull, gray surface susceptible to corrosion. |
| Bruises, chips, dents, etc. | Bruises, chips, dents, etc. | These defects may occur as a result of improper processing or transport of parts. |
| Contamination by coolant, lubricants, grease, etc. | Burns, blackening of details | Oil can pyrolyze, forming soot (elemental carbon) on the part surface. This soot acts as a barrier, locally blocking the absorption of carbon atoms from the carburizing gas. |
| None | Deformations | Rapid cooling (hardening) leads to large temperature differences between the surface and the core of the part. The surface that comes into contact with the coolant contracts first, generating tensile stresses within and compressive stresses on the surface. Incorrect detail geometry resulting from stresses, requiring straightening or machining. |

**Fig. 7** Detail before (a) and after (b) heat treatment – acceptable appearance.

Also, currently, the first experimental work has been conducted regarding the carburizing process of samples with identified quality defects prior to heat treatment (Fig. 8). Before the process, parts should be silver in color with a characteristic sheen, achieved through machining (turning, milling, etc.). Additionally, they should be dry and free of contaminants, which can affect the quality of the parts after treatment and the hygiene of the furnace. Lubricants and machining fluids present on the parts during the process evaporate and then settle on the cold components of the furnace and pump system, causing their degradation.

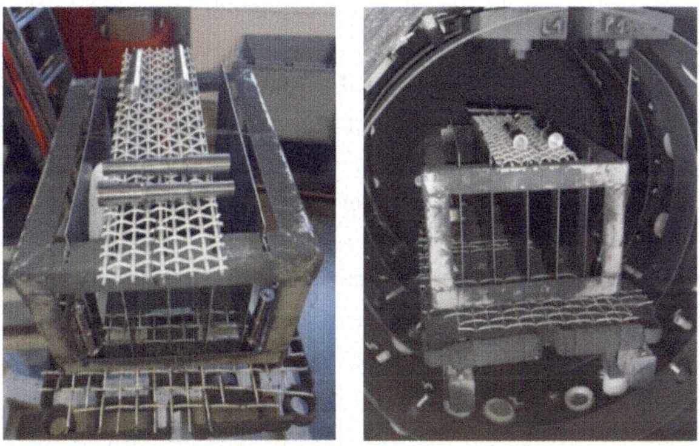

**Fig. 8** Rollers prepared for heat treatment.

Furthermore, after heat treatment in a vacuum with gas quenching, the appearance of parts should be characterized by uniform color throughout the entire furnace workspace, both within a single part and across the entire furnace workspace. As a result, the parts should become slightly dull and lose their gloss. This loss of gloss, along with the absence of discoloration and color uniformity, can indicate proper preparation of the parts for heat treatment (machining, cleaning) and the quality of the process (furnace tightness, uniform temperature and carburizing gas flow throughout the entire batch, and the purity of the process gases used).

## 4 Discussion and Conclusions

Heat treatment is one of the least automated areas of production, excluding the thermal process itself, which takes place in an industrial furnace. A human operator is responsible for receiving parts into the heat-treatment cell and verifying the delivery. A human is also responsible for assessing the conformity of the delivery with the order, as well as transporting batches to the appropriate locations within the heat-treatment cell, such as the washing area, if necessary, and arranging the parts into the so-called process load that will undergo the industrial thermal process.

In order to improve the quality control process, it is necessary to introduce automation into this process. Therefore, cyber-physical systems (CPSs) in line with the concept of Industry 4.0, which support correct responses to requests for decisions from the manufacturing process [12], based on methods for data processing [13] should be integrated with AI-based solutions. Due to the need to improve the efficiency of the inspection process and product quality, the implementation of AI is growing in popularity due to the possibility of increasing the accuracy and speed of the process [14]. Examining the possibilities of finding an answer to RQ3: What are the future challenges and directions for improving quality control in the heat treatment process? it can be stated that a very important aspect of further work. Current optical systems used for quality control are mainly used for geometric and dimensional inspection or surface detection [15]. However, apart from the issue of appropriate methods, recent years have also seen, to a large extent, the accuracy of the AI-based model that will be used for automatic quality defect recognition. Currently, laboratory work at the Institute of Mechanical Engineering of University of Zielona Góra is being conducted to build the images database for defects occurring prior to the carburizing process. Further work includes developing a tool to support automatic defect recognition and verifying this tool in practice.

The production of defective products is associated not only with costs but also with maintaining customer trust and building a strong brand. Efficiently automated optical and geometric inspection of all parts will certainly contribute to saving time and resources required for this essential activity, especially where the heat-treatment cell operates within a production chain as a subcontractor, for example, a commercial hardening plant. Fast and methodically effective actions will also allow for almost immediate conclusions to be drawn regarding the quality and uniformity of the parameters of the industrial heat-treatment furnace, and for a response in the form of halting production until the problem is resolved.

**Acknowledgments.** This work is supported by a programme of the Polish National Centre for Research and Development under the name of INNOGLOBO project no. INNOGLOBO/3/87/GRIP/2024/1 (5/2024/12/833).

## References

1. Kim, S., Nguyen, T.P., Yoon, J.: A systematic review of machine vision applications in factory and manufacturing processes: From quality control to predictive diagnostics. Int. J. Precis. Eng. Manuf.-Green Technol. (2025). https://doi.org/10.1007/s40684-025-00769-2

2. World Manufacturing Foundation: 2022 World Manufacturing Report: Redesigning Supply Chains in the New Era of Manufacturing. World Manufacturing Foundation (2022).
3. Sader, S., Husty, I., Daróczi, M.: Quality management practices in the era of Industry 4.0. Sci. J. Czest. Univ. Technol. – Manag. **35**(1), 117–126 (2019)
4. Foidl, H., Felderer, M.: Research challenges of Industry 4.0 for quality management. In: Innovations in Enterprise Information Systems Management and Engineering. 4th International Conference, ERP Future 2015, Munich, Germany, Revised Papers, November 16–17, 2015 (2016).
5. Alkhatib, F., Allam, M., Swarnakar, V., Alsadi, J., Maalouf, M.: Implementing and evaluating the quality 4.0 PMQ framework for process monitoring in automotive manufacturing. Sci. Rep. **15**, 24742 (2025)
6. Beneš, P., Vítovec, V.: Selected NDT techniques in heat treatment control. Manuf. Technol. **21**(1), 20–28 (2021). https://doi.org/10.21062/mft.2021.007
7. Khdoudi, A., Barka, N., Masrour, T., El-Hassani, I., El Mazgualdi, C.: Online prediction of automotive tempered glass quality using machine learning. Int. J. Adv. Manuf. Technol. **125**, 1577–1602 (2023). https://doi.org/10.1007/s00170-022-10649-7
8. Kuo, C.J., Ting, K.C., Chen, H.M.: Automatic machine status prediction in the era of Industry 4.0: Case study of machines in a spring factory. J. Syst. Archit. **81**, 44–53 (2017)
9. Chen, M., Wei, Z., Zhang, K.: Edge computing-based proactive control method for industrial product manufacturing quality prediction. Sci. Rep. **14**, 1288 (2024)
10. Morsi, A., Abbas, H.S., Ahmed, S.M., Mohamed, A.M.: Model predictive control based on linear parameter-varying models of active magnetic bearing systems. IEEE Access. **9**, 23633–23647 (2021)
11. Ismail, M., Mostafa, N., El-Assal, A.: Quality monitoring in multistage manufacturing systems by using machine learning techniques. J. Intell. Manuf. **33**(8), 2471–2486 (2022). https://doi.org/10.1007/s10845-021-01792-1
12. Bocewicz, G., Bożejko, W., Wójcik, R., Banaszak, Z.: Routing and scheduling subject to a trade-off between vehicle fleet size and storage capacity. Manag. Prod. Eng. Rev. **10**(3), 41–53 (2019). https://doi.org/10.24425/mper.2019.129597
13. Poczekajło, P., Suszyński, R.: Modern computing methods for digital signal processing engineering systems. Procedia Comput. Sci. **192**, 3534–3541 (2021). https://doi.org/10.1016/j.procs.2021.09.126
14. Kubik, C., Molitor, D.A., Varchmin, S., Leininger, D., Ohrenberg, J., Groche, P.: Image-based feature extraction for inline quality assurance and wear classification in high-speed blanking processes. Int. J. Adv. Manuf. Technol. **129**, 4883–4897 (2023). https://doi.org/10.1007/s00170-023-12653-x
15. Nyabadza, A., Azoulay-Younes, L., Vázquez, M., Brabazon, D.: Machine learning-based process quality control of screen-printed titanium dioxide electrodes. Results Mater. **26**, 100692 (2025)

# Risks of AI Over-Reliance in Digital Project Management: Implications for Intelligent Adaptive Systems

Katarzyna Ragin-Skorecka(✉), Marta Banach, Daria Motała, and Agnieszka Krugiełka

Faculty of Engineering Management, Poznan University of Technology, Poznan, Poland
katarzyna.ragin-skorecka@put.poznan.pl

**Abstract.** The rapid digitalization of organizational processes has intensified the integration of artificial intelligence into project management, positioning it as a key enabler of automation, predictive analytics, and operational efficiency. However, the growing reliance on AI-based decision support carries significant risks that can compromise project outcomes and the sustainability of intelligent adaptive systems. This article examines the risks associated with overreliance on AI in project management and discusses its implications for adaptive systems in production and organizational environments. Based on a survey of 100 IT project practitioners and qualitative content analysis of open-ended responses, the study empirically confirms four dominant risk areas: data security and privacy vulnerabilities, algorithmic errors related to data quality, unclear legal and ethical frameworks, and overreliance on AI-generated results. Despite the increasing maturity of AI tools, the results show that the majority of respondents (78%) believe that AI cannot replace project managers due to the irreplaceable role of human judgment, creativity, and interpersonal skills, capabilities that remain crucial in complex and adaptive management ecosystems. Rather than replacing human decision-makers, the results highlight the need for hybrid models of human-AI collaboration, robust risk mitigation strategies, and effective governance mechanisms to ensure that intelligent adaptive systems support and complement, rather than replace, human decision-making in project environments.

**Keywords:** Artificial intelligence in project management · Adaptive production systems · AI risk assessment

## 1 Introduction

The digitalization of organizational and production processes has become one of the key development directions for modern enterprises [1–3]. This transformation enables deeper process integration, improves operational efficiency, and facilitates access to new markets, especially in dynamic and data-intensive environments [4–8]. In this context, projects are increasingly becoming the primary mechanism for implementing digital solutions, including technologies supporting the development of intelligent and adaptive production management systems.


A. Hamrol et al. (Eds.): MANUFACTURING 2026, LNME, pp. 30–42, 2026.
https://doi.org/10.1007/978-3-032-21701-1_3

At the same time, project management is undergoing a significant transformation, evolving towards a more holistic and systemic approach supported by Industry 4.0 technologies [9–12]. Artificial intelligence (AI) has emerged as one of the keys enabling technologies, supporting automation, predictive analytics, and data-driven decision-making in projects [13–16]. As previous research has shown, AI offers significant potential for increasing planning accuracy, streamlining monitoring, and strengthening risk management processes, particularly in complex and adaptive project environments [10, 17–19].

In addition to its benefits, the literature consistently highlights significant risks associated with the use of AI in project management. These risks include issues related to data security and privacy, algorithmic biases, and the lack of clear and comprehensive legal regulations governing AI-based solutions [15, 20–23]. Despite a growing body of research highlighting the benefits of implementing artificial intelligence [24, 25], empirical research focusing on the risks of overreliance on AI remains limited, particularly in the context of intelligent adaptive systems operating in organizations and production environments.

This identified research gap forms the basis for this study, which aims to analyse the risks associated with the use of AI in project management and assess its impact on the performance and resilience of adaptive organizational and production systems.

## 2 Literature Review

For many years, the digitalization of enterprises has been described as a driving force for organizational change, influencing both the functioning of internal processes and the competitive position of companies in the market. Researchers indicate that digital transformation enables not only improved efficiency but also broader process integration and the opening of organizations to new business models [1–3]. The literature also emphasizes that digitalization promotes better resource utilization, contributes to the creation of collaboration networks, and enables entry into new markets, as confirmed by numerous empirical and review studies [4–8]. As a result, projects as temporary structures are becoming the main mechanism for implementing technological and organizational innovations.

As digitalization progresses, the very understanding of project management is changing. The literature suggests a shift from a narrow, tool-based approach to a systems approach that encompasses both technical and behavioural competencies, as well as the organizational context [11, 12]. This evolution is closely linked to the development of Industry 4.0 technologies, which impact productivity, efficiency, and sustainable resource use [9, 10, 26]. As a result of these changes, the importance of IT systems supporting project management is growing, integrating data, automating workflows, and streamlining decision-making [15, 27–31].

The literature increasingly emphasizes that intelligent technologies such as artificial intelligence, data analytics, blockchain, and the Internet of Things will shape the future of project management in the coming years [10, 19, 24, 32, 33]. AI stands out from other technologies because it combines the ability to process large data sets, adaptive learning, and decision support in dynamic project environments.

According to literature reviews and empirical studies, AI systems can automate administrative tasks, streamline scheduling, predict risks, and support project managers in real-time data analysis [13–16]. The effectiveness of such tools is confirmed by the research of Barcaui and Monat [34] and works on predictive models used in project risk assessment [17, 18]. Moreover, some authors suggest that AI may gradually take over tasks previously performed by PMOs and even partially replace the role of project managers [10, 19].

Despite its numerous benefits, the literature also extensively describes the limitations and risks associated with the use of AI. These include data security, model quality, algorithmic errors, lack of transparency, and unclear legal regulations [20–23]. Authors also point to the risks of overreliance on AI, the reduction of the human factor, and the social consequences of decision automation [35, 36]. Mangal [24] and Müller et al. [25] further emphasize the need to create ethical and organizational frameworks for the use of AI that ensure control, accountability, and proper data management.

In summary, the literature presents a broad spectrum of research on the use of AI in projects, from operational efficiency and predictive models to ethical and organizational risks. At the same time, the number of empirical studies examining the risks of overreliance on AI in the context of projects implemented in intelligent, adaptive manufacturing systems environments remains limited. This research gap provides a direct justification for undertaking further analyses presented in this article.

## 3 Materials and Methods

The main objective of the study was to identify risks associated with the use of artificial intelligence in project management, with particular emphasis on the problem of overreliance on AI, including its potential replacement of selected aspects of the project manager's role. To achieve this objective, the following research questions were formulated:

**RQ1:** What areas of application of IT technologies, including artificial intelligence, in project management are described in the literature?

**RQ2:** What risks related to the use of artificial intelligence in project management are indicated in the literature?

**RQ3:** What are the key risks associated with using AI in project management?

**RQ4:** In what areas can artificial intelligence replace the role of a project manager?

The research outline, illustrating the sequence of research activities, is presented in Fig. 1. The first step was a literature review, which allowed us to identify the main areas of application of IT solutions, including artificial intelligence, supporting project management (RQ1). This step also allowed us to outline the current state of research and the dominant risk categories discussed in the literature (RQ2). The next step involved conducting a survey, and the responses to open-ended questions were analysed using qualitative content analysis (RQ3, RQ4). The final step involved discussing the results and comparing them with those presented in previous studies.

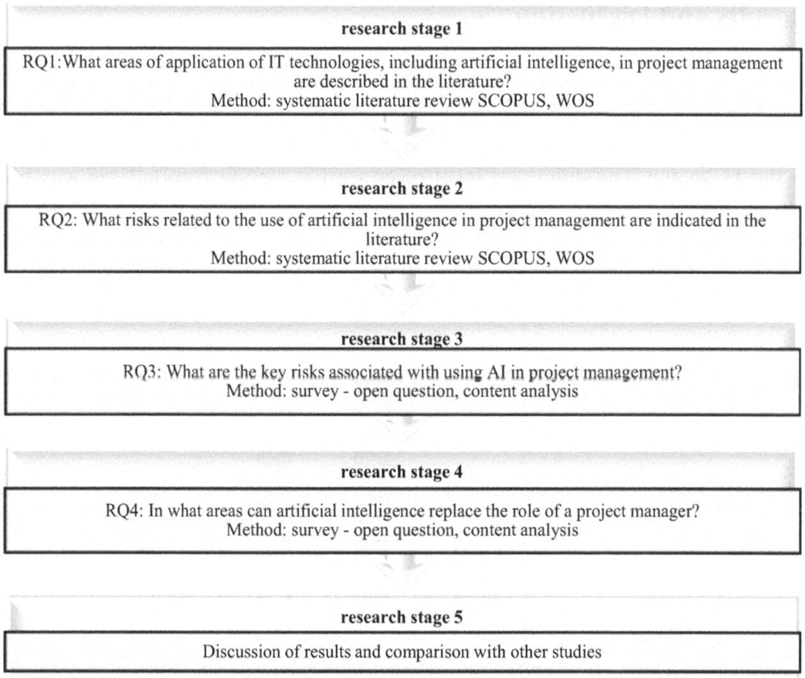


**Fig. 1** Research framework.

The chosen research method, a survey based on open-ended questions, was chosen due to the exploratory nature of the study and its emphasis on capturing practitioners' experiences and perceptions. Although this method does not allow for dynamic interaction between the researcher and respondents, it does enable the collection of rich qualitative insights regarding perceived risks and attitudes toward the use of AI in project management, which is essential to answering the research questions.

The geographic scope of the study included respondents employed at enterprises located in Poland. Participants were selected purposively, as they included project managers and team members involved in IT project implementation. All responses were collected anonymously and did not require additional anonymization procedures. The study was conducted in the second quarter of 2024. Limitations of the study include the unrepresentative nature of the sample, reliance on self-report data, subjective interpretation of open-ended questions and responses, and lack of control over external factors influencing respondents at the time of the survey. Furthermore, the lack of triangulation and validation methods should be considered when interpreting the results, as the study aims to provide exploratory observations rather than statistical conclusions [37].

## 4 Results and Discussion

The survey involved 100 people. The majority of respondents are young people aged 20–30 (83%) and 30–40 (15%), which is partly related to the recruitment method based on social media channels, as well as the relatively high representation of younger employees in the IT sector. The dominant level of education among respondents was higher education (75%). In terms of experience in IT project implementation, 57% of respondents reported between 2 and 5 years of experience, while 18% had more than 5 years of professional experience.

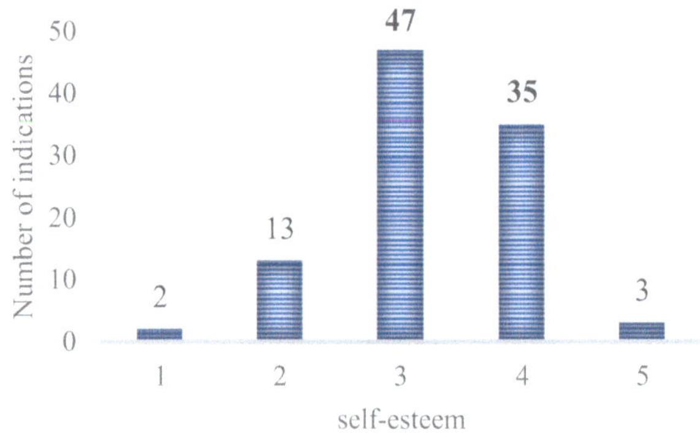


**Fig. 2** Respondents' self-assessment of knowledge about artificial intelligence.

An important element of the study was the respondents' self-assessment of their knowledge of artificial intelligence. The results are presented in Fig. 2. Most respondents rated their knowledge of artificial intelligence as average or above. This result suggests that respondents have a basic or intermediate understanding of AI-related topics, confirming the accuracy of their perceptions regarding the risks associated with AI in project management. It should be emphasized, however, that this assessment is based on self-assessment data and does not constitute an objective measure of expertise, which is a limitation of the study.

Approximately half of the respondents (48%) indicated they perceived risks associated with the use of AI in project management (RQ3). To further explore these concerns, an open-ended question was used to collect qualitative data on specific risks. Responses were analysed using content analysis [38], and the results are presented in Table 1. The analysis revealed four dominant risk categories: security and privacy, mistakes, legal aspects, and overreliance on AI solutions. These results are consistent with existing research and largely confirm previously identified risk categories, as also noted by Prifti [15].

Regarding security, the author points out that AI technology may not adhere to organizational security standards, which could impact human safety within a project. Linked to this is the risk of privacy breaches, as AI may not adequately distinguish between approved and restricted data. Legal aspects related to these two areas remain unclear, and the ethical use of AI has yet to be fully addressed. Artificial intelligence also

**Table 1** Risks associated with the use of artificial intelligence.

| Separated categories | Selected answers |
|---|---|
| Security and privacy | Security and privacy risks, quality of tasks performed. |
| | If it is misconfigured and has access to data, it may cause unintentional data leakage. |
| | Privacy threat. |
| | Data theft. |
| Mistakes | Too much "reliance" on AI solutions and considering these tools as infallible. Currently, I believe that artificial intelligence is fallible and still makes a lot of mistakes. Another threat is the leakage of confidential data/information. |
| | Errors and incorrect conclusions, sensitivity to attacks and manipulation. |
| | AI can make mistakes that, if not caught in time, can cause serious problems and costs. |
| | Lack of control, errors, incorrect content |
| Legal aspects | Lack of appropriate knowledge, lack of openness to new solutions, lack of specific regulations in law. |
| | Too much reliance on AI solutions and lack of verification of what is created by it. Undefined legal issues when it comes to the use of AI. Possible risk when it comes to data privacy. |
| | The issue of legal aspects – it is currently not 100% defined what the law on using elements generated by AI looks like. |
| Overreliance on AI solutions | Security of personal data, questionable legal issues with the use of AI-derived creations, too little verification of work done by AI, too much reliance on AI work results alone – lack of own creativity, too little knowledge of project teams. |
| | Over-reliance (blindness) on artificial intelligence, lack of doing own research. |
| | Lack of verification of data coming out of AI. Lack of own creative invention. Legal issue of using elements taken from AI. |

makes errors due to incomplete or missing data, which are used to generate forecasts or reports that serve as the basis for project decision-making. Therefore, it is crucial to ensure data quality and availability. Over-reliance on AI solutions and treating it as an autonomous entity may lead to situations where AI starts acting independently, causing project managers to lose control over what, how, and when things happen. An additional risk category described by Prifti [15] concerns employees. AI can eliminate repetitive and low-skilled positions, but it cannot always assign tasks to appropriate individuals due to a lack of human social skills and understanding. Furthermore, AI cannot be considered an autonomous and infallible entity, requiring human oversight. Similar observations can be found in the works of Alshaikhi et al. [20], Karamthulla et al. [22] and Peddisetty [39].

**Table 2** Replacing the role of the project manager with artificial intelligence.

| Separated categories | Selected answers |
|---|---|
| No human factor | A human factor is needed, a human assessment in a given situation. AI adaptation to changing conditions may take more time than MP intervention in specific cases. |
| | I believe that a human factor is necessary in every project. Artificial intelligence can help us in some areas, but not completely replace it. |
| | No, because if we still take into account the human factor as a design team, then such features as empathy and understanding cannot be replaced by AI (or at least probably not). |
| | In project management, the human factor is important, analysis of the situation, feeling emotions, empathy, context – AI only generates answers, ways to do a given thing, but it is the human factor that is important in managing the project, people and the entire business |
| | A Project Manager deals not only with hard data and process design (which artificial intelligence could actually replace or strongly support), but this job also involves a lot of soft skills: assessing the potential, talents and skills of the team, flexibility in terms of communication and organization in terms of private conditions, and a high level of personalization of the message in terms of personal circumstances. |
| | Artificial intelligence is just a tool that needs to be supported. |
| | No, because in the role of a project manager, interpersonal communication is important, which AI will never be able to truly replace and gain the same benefits. |
| | AI can significantly facilitate and streamline the work of PM, but in working with entire project teams, aspects of empathy and conflict resolution and a 'human' approach to the team are still important. The key is to learn AI solutions or new technologies in general so that they can be used correctly. |
| Limitations of Artificial Intelligence | In the current situation, this is possible, but AI still has its problems and needs huge amounts of data that it will collect over time. |

(*continued*)

**Table 2** (*continued*)

| Separated categories | Selected answers |
|---|---|
| | No, because only a human can go beyond the pattern. Artificial intelligence operates based on x-y frames and will not go beyond them (e.g. combining threads to find a solution to two problems with one solution). |
| | There is no exact knowledge. |
| | Not at the moment, because it is not that developed yet, but it is possible in the future |
| | It seems to me that at the moment, AI is not developed enough to properly perform this type of role. I have often noticed errors, e.g. in the answers provided by AI or its lack of understanding of the question asked. |
| | It should not completely replace the role of a project manager, because it is artificial intelligence and can make mistakes and errors, it must be supervised. |
| | AI helps, generates general content, ideas, the human role is to adapt it to the specifics of the company/project. |
| | Robots cannot replace humans completely. |
| | Definitely not. Technology is just a tool that people need to learn to use, but it cannot replace us completely. |

Regarding the potential replacement of the project manager role by AI (RQ4), the majority of respondents indicated that AI cannot replace the project manager (78%), while 17% indicated "maybe". Only 5% of respondents believed that AI could fully replace the project manager. A qualitative analysis of open-ended responses [38], summarized in Table 2, shows that respondents most frequently emphasized the importance of the human factor in project management, particularly in relation to interpersonal communication, contextual assessment, and relationship building.

Respondents also noted the current limitations of AI technology, which prevent it from completely replacing the project manager role. Respondents who suggested that AI could eventually replace project managers (a significant minority) typically viewed this as a long-term possibility, not an immediate outcome dependent on future technological advancements. Gupta [27] presented similar findings, emphasizing that despite the growing role of AI in project management, human competencies, especially emotional intelligence, remain crucial for motivating teams and managing complex social interactions. While AI effectively supports decision-making and resource allocation, it is not yet capable of fully assuming responsibility for team leadership and adaptive response to dynamic project conditions.

Alshaikhi et al. [20] cite research indicating that 73% of respondents believe technology will never replace the human mind. Human creativity and social interaction remain fundamental elements of effective project management [40]. The results of this study

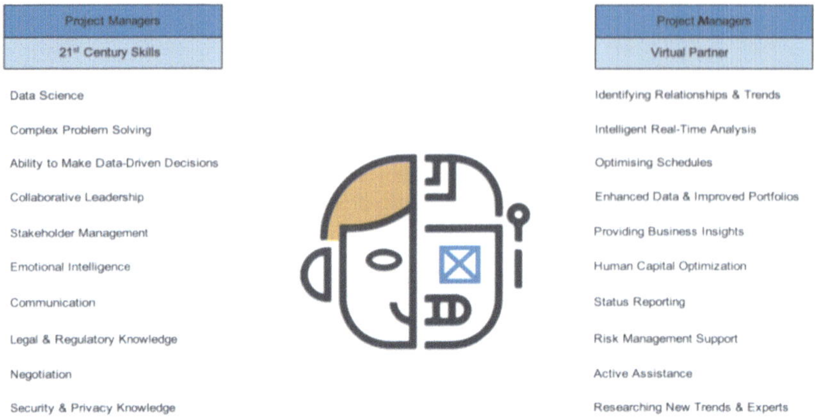


**Fig. 3** Modern Project Manager Skills and AI Support [20].

further support the view that AI and project managers should be viewed as interdependent, not competing, entities. This interdependence is reflected in the complementary division of competencies between project managers and AI-based tools, as illustrated in Fig. 3. Project managers must continually develop competencies that enable them to manage and oversee AI-based tools, while AI systems require effective human oversight to generate value through high-quality data and informed interpretation of results. Therefore, the effective use of AI in project management requires a diverse and flexible set of skills. Although many routine project management tasks can be automated, the core value of project management still lies in advanced interpersonal skills, understanding context, and the ability to adapt to new technologies [20, 35, 41, 42].

## 5 Conclusions

Digitalization and technological advancements are playing a key role in transforming modern project management, especially in environments based on integrated and intelligent adaptive systems. The use of AI offers significant benefits, such as the automation of repetitive tasks, improved resource allocation, improved risk management, and more accurate outcome forecasting. The results of this study empirically confirm the findings presented in previous literature, demonstrating that despite these advantages, AI cannot fully replace project managers, as key elements such as creativity, empathy, contextual judgment, and interpersonal skills remain inherently human. These findings clearly indicate that AI delivers the greatest value when used as a decision-supporting and facilitating tool, enabling project managers to focus on higher-value, strategic, and relational activities rather than routine operational tasks.

One of the main challenges of integrating AI into project management is ensuring high-quality data and adequately preparing employees to effectively use these technologies. At the same time, the results highlight the need for clearly defined governance, ethics, and organizational frameworks governing the use of AI to minimize risks related to data privacy, algorithmic bias, and loss of human oversight. These challenges are particularly relevant in intelligent adaptive systems, where project-level decisions directly impact the stability, flexibility, and resilience of organizational and production processes. Although AI has the potential to significantly increase efficiency and innovation in project management, the study confirms that it cannot and should not completely replace human input, but rather operates within hybrid models of human-AI collaboration.

When considering future research directions, it is worth referring to the work of Wynsberghe [43], who addressed the issue of AI for sustainability and the sustainable development of AI. This perspective represents a significant extension of the present study, as it connects AI-enabled project management with broader organizational and social responsibility. This approach is relevant both for IT projects implementing AI solutions and for project management practices that rely on AI-based tools. Key directions for future research include the development of governance mechanisms for the sustainable use of AI, the empirical evaluation of hybrid human-AI decision-making models, and the assessment of the long-term impact of AI on adaptive organizational and production systems. Key directions for the sustainable implementation of AI in project management include optimizing resource consumption, minimizing energy-intensive activities, and ensuring that projects support long-term ecological and social goals. Striking a balance between technological innovation, human oversight, and sustainability emerges as a critical condition for the ethical and responsible integration of AI into project and production management.

**Acknowledgments.** This research was funded by the Poznan University of Technology, Faculty of Engineering Management – project No. 0812/SBAD/4240.

## References

1. Begnini, S., Oro, I.M., Tonial, G., Dalbosco, I.B.: The relationship between the use of technologies and digitalization strategies for digital transformation in family businesses. J. Fam. Bus. Manag. **14**(4), 710–726 (2023). https://doi.org/10.1108/JFBM-06-2023-0087
2. Szopa, Ł., Cyplik, P.: The concept of building a digital transformation model for enterprises from the SME sector. LogForum. **16**(4), 593–601 (2020). https://doi.org/10.17270/J.LOG.2020.497
3. Vărzaru, A.A.: An empirical framework for assessing the digital technologies users' acceptance in project management. Electronics. **11**(23), 3872 (2022). https://doi.org/10.3390/electronics11233872
4. Bresciani, S., Huarng, K.-H., Malhotra, A., Ferraris, A.: Digital transformation as a springboard for product, process and business model innovation. J. Bus. Res. **128**, 204–210 (2021). https://doi.org/10.1016/j.jbusres.2021.02.003
5. Mitrega, M., Spacil, V., Pfajfar, G.: Co-creating value in post-communists' contexts: Capability perspective. J. Serv. Mark. **35**(2), 169–181 (2021). https://doi.org/10.1108/JSM-03-2019-0114
6. Vial, G.: Understanding digital transformation: A review and a research agenda. Manag. Digit. Transform. **13–66** (2021). https://doi.org/10.1016/j.jsis.2019.01.003
7. Piechowski, M., Meller, A.: Use of KPI indicators to optimize production process performance – case study. In: International Symposium on Distributed Computing and Artificial Intelligence, pp. 3–12. Springer, Cham (2023). https://doi.org/10.1007/978-3-031-38318-2_1

8. Vuchkovski, D., Zalaznik, M., Mitręga, M., Pfajfar, G.: A look at the future of work: The digital transformation of teams from conventional to virtual. J. Bus. Res. **163**, 113912 (2023). https://doi.org/10.1016/j.jbusres.2023.113912
9. Adegbite, A.O., Adefemi, A., Ukpoju, E.A., Abatan, A., Adekoya, O., Obaedo, B.O.: Innovations in project management: Trends and best practices. Eng. Sci. Technol. J. **4**(6), 509–532 (2023). https://doi.org/10.51594/estj.v4i6.670
10. Aliu, J., et al.: Towards a new paradigm of project management: A bibliometric review. Sustainability. **15**(13), 9967 (2023). https://doi.org/10.3390/su15139967
11. Gemünden, H.G., Schoper, Y.: Future trends in project management. In: Proceedings of the International Research Network on Organizing by Projects (IRNOP) Conference, London, UK, vol. 13 (2015).
12. McGrath, J., Kostalova, J.: Project management trends and new challenges. In: Proceedings of the 13th Annual International Conference on Project Management (2020). https://doi.org/10.36689/UHK/HED/2020-01-061
13. Babu, P.P., Vasumathi, A.: Role of artificial intelligence in project efficiency mediating with perceived organizational support in the Indian IT sector. Indian J. Inf. Sources Serv. **13**(2), 39–45 (2023). https://doi.org/10.51983/ijiss-2023.13.2.3786
14. Nuhn, H., Oswald, A., Flore, A., Lang, R.: AI-supported natural language processing in project management: Capabilities and research agenda. In: IPMA Research Conference Proceedings (2022). doi:10.56889/nrkr7690.
15. Prifti, V.: Optimizing project management using artificial intelligence. Eur. J. Form. Sci. Eng. **5**(1), 30–38 (2022). https://doi.org/10.26417/667hri67
16. Weng, J.C.: Putting Intellectual Robots to Work: Implementing Generative AI Tools in Project Management. NYU SPS Applied Analytics Laboratory (2023)
17. Afzal, F., Yunfei, S., Nazir, M., Bhatti, S.M.: A review of artificial intelligence-based risk assessment methods for capturing complexity-risk interdependencies: Cost overrun in construction projects. Int. J. Manag. Proj. Bus. **14**(2), 300–328 (2021). https://doi.org/10.1108/IJMPB-02-2019-0047
18. Nyqvist, R., Peltokorpi, A., Seppänen, O.: Can ChatGPT exceed humans in construction project risk management? Eng. Constr. Archit. Manag. **31**(13), 223–243 (2024). https://doi.org/10.1108/ECAM-08-2023-0819
19. Steen, J., Klein, G., Potts, J.: 21st-century general-purpose technologies and the future of project management. Proj. Manag. J. **53**(5), 435–437 (2022). https://doi.org/10.1177/87569728221125095
20. Alshaikhi, A., Khayyat, M.: An investigation into the impact of artificial intelligence on the future of project management. In: 2021 International Conference of Women in Data Science at Taif University (WiDSTaif), pp. 1–4. IEEE (2021). https://doi.org/10.1109/WiDSTaif52235.2021.9430234
21. Elmousalami, H.H.: Comparison of artificial intelligence techniques for project conceptual cost prediction: A case study and comparative analysis. IEEE Trans. Eng. Manag. **68**(1), 183–196 (2021). https://doi.org/10.1109/TEM.2020.2972078
22. Karamthulla, M.J., Muthusubramanian, M., Tadimarri, A., Tillu, R.: Navigating the future: AI-driven project management in the digital era. Int. J. Multidiscip. Res. **6**(2), 1–11 (2024). https://doi.org/10.36948/ijfmr.2024.v06i02.15295
23. Wang, Q.: How to apply AI technology in project management. PM World J. **8**(3) (2019)
24. Mangal, A.: Revolutionizing project management with generative AI. ESP J. Eng. Technol. Adv. **3**(4), 53–60 (2023). https://doi.org/10.56472/25832646/JETA-V3I8P106
25. Müller, R., Locatelli, G., Holzmann, V., Nilsson, M., Sagay, T.: Artificial intelligence and project management: Empirical overview, state of the art, and guidelines for future research. Proj. Manag. J. **55**(1), 9–15 (2024). https://doi.org/10.1177/87569728231225198

26. Torre, D., Colapinto, C., Durosini, I., Triberti, S.: Team formation for human-artificial intelligence collaboration in the workplace: A goal programming model to foster organizational change. IEEE Trans. Eng. Manag. **70**(5), 1966–1976 (2023). https://doi.org/10.1109/TEM.2021.3077195
27. Gupta, S.: Artificial intelligence, analytics and agile: Transforming project management in the 21st century. Int. J. Recent Technol. Eng. **11**, 1–8 (2022). https://doi.org/10.9734/bpi/rpst/v1/4843A
28. Maruthi, S., Dodda, S.B., Yellu, R.R., Thuniki, P., Reddy, S.R.B.: Automated planning and scheduling in AI: Studying automated planning and scheduling techniques for efficient decision-making in artificial intelligence. Afr. J. Artif. Intell. Sustain. Dev. **2**(2), 14–25 (2022)
29. Zadeh, E.K., Khoulenjani, A.B., Safaei, M.: Integrating AI for agile project management: Innovations, challenges, and benefits. Int. J. Ind. Eng. Constr. Manag. **1**(1), 1–10 (2024)
30. Piechowski, M., Kudelska, I.: Integration of IT systems using QR codes and RPA through automation of data registration processes. In: International Conference on Intelligent Systems in Production Engineering and Maintenance, pp. 409–422. Springer, Cham (2025). https://doi.org/10.1007/978-3-032-01517-4_30
31. Loska, A., Paszkowski, W., Waszkowski, R.: The scenario approach to the concept of maintenance of technical systems of urban engineering. In: International Scientific-Technical Conference Manufacturing, pp. 84–97. Springer, Cham (2024). https://doi.org/10.1007/978-3-031-56444-4_7
32. Ezeigweneme, C.A., Umoh, A.A., Ilojianya, V.I., Adegbite, A.O.: Telecom project management: Lessons learned and best practices: A review from Africa to the USA. World J. Adv. Res. Rev. **20**(3), 1713–1730 (2023). https://doi.org/10.30574/wjarr.2023.20.3.2707
33. Zaman, U., Nawaz, S., Nadeem, R.D.: Navigating innovation success through projects: Role of CEO transformational leadership, project management best practices, and project management technology quotient. J. Open Innov.: Technol. Mark. Complex. **6**(4), 168 (2020). https://doi.org/10.3390/joitmc6040168
34. Barcaui, A., Monat, A.: Who is better in project planning? Generative artificial intelligence or project managers? Proj. Leadersh. Soc. **4**, 100101 (2023). https://doi.org/10.1016/j.plas.2023.100101
35. Sahadevan, S.: Project management in the era of artificial intelligence. Eur. J. Theor. Appl. Sci. **1**(3), 349–359 (2023). https://doi.org/10.59324/ejtas.2023.1(3).35
36. Wamba-Taguimdje, S.L., Wamba, S.F., Kamdjoug, J.R.K., Wanko, C.E.T.: Influence of artificial intelligence (AI) on firm performance: The business value of AI-based transformation projects. Bus. Process. Manag. J. **26**(7), 1893–1924 (2020). https://doi.org/10.1108/BPMJ-10-2019-0411
37. Ragin-Skorecka, K., Hadaś, Ł.: Sustainable e-procurement: Key factors influencing user satisfaction and dissatisfaction. Sustainability. **16**(13), 5649 (2024). https://doi.org/10.3390/su16135649
38. Kleinheksel, A.J., Rockich-Winston, N., Tawfik, H., Wyatt, T.R.: Demystifying content analysis. Am. J. Pharm. Educ. **84**(1), 7113 (2020). https://doi.org/10.5688/ajpe7113
39. Peddisetty, N., Reddy, A.K.: Leveraging artificial intelligence for predictive change management in information systems projects. Distrib. Learn. Broad Appl. Sci. Res. **10**, 88–94 (2024)
40. Felicetti, A.M., Cimino, A., Mazzoleni, A., Ammirato, S.: Artificial intelligence and project management: An empirical investigation on the appropriation of generative chatbots by project managers. J. Innov. Knowl. **9**(3), 100545 (2024). https://doi.org/10.1016/j.jik.2024.100545
41. Minelle, F., Stolfi, F.: AI to support PM: A ChatGPT quality assessment (β test). PM World J. **12**(5) (2023)

42. Muda, I., Almahairah, M.S., Jaiswal, R., Kanike, U.K., Arshad, M.W., Bhattacharya, S.: Role of AI in decision making and its socio-psycho impact on jobs, project management and business of employees. J. ReAtt. Ther. Dev. Divers. **6**(5s), 517–523 (2023)
43. van Wynsberghe, A.: Sustainable AI: AI for sustainability and the sustainability of AI. AI and Ethics. **1**(3), 213–218 (2021). https://doi.org/10.1007/s43681-021-00043-6

# Suggestion Platform to Develop Quality and Sustainability - New Possibilities and Solutions

Anna Wencek[1](✉), Agnieszka Misztal[1], Laurentiu Ionescu[2], Kinga Ratajszczak[1], and Nadia Ionescu[2]

[1] Faculty of Engineering Management, Poznan University of Technology, Poznan, Poland
anna.wencek@doctorate.put.poznan.pl

[2] National University of Science and Technology Politehnica Bucharest, Bucharest, Romania

**Abstract.** Constant quality improvement and sustainable development of an organization rely on the involvement of all its members, regardless of their roles. Large organizations typically establish systems to facilitate the efficacy of their initiatives. Employees are encouraged to voice problems, share ideas, and suggest improvements. However, unresolved issues can lead to employee discouragement. Therefore, management commitment is essential to model desired behaviors and provide necessary tools and support. In this paper example of such instrument has been presented to show essential features the system should have to bring real benefit and ensure sustained engagement of workers. The location-aware feedback platform designed to enhance communication between management and employees in industrial environments has been presented. The system enables users to issue suggestions as tokens linked to specific physical locations. These tokens, accessible via mobile devices, include content such as text, images, and discussion channels. Employees receive notifications when near a token and can provide feedback. The platform bridges the digital and physical workspace, supporting real-time, context-sensitive communication without disrupting production processes. The purpose of the article is to present a location-based communication system that enhances employee engagement and supports continuous quality improvement and sustainable development in industrial environments.

**Keywords:** Mobile feedback platform · Interactive analysis · Real-time communication · Quality improvement · Sustainable development · Suggestion systems

## 1 Introduction

From the early stages of scientific inquiry into quality or sustainability as a distinct aspect of production companies, it was recognized that they cannot be considered in isolation. At the same time, its deep connection to the broader organizational framework became increasingly evident. Over time, it became clear that quality and sustainability are fundamentally tied to the commitment and active involvement of the entire organization. This includes the collective efforts of all employees, regardless of their specific roles or positions within the corporate hierarchy.


A. Hamrol et al. (Eds.): MANUFACTURING 2026, LNME, pp. 43–58, 2026.
https://doi.org/10.1007/978-3-032-21701-1_4

There are many recommendations, insightful strategies, beneficial concepts, and lessons learned in literature and online forums. It is likely that almost every organization has undertaken various improvement initiatives, initially encountering some mistakes, accumulating experience over time, and ultimately, through this iterative process, achieving significant progress [1–4].

The aim of individual employee suggestions regarding quality and sustainability can be multifaceted: process improvement by reducing defects and variability, shortening cycle times, eliminating downtime and errors, optimizing resource or material use, improving ergonomics by adapting tools, machines, spaces and working conditions for employees comfort, reducing energy consumption, etc. The proposed location-based token platform directly contributes to organizational ergonomics by reducing unnecessary operator movements, minimizing cognitive load associated with information searching and eliminating interruptions caused by non-contextual communication. By delivering instructions and feedback exactly at the point of work, the system reduces ergonomic waste related to motion, waiting, and overprocessing, which are key components of Lean ergonomics.

Eliminating all waste within whole organization and standardizing activities can not only improve efficiency but also create more sustainable processes that not only save resources but also minimize the carbon footprint [5].

This article aims to introduce a location-based communication system that increase employee engagement and facilitates ongoing quality enhancement in industrial settings. Considering that it is also worth asking the question of how to make this system as effective as possible. What facilitates people engagement in the improvement process?

## 2 Literature Review Regarding Employee Suggestion and Idea Management Systems

Implementing employee suggestion systems have progressed significantly as it is presented on fig. 1 - from basic suggestion boxes (with the first recorded suggestion system originated in 1721 when Shogun Yoshimune Tokugawa implemented a suggestion box at Edo Castle) by other solutions created throughout $19^{th}$ century like the one in Bristish Navy or Scottish shipbuilding firm, by japanese Kaizens to advanced digital platforms that are used nowadays. In time, there has been also formalized systems developed to support the process of collecting employees ideas, like: EIA (The Employment Idea Association), IAP2 (Internation Association for Public Participation), etc. Development of suggestion systems has been shaped by several key factors [6, 7]:

- Employee Engagement - involving staff in decision-making enhances problem-solving and can lead to up to 48% higher engagement,
- Training and Development - skills training and a culture of learning encourage meaningful contributions,
- Communication - promoting the system effectively raises awareness and boosts participation,
- Incentives - recognition and rewards motivate employees to share valuable ideas,

- Ownership and Confidentiality - clear policies protect both the employee's input and the organization's interests, outlines key milestones in the evolution of employee suggestion programs.

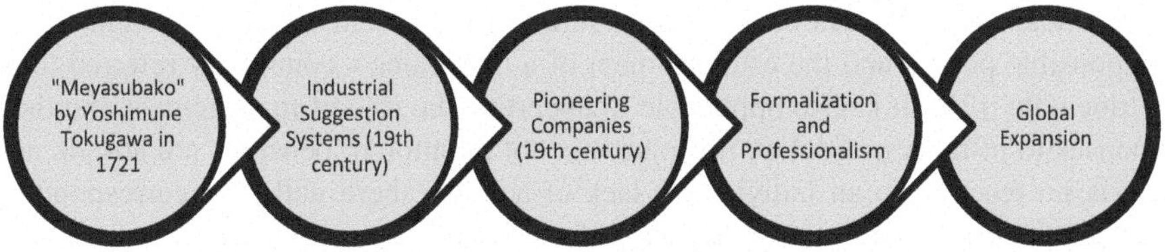


**Fig. 1** Key milestones in the evolution of employee suggestion schemes [6].

The success of any strategic initiative fundamentally relies on the commitment of upper management, whose leadership is pivotal in guiding the organization towards its objectives. Managers must set targets, indicate pathways, and formulate objectives; and, most significantly, they accomplish this by exemplifying a focused commitment to these principles and demonstrating personal involvement. Another vital aspect is the provision of ongoing feedback to employees regarding the progress of initiatives, thereby acknowledging their role as key contributors to the transformation process. It is imperative that employees are kept informed and recognized as essential stakeholders, given their role as creators and end users of the ideas proposed. Employees must trust their managers, believing that their concerns will not be disregarded [8–10].

Many organizations perceive improvement strategies as isolated events or episodic initiatives, addressing them only during designated brainstorming sessions. Conversely, a significant number of organizations have embedded improvement within their structures, adopting KAIZEN ideology, and nurturing a culture of continuous enhancement [11–15].

Transparency is an aspect appreciated by employees; comprehending the organization's current state alongside a well-articulated strategy can effectively secure employee support, as it clarifies their direction and motivations. This is often accomplished through Town Hall meetings, yet due to their infrequent nature, it is advisable to disseminate this information regularly in daily meetings [15–17].

Moreover, it is imperative to communicate explicit expectations for individuals, as well as to provide clearly defined goals to align their efforts with the organization's mission. The engagement and ongoing motivation of employees necessitate an understanding of the factors influencing their commitment. Recognizing quality and sustainability as a critical factor alongside safety and productivity helps employees understand prioritization areas for optimal performance [18–20].

Thus, it is crucial to highlight the importance of this theme and explain our strategy for enhancing competitiveness through high-quality product delivery that meets or exceeds customer expectations [21, 22].

Moreover, articulating this strategy aids employees in comprehending their roles within the broader framework of the company's objectives. By effectively communicating these goals and the rationale behind our competitive strategy, we can cultivate an environment where employees feel empowered to contribute significantly to their teams [23].

Such transparency and open communication can significantly enhance individuals' commitment to the organization. Additionally, it is vital to acknowledge and celebrate even minor achievements that advance us toward our goals [24, 25].

A fundamental element of the success of every minor activity is the assignment of a responsible person and the establishment of a timeframe - commonly referred to as the 'due date' [26]. It is also applicable to underline the role of management and their responses to instances of delay; if progress is not monitored, penalties are absent, and there is no reaction to an individual's lack of activity, there will be a corresponding absence of commitment, improvements, and a sense of responsibility.

Encouraging participation of employees can be executed in many ways. People need to have possibility to speak out without embarrassment or the threat of dissatisfaction from the superior like for example on some quality circles or brainstorming sessions. Empowering employees to give opinions and force change is building strong feeling of responsibility [27]. However, just giving them voice is sometimes not enough to encourage. Some researchers argue that implementing suggestion systems in organizations that allow employees to anonymously submit ideas promotes participation and can yield innovative solutions to improve processes [1].

On the other hand, an alternative perspective is that if such a mechanism does not guarantee anonymity, it can yield even greater benefits. Associating individuals with their ideas can foster a sense of responsibility for quality improvement, encourage ownership, and demonstrate commitment and participation among coworkers. However, this approach entails a potential risk: if such commitment is not acknowledged, valued, or utilized, individuals may become disengaged, cease contributing input, and withdraw from efforts toward continuous improvement [28].

Another approach to motivate employees involves a more explicit empowerment strategy, wherein individuals are granted real authority. For example, employees can be designated as problem-solving specialists who determine optimal solutions from a range of alternatives or be given the autonomy to stop production when irregularities are identified. Additionally, the formation of cross-functional teams enables interdisciplinary problem-solving through collaboration among experts from various departments. Importantly, it is essential to include shop-floor operators in these teams, as their involvement is critical. No stakeholders are likely to demonstrate greater engagement or commitment to process improvements than the operators themselves, who spend the most time directly interacting with production systems [29–31].

The practice of celebrating success is contingent upon manager's insight. Managers should possess the ability to recognize when an individual or team deserves acknowledgment and to express this appreciation in an authentic and spontaneous manner. The perception of sincerity in such recognition is paramount; employees need to feel that expressions of gratitude are genuinely reflective of their contributions to the organization. Some argue that formalizing procedures regarding recognition contradicts the naturality of appreciation. However, many individuals prefer clarity regarding the circumstances and criteria under which they will be recognized and rewarded.

There are a variety of incentive mechanisms that can be supplied to appreciate accomplished work:

- Monetary incentives (bonuses, commendations, gifts),
- Non-monetary incentives (certificates, opportunities for professional development).

According to Lewiński research more than 50% of respondents, which means the most popular form of award for suggestions was the financial reason [32].

Modern IT solutions already available on the market allow for the combined use of many of the above-mentioned methods of achieving employee motivation. Their basic and primary function is to collect employees' ideas. Off-shelf solutions offer also possibility of interaction between employees by ranking the ideas, sharing own experience and also offering other solutions [33–35]. These systems not only support the generation and exchange of ideas but also facilitate the structured evaluation and implementation of employee initiatives. At this stage, a process for assessing and implementing employee initiatives using a suggestion system was developed. Starting with Issues reported by employee in the system through to Assessment of whether the action brought improvement in the last step (Fig. 2).

**Fig. 2** The process of evaluating and implementing employee initiatives using the suggestion system (own work).

The system might be available for every employee by logging onto a website. Each reported issue should be ranked and enable employee to gain points. Of course, if the solution is also proposed, number of acquired points should be much higher. In some cases, it can be possible to solve the problem rapidly by operator himself/herself. Most probably the case then was also of not a real, importance, but even then, such initiative and self-solving should be highly appreciated.

Based on the experience of implementing such a ready-made application for the suggestion system available on the market, it can be stated that its usefulness can be surprising. It seems that organizations are open to changes, that they are introducing Kaizen, that employees have the opportunity to report improvements.

However, this system offers additional benefits which motivate employees - a clear remuneration system and the ability to track their own idea.

The number of ideas reported at the beginning of the implementation can be alarming. As it turns out, this system is ultimately a box for complaints and grievances. Aspects reported to leaders so far have been forgotten and not implemented. Some employees, on the other hand, were afraid that their ideas would be presented to management by someone else as their own. No one returned to this employee with feedback on whether the comment was correct or the idea was good. There will definitely be some feedback in the application. And definitely a reward if the idea turns into an improvement.

# 3 Results

## 3.1 Platform Presentation

The block diagram below (Fig. 3) illustrates the architecture and key functionalities of the proposed communication platform, designed to support employee engagement, idea management, and seamless information flow between shop floor operators and management. The platform is web-based and aims to enhance communication between management and employees, following the principles previously outlined. The choice of web technology is justified by its ease of integration into existing IT infrastructure (network, computers, tablets, smartphones) - installation efforts are minimal and limited to the notification app for mobile tablet devices. All other access is done directly through a web browser.

The server is integrated into the local network and runs an application that supports two types of users:

1) The management team, acting as content creators, accesses the application to create a "token." This token represents a web page that can include information such as a suggestion, a proposed workflow, or a reward directive. The content can be in the form of text, images, or even video - the application's input interface allows media files to be uploaded (within a predefined size limit) and displayed on the page. In addition, the page includes a feedback form where users who access the page can submit their responses. During the token creation process, the location where the token will apply and the area affected by it - where notifications can be received - are also defined. Furthermore, specific target users can be assigned to each token, identifying the employees for whom the content is intended.
2) Employees from various departments within the institution - acting as notification recipients and feedback contributors - have implicit access to the application through

a notification mechanism. When they are logged into the application and present in the area where a token has been placed, they receive a notification alerting them of the token's presence. With a simple click, they can access the page containing the management's suggestion. Depending on how the page is configured, their feedback can be submitted either anonymously (public access) or with their identity (signed). Additionally, based on the page settings, employee submissions can be made visible to all visitors of the page, to logged-in target users, or exclusively to the page's initiator.

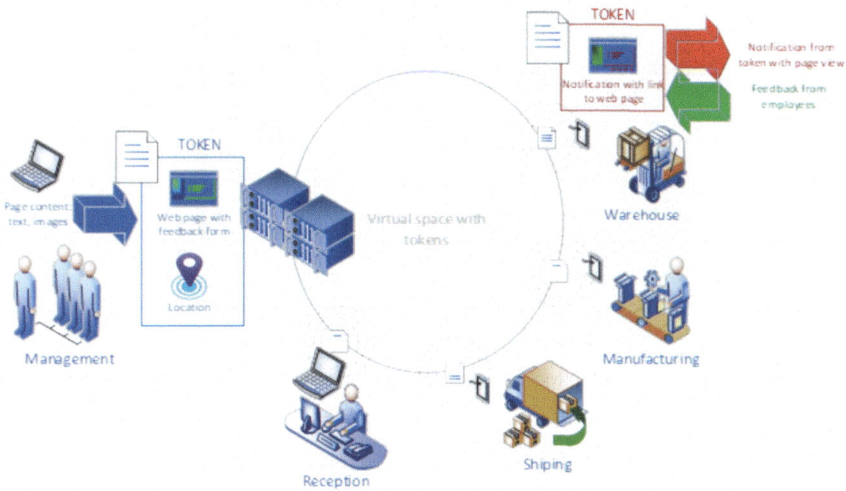


**Fig. 3** Platform block diagram (own study).

The token is therefore a virtual entity stored on the server and defined by a specific location. On the mobile devices used by employees in various sectors (such as the warehouse, production area, delivery zone, etc.), high-precision location - a standard feature on modern mobile devices - are enabled, along with communication through the notification app linked to the server. When a target user enters the token's notification area, their device sends its location to the server, triggering a notification alert.

The user can then access the token's page and submit feedback. This feedback can take the form of a quick response via radio buttons ("yes", "no" - single choice), checkboxes (multiple selections), or a text entry - all designed to allow fast and efficient feedback collection, facilitating both reaction gathering and idea submission from employees.

The localization feature on the management side works by uploading a scaled floor plan and placing the token's location directly on it, as illustrated in the Fig. 4.

In this way, the token's position is defined relative to the coordinates (0,0), which represent the location of the manager setting up the tokens (the origin coordinates can be adjusted as needed). Additionally, the area surrounding the token where notifications can take effect is also specified.

Multiple tokens can be managed simultaneously, and a color-coding system allows for visual feedback tracking on the map (e.g., red indicates feedback received).

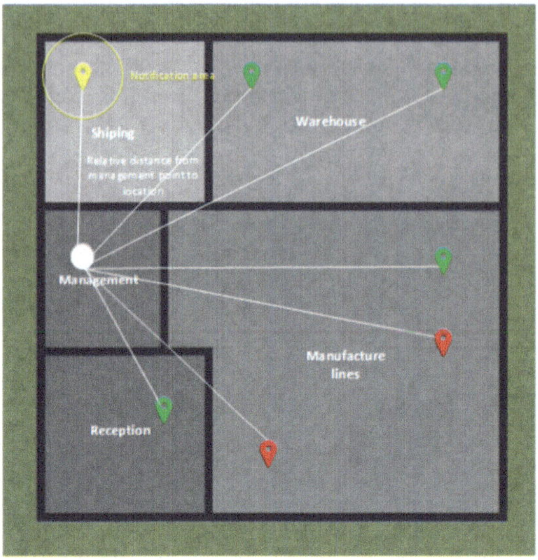


**Fig. 4** Placement of the location and notification area for a token (own study).

All responses are centralized and stored, and a dashboard engine processes them to generate statistics based on response analysis.

### 3.2 Suggestions Recorded

An illustrative example of a Polish factory operating within the metal industry gives an example of introduction of such system. Specializing in the production of steel components for the automotive, railway, mining, and agricultural sectors, the facility employs approximately 800 people, with around half directly engaged in production activities. Following its acquisition by a large international corporation, the factory retained its existing infrastructure, workforce, machinery, and production lines. Since the takeover, the company has been systematically implementing innovative tools aimed at enhancing employee involvement, strengthening quality orientation, and driving continuous improvement in both processes and products.

In the example enterprise implementation of the system was firstly performed on a singular production line as a pilot. There was no formal solution for suggestion/idea handling implemented in the organization before. Things were reported by employees only orally or by emails. Most of them were not followed or forgotten, therefore no statistics are known from that time.

Solution chosen by company was a software application offered by a local company customized slightly to organization's specific needs. After starting up a new tool an impressive number of 282 suggestions from this one pilot channel has been recorded within the initial three-day period. This involved a workforce comprising 36 individuals: operators functioning within a three-shift system, quality inspectors, engineers, team leaders, others. Consequently, the mean number of reported notifications per individual was approximately 7.3, which is a substantial figure, however such result was also anticipated during the initiation phase. Potential explanations for this phenomenon may include the likelihood that employees have documented issues that have persisted over an extended duration without resolution from management. An additional rationale could be the pursuit of incentives. Even though some issues were known to employees, they refrained from reporting them in a conventional manner, opting instead to await the system's introduction for the accumulation of points. During next two weeks the average number of notifications per employee declined to 3.8.

After one month of operation, the statistical data revealed the following: the cumulative total of all improvement suggestions submitted to the system reached 512, resulting in an average of 14 notifications per employee. However, a significant proportion of these notifications were dismissed by management for various justifiable reasons – 409 rejections with only 103 records approved for further processing. Majority of the records that were rejected were recurrent from other notifications. In such situations, the notification of the issue that was registered as first in the system was ranked. Nevertheless, it is also likely That notifications reported afterwards may consist valuable methodologies for addressing the problem.

It is not uncommon for only a limited number of individuals to exhibit a high level of enthusiasm regarding innovative solutions such as this suggestion system, thereby actively engaging in the reporting of improvements. Furthermore, there is a substantial group of employees, slightly exceeding the number of those who are enthusiastic, who do not submit any suggestions for improvement. This phenomenon may be attributable to various factors, including:

- Inadequate training,
- Resistance to the adoption of new technologies (age, language proficiency, etc.),
- Unclarity in instructions,
- Inability to utilize the application on personal devices.

It is imperative to acknowledge these reasons, as these employees may possess valuable insights yet lack the means to communicate them effectively. Figure 5 illustrates employees' activity in submitting improvement ideas through the suggestion system during its first month of operation. The figure shows the level of employee participation and reflects early engagement with the system as staff began contributing ideas for process and performance improvements.

In following months, a gradual decline in engagement with the system and the generation of new entries could be observed (Fig. 6)

It is highly probable that the primary contributing factors were:

- the previously noted substantial number of rejected proposals, mostly being caused by the redundancy of the reported notifications. As the quantity of records in the system escalated, the duration required to analyze whether a particular idea had already been documented also increased.

**Fig. 5** Employees activity to report improvement ideas in the suggestion system in first month of operating (own study).

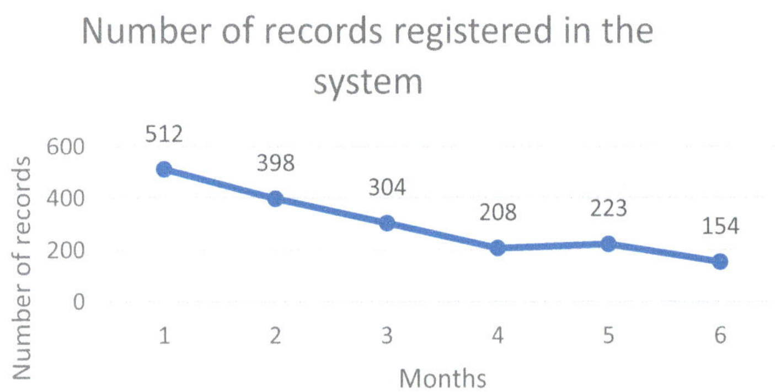


**Fig. 6** Number of records registered in the system in months (own study).

- extended application processing times – time from registering an idea to its consideration (decision to implement or reject it). Managers, overwhelmed by the number of records within the system, found themselves unable to conduct ongoing analyses. The average processing duration lengthened progressively with each passing week, resulting in a diminishment of employees' initial enthusiasm (Fig. 7).

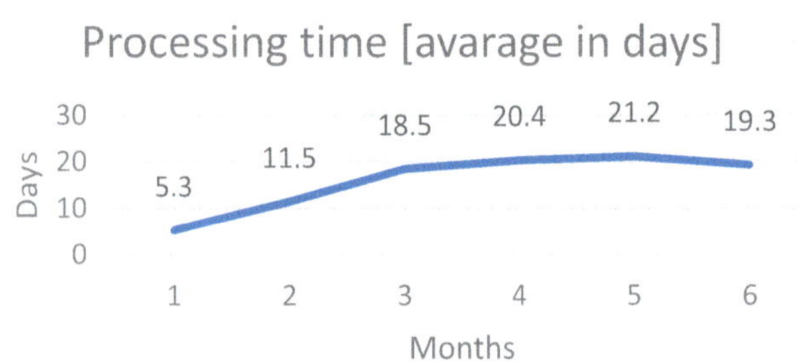


**Fig. 7** Processing time (own study).

An example of proposed and implemented suggestion is illustrated on Fig. 8

**Fig. 8** Example of improvement processed by suggestion system (own study).

The total processing time of the record described in the example presented in Fig. 8 was 19 days, which can be considered satisfactory given the level of investment required. The longer processing time was also influenced by the fact that the issue ultimately involved more than a single channel.

Based on this example, it can be concluded that these methods are effective. However, it is essential to acknowledge that not all employees may respond positively to engagement initiatives. Some may resist change due to fear of the unknown or a lack of confidence in their abilities, which highlights the need for tailored approaches that address diverse employee needs.

There is a clear consensus across the different methods examined: to encourage shop-floor operators to initiate and communicate improvement ideas, organizations must prioritize communication and trust, demonstrate exemplary leadership, and allocate resources to training and cross-functional collaboration. To support these principles effectively in practice, dedicated technological tools can play a crucial role. One such solution is the implementation of a digital platform specifically designed to enhance interaction between management and employees.

## 4 Discussion of Results

Table 1 below presents the results obtained from the pilot study.

**Table 1** Results presentation for pilot testing (own study).

| | |
|---|---|
| Evaluation time | 6 months (180 days) |
| Users | 36 |
| Suggestions | 1788 |
| Accepted/Rejected Rate | ¼ |
| Average response time (for each suggestion) | 15 days |

For each suggestion placed in a specific location, a response time was ranged from 5 to 24 days (number of working days per week and number of working days per month). However, the time in which responses to suggestions (tokens) were given was shorter – this indicates an interest of the target group in the first days in which the suggestion is published and then a decrease in interest. It is possible that the decrease in reactions to a token to 0 before the time expires is also caused by the number of users – probably an increase would lead to an increase in the reaction time. However, it was found that a duration of 24 days (number of working days per month) is too long. On the other hand, a record with a short duration risks not being accessed by all users. The average time of 15 days is, from our point of view, recommended life cycle of a token.

Based on experience of implementation suggestion system in the company there are numerous doubts and problems worth considering and for sure are points for further discussion:

(1) Time spent on reviewing the records - managers who usually are evaluators of the ideas need to spend a lot time assessing the records - by using the centralized system for collecting employee responses, the effort required for communication and feedback processing is significantly reduced [36–38].

(2) if employees use the app on their mobile phones while many companies forbid to use them on production areas as they bring disturbance to operators - the proposed solution encourages the use of mobile devices (smartphones and tablets) without impacting employee performance - it enables quick responses or the simple submission of suggestions or ideas [39]. The implementation of mobile technology in manufacturing environments presents several challenges, including operator distraction, security concerns, and work-life balance issues. Distractions can be minimized through purpose-built apps tailored to specific user needs. Security risks may be addressed by using shared device pools and conducting compliance assessments. To maintain a healthy work-life balance, companies should establish clear policies governing mobile device usage [40–42].

(3) Ranking the ideas by other employees can be more involved with personal sympathies rather than real assessment of the issue – by localizing the token and, if necessary, restricting access to it, the system enables the selection of a targeted group of employees involved in the specific process, based on their individual competencies [43].

(4) Number of rejected ideas can be discouraging for employees – the presented platform establishes a framework that demonstrates to employees that management is open

to bottom-up suggestions. Through long-term use, it fosters best practices among both employees and the management team, promoting a culture of receptiveness to individual concerns and ideas [44].

(5) Risk of "loop" - one employee reports an issue of for example "uncomfortable place of storing documents, too far away from the device". After change another worker reports the same thing about new place claiming it was better before - the public display of reactions can, on a small scale, create a discussion forum where the final solution is reached through consensus among all parties involved. The implementation of a suggestion system in a manufacturing setting reveals several practical challenges. Managers often spend significant time reviewing input, but a centralized platform reduces this burden. Mobile device use, while often restricted on production lines, can be optimized through purpose-built apps that address security and work-life balance concerns. To ensure objective evaluation of ideas, access can be limited to relevant personnel. The system also helps mitigate discouragement from idea rejection by promoting a culture of openness and feedback. Lastly, it enables transparent discussions to resolve conflicting feedback through consensus-building.

(6) Quality of submitted ideas – how to make rules or templates in the system that would filter submitted ideas.

(7) Location-based solution proposed in the system may inhibit employees by being recognized and potentially monitored. This exposes certain workplace irregularities, which can lead to excessive attention from management. Therefore, location-based aspect of the solution brings risks and ethical -aspects that should be mitigate. Users may be afraid of privacy implications, as they agree to use their private mobile devices, which may concern private data security, risk of surveillance. This may impact people engagement, which is why it needs to be clearly communicated, analyzed and assessed.

## 5 Conclusions and Prospects for Future Research

The article achieves its stated objective by presenting an innovative location-based communication platform designed to enhance employee engagement while simultaneously streamlining quality improvement processes. In addition, it identifies key success factors and methods for increasing employee involvement in improvement activities, drawing on evidence from a real-life implementation case study.

The proposed platform demonstrates clear advantages over traditional suggestion systems, such as physical suggestion boxes or paper-based forms. Its distinct features include:

- Context-aware, location-based feedback - instead of relying on passive suggestion mechanisms, the platform uses tokens assigned to specific physical locations (e.g., workstations). Employees receive notifications only when they are present in relevant areas, which improves the accuracy and timeliness of feedback while reducing informational noise.
- Bidirectional and personalized communication - managers can send targeted messages, suggestions, or inquiries to selected employee groups, while employees can respond directly through the application, either anonymously or with their identity disclosed.

– Minimal disruption and ease of integration - the platform is web-based and accessible through a browser, without the need to install software on workstations. Only a lightweight notification application is required on mobile devices (e.g., tablets). As a result, system operation does not interfere with production processes, unlike traditional approaches that are often time-consuming.
– Instant and measurable feedback - employee responses are collected and analyzed in real time. The system provides management dashboards with intuitive visualizations, such as color-coded factory maps, enabling faster and more informed decision-making.
– Personalization and targeted evaluation - tokens can be assigned to specific employees, and notification zones can be precisely defined. This reduces irrelevant or inaccurate feedback and allows suggestions to be assessed by competent personnel, mitigating biases commonly observed in open-ranking systems.

Despite these advantages, the platform constitutes a subject for further research. Future studies should focus on employees' ability to use mobile devices across different organizational levels, the automation of initial suggestion classification (e.g., through artificial intelligence), the prioritization and clustering of similar suggestions, as well as efficient data processing and archiving mechanisms. Overall, the presented system illustrates how digital technologies can be effectively integrated into everyday industrial operations, delivering tangible value for both organizations and their workforce.

## References


1. Galeazzo, A., Furlan, A., Vinelli, A.: The role of employees' participation and managers' authority on continuous improvement and performance. Int. J. Oper. Prod. Manag. **41**(13), 34–64 (2021)
2. Rowe, S., Sikes, S.: Lessons learned: Taking it to the next level. In: PMI® Global Congress 2006, North America, Seattle (2006).
3. McIntyre, S.G.: Learning lessons for organizational learning, process improvement, and innovation. In: Utilizing Evidence-Based Lessons Learned for Enhanced Organizational Innovation and Change, pp. 24–40. IGI Global (2015)
4. Romano, P., Danese, P., Beraldin, A.: Employee involvement for continuous improvement and production repetitiveness: A contingency perspective for achieving organisational outcomes. Prod. Plan. Control. **33**(4), 323–339 (2020)
5. Misztal, A., Ratajszczak, K.: Possibilities of using contemporary quality management methods and tools for sustainable development of the organization. Sustainability. **17**, 617 (2025). https://doi.org/10.3390/su17020617
6. Angyal, V.: The historical evolution of employee idea management: A comprehensive review. In: International Scientific Conference on the Occasion of the Hungarian Science Festival (2024).
7. Misztal, A., Butlewski, M., Belu, I., Ionescu, L.: Creating involvement of production workers by reliable technical maintenance. In: Proceedings of the 2014 International Conference on Production Research – Regional Conference Africa, Europe and the Middle East, pp. 331–336 (2014).
8. CrosQ: 10 ways to boost employee involvement and buy-in for your quality improvement programme. https://qfactor.crosq.org/2019/11/01/10-ways-to-boost-employee-involvement-and-buy-in-for-your-quality-improvement-programme/. Accessed 20 May 2025.

9. Morgan, D., Rachid, Z.: Employee involvement, organizational change and trust in management. Int. J. Hum. Resour. Manag. **14**(1), 55–75 (2003)
10. LaMarsh Global: The importance of employee involvement in change management. https://lamarsh.com/the-importance-of-employee-involvement-in-change-management/. Accessed 20 May 2025.
11. Saleem, M., Khan, N., Hameed, S., Abbas, C.M.: An analysis of relationship between total quality management and Kaizen. Life Sci. J. **9**(3), 31–40 (2012)
12. Mendez, J., Vila-Alonso, M.: Three-dimensional sustainability of Kaizen. TQM J. **30**(4), 391–408 (2018)
13. Singh, J., Singh, H.: Kaizen philosophy: A review of literature. IUP J. Oper. Manag. **8**(2) (2009)
14. Continuous Improvement Institute: Continuous improvement plan using Kaizen. https://ccitracc.com/blog/continuous-improvement-plan-using-kaizen/. Accessed 20 May 2025.
15. Stranzl, J., Ruppel, C., Einwiller, S.: Examining the role of transparent organizational communication for employees' job engagement and disengagement during the COVID-19 pandemic in Austria. J. Int. Crisis Risk Commun, Res. **4**(2), 271–308 (2021)
16. Men, L., O'Neil, J., Ewing, M.: Examining the effects of internal social media usage on employee engagement. Public Relat. Rev. **46**(2), 101880 (2020)
17. Gibson, N., King, A.: The promise and peril of constituent interactions: The determinants of town hall participation among members of Congress. Legis. Stud. Q. **49**(4), 745–772 (2024)
18. Schroeder, J., Fishbach, A.: How to motivate yourself and others? Intended and unintended consequences. Res. Organ. Behav. **35**, 123–141 (2015)
19. Cheng, T., Zhou, H.: Same-rank versus multiple-rank employee recognition: The effect on subsequent motivation of unrecognized employees. J. Manag. Account. Res. **37**(1), 1–19 (2025)
20. Boikanyo, D., Naidoo, M.: The influence of organisational support, advancement, meaningfulness and psychological safety on employee engagement in a petrochemical organisation. WSEAS Trans. Bus. Econ. **20**, 1100–1109 (2023)
21. Sanchez-Lizarraga, M., Limon-Romero, J., Tlapa, D., Baez-Lopez, Y., Puerta-Sierra, L., Maciel-Monteon, M.: Enablers and barriers for a quality management system implementation in Mexico: An exploratory analysis. In: Trends in Industrial Engineering Applications to Manufacturing Process (2021).
22. Helgesen, Ø.: Are loyal customers profitable? Customer satisfaction, customer (action) loyalty and customer profitability at the individual level. J. Mark. Manag. **22**, 245–266 (2006)
23. Project Management Institute: A Guide to the Project Management Body of Knowledge. Project Management Institute, Newtown Square, PA (2008)
24. Hadziahmetovic, N., Salihovic, N.: The role of transparent communication and leadership in employee engagement. Int. J. Acad. Res. Econ. Manag. Sci. **11**(2) (2022)
25. Nadkarni, A., Levy-Carrick, N., Kroll, D.S., Gitlin, D., Silbersweig, D.: Communication and Transparency as a Means to Strengthening Workplace Culture During COVID-19. NAM Perspectives (2021)
26. Szydło, J., Szpilko, D., Glińska, E., Kobylińska, U., Rollnik-Sadowska, E., Ryciuk, U.: Theoretical and Practical Aspects of Business Activity. Starting a business. Publishing House of Bialystok University of Technology (2022) (in Polish)
27. Modise, J.: The impacts of employee workplace empowerment, effective commitment and performance: An organizational systematic review. Int. J. Innov. Sci. Res. Technol. **8**(7), 25–30 (2023)
28. Sutcliffe, K., Christianson, M., Ashford, S.: Peaking up and speaking out: The leadership dynamics of voice in organizations. In: Voice and Silence in Organizations, pp. 175–202 (2009).

29. Dhillon, G.: The power of empowerment. J. Inf. Syst. Secur. **20**(3), 143–145 (2024)
30. Monje-Amor, A.: Empowerment. In: Elgar Encyclopedia of Organizational Psychology (2024).
31. Zhou, Q., Li, Q., Gong, S.: How job autonomy promotes employees' sustainable development? A moderated mediation model. Sustainability (2019).
32. Lewiński, A.: Employee suggestion program as a way to involve employees in improving business processes. Bridge Knowl. **18–28** (2016) (in Polish)
33. Kagan, R., Krzos, G.: Economic efficiency of employee suggestion systems in manufacturing enterprises. IBIMA Business Review.
34. Schweinitz, R.: Training and qualification: Team development at a flexible packaging factory. In: Handbook of Human Resources Management (2016).
35. Vel, V., Park, I., Liu, J.: The effect of enterprise crowdsourcing systems on employees' innovative behavior and job performance. In: Proceedings of the Annual Hawaii International Conference on System Sciences (2018).
36. Tummala, J.: Six strategic steps to make systems integration seamless. Plant Eng. (2022)
37. Valcik, N., Sabharwal, M., Benavides, T.: The functional areas utilizing HRIS in a public organization. Management for Professionals (2021).
38. Yusuf, S., Adekoya, O.: Trends in contemporary record management. In: Handbook of Research on Information and Records Management in the Fourth Industrial Revolution (2021).
39. Alfredsson, F., Brönmark, J., Dahlstedt, P., Larson, M., Vartiainen, E.: Always in touch: Adapting mobile devices to industrial applications. ABB Rev., 46–51 (2012)
40. Schönig, S., Ermer, A., Silva, A., Reger, A.: Process monitoring and control with smart services – An implementation approach. Prod. Manag. (2017).
41. Agasi, O.: Encapsulating mobile security. Comput. Fraud Secur. (6), 10–12 (2015)
42. Marett, K., Xiao, S., Kim, S.: Security compliance and work-issued mobile devices: Out of sight, out of mind? IseB. **21**(4), 913–925 (2023)
43. Marion, S., Worthen, M., Evans, C.: How Systems of Assessments Aligned with Competency-Based Education can Support Equity. Aurora Institute and Center for Assessment, Vienna (2020)
44. Gerlach, S., Brem, A.: Idea management revisited: A review of the literature and guide for implementation. Int. J. Innov. Stud. **1**(2), 144–161 (2017)

# Automation of Quality Data Management in a Food Industry Company

Ewa Dostatni[1], Anna Dudkowiak[1(✉)], Anna Kończal[1], and Nataša Náprstková[2]

[1] Poznan University of Technology, Pl. Marii Skłodowskiej-Curie 5, 60-965 Poznan, Poland
anna.dudkowiak@put.poznan.pl

[2] University of Jan Evangelista in Ústí nad Labem, Pasteurova 3334/7, 400 01 Usti nad Labem, Czech Republic

**Abstract.** Modern organizations operate in a rapidly changing environment, where the growing amount of data and the need for its rapid processing necessitate the use of advanced IT tools. Data analysis and process optimization have become key elements of effective business management. The automation of data management processes often involves the use of various IT tools, including integrated IT systems. However, these are expensive and often require a lengthy pre-implementation analysis process. A temporary alternative are tools that are flexible and can be adapted to the current situation in the company e.g. Excel and Access. The main objective of this work is to automate the management of quality data for the storage sample archiving process in a food industry company using Excel. The article shows that automation has improved the quality of work and efficiency and saved time that can be spent on controlling current production processes.

**Keywords:** Process Automation · Process Analysis · SWOT Analysis · Food Industry · MS Excel

## 1 Introduction

Nowadays, organizations operate in a fast-changing environment where the growing amount of data and the need for quick analysis make it necessary to use advanced IT tools. Data processing and process optimization have become key elements of effective business management.

Often, companies can identify processes that take longer to complete due to their complexity. That is why these areas are most often analyzed in terms of potential optimization. In the company under analysis, such an area is the process of archiving storage samples. Archiving storage trials in food sector companies is a fundamental element of quality assurance systems, playing a key role in monitoring product stability and compliance with declared parameters. This process enables systematic quality assessments to be carried out, including analyses of the shelf life of the finished product at each stage of its shelf life. Storage tests include the examination of physical, chemical, microbiological, and organoleptic changes that may occur during the storage of a product in various


A. Hamrol et al. (Eds.): MANUFACTURING 2026, LNME, pp. 59–75, 2026.
https://doi.org/10.1007/978-3-032-21701-1_5

environmental conditions. The aim of the tests is to optimize storage conditions, determine the actual shelf life, and identify potential quality risks. The results obtained from archiving samples are used both to improve technological processes and in preventive measures to minimize the risk of consumer complaints. In addition, archived samples play an important role in complaint investigation processes, providing an objective reference point for analyzing potential non-conformities. The archiving of storage samples also plays a role in the development of new recipes and the improvement of production processes. It is an indispensable tool in assessing the impact of the technologies and raw materials used on the quality of the final product. This approach ensures food safety and durability, which is crucial in building consumer confidence and the company's competitive position in the market. The traditional method of archiving samples, without the support of advanced IT programs, is feasible, but it involves many difficulties. In a production plant with a range of over 400 finished products, managing such a complex process becomes a challenge.

The main objective of this paper is to automate the management of quality data for the archiving of storage samples in a food industry company. Optimization concepts were analyzed, and it was decided to implement a label printer and an improved MS Excel spreadsheet with a barcode reader. The article shows that automation has improved the quality of work and efficiency, and saved time that can be spent on monitoring current production processes.

The article consists of six chapters. The second section describes the literature review. The third chapter presents a research methodology and an analysis of the company, its current status, along with a calculation of the duration of operations at the analyzed workstation. The fourth chapter presents the optimization concept and describes its implementation. The fifth chapter discusses the results, and the sixth chapter formulates the final conclusions.

## 2 Literature Review

Process optimization is an integral part of operational management. George [1] presents the Lean Six Sigma concept, combining waste elimination (Lean) with a statistical approach (Six Sigma), which enables the identification of sources of inefficiency and their elimination through continuous improvement. Del Castillo [2] describes a statistical approach to optimization based on design of experiments (DoE), analysis of variance, and response function modeling, which is particularly useful in manufacturing environments. Dumas et al. [3] highlight the importance of BPM (Business Process Management) systems for modeling, simulating, and monitoring processes in real time, while Duczmal [4] emphasizes the role of mathematical and probabilistic methods, such as linear programming and simulations. Excel is also used in process optimization, e.g., in assembly planning [5] or the matching of precision components taking into account dimensional tolerances [6].

One of the basic solutions used in companies to increase operational efficiency is the implementation of a ready-made IT system, either integrated or supporting specific activities in the company. However, such solutions are usually expensive and require the involvement of many resources to be implemented effectively. In addition, there are publicly available tools that facilitate everyday work with their functionality. These are tools used in many companies, such as Microsoft Office programs, including MS Excel.

For many years, MS Excel has served as the primary analytical tool in organizations due to its simplicity, wide range of mathematical and statistical functions, and ability to integrate with other systems. Kosny [7] emphasizes that a significant advantage of Excel is the ability to program in VBA (Visual Basic for Applications), which allows for the creation of automated data processing procedures. The literature points to both the role of Excel in data management and its limitations in processing qualitative information. Bissett [8] highlights Excel's potential for automated data analysis, emphasizing the importance of rapid prototyping tools. Alexander and Kusleika [9] as well as Clough [10] point out that VBA functionality allows for the automation of repetitive tasks in spreadsheets, including the creation of user interfaces, reporting modules, data validation, and notification systems [11].

Extensions such as Power Query and Power Pivot significantly increase Excel's capabilities in processing large and complex data. Power Query enables intuitive downloading, filtering, and transformation of data from various sources, including databases, CSV files, XML, and the Internet, without the need for advanced coding [7]. Power Pivot allows you to create relational data models that can be analyzed using DAX (Data Analysis Expressions), which enables the calculation of indicators, cumulative values, and advanced analytical measures [12]. The combination of VBA, Power Query, and Power Pivot allows you to build complex analytical environments without having to invest in expensive ERP or BI systems. Examples of practical applications can be found, among others, in the research by Brzozowska and Gola [5] on supporting assembly planning in Excel, as well as in the works of Opara [13] and Blayney and Sun [14], which show that Excel with VBA can be a low-budget tool to support preliminary analyses in big data research.

In the context of large data resources, organizations are increasingly using BI tools that support strategic decisions by integrating data from various sources, creating reports, and dynamic dashboards [15]. Excel is used in preliminary big data analyses, enabling data cleansing, creating summaries, and identifying key indicators [13, 14]. Alternatives to Excel include platforms such as Power BI and Access – Power BI allows for advanced data integration, visualizations, and strategic analysis, while Access serves as a relational database for small and medium-sized enterprises, supporting operational management. The choice of tool depends on the nature of the analysis, available resources, and the scale of data processing.

In summary, modern IT tools, especially MS Excel with Power Query, Power Pivot, and VBA extensions, support data processing and business process optimization. Their use allows for the automation of activities, improvement of data quality, and effective support for management decisions. These are complemented by BI systems and statistical methods, which, in combination with Lean and Six Sigma practices, create a comprehensive system for organizational improvement.

# 3 Materials and Methods

## 3.1 Research Methodology

The research was conducted based on a sequential approach to process analysis and optimization. The adopted methodology assumes a gradual transition through successive stages of the procedure, which enables both a reliable assessment of the current state and the formulation of justified proposals for improvements. The entire research process is presented in the form of a flowchart illustrating the logic and relationships between the individual steps (Fig. 1).

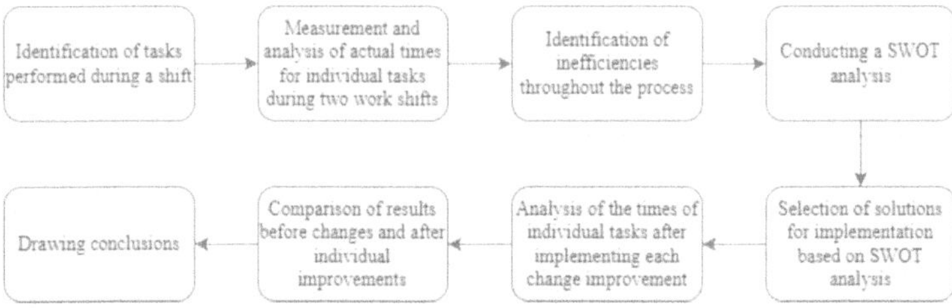


**Fig. 1.** Research methodology scheme.

In the first stage, the scope of the analyzed process was defined and its key elements and limitations were identified. Next, the data necessary to evaluate its functioning was collected through observation of the process. Next, the results obtained were analyzed to identify areas for improvement. The identified problems were the starting point for developing optimization options, which were then subjected to SWOT analysis. Improvements were implemented and their impact on process times was verified.

## 3.2 Company Description

The analyzed food industry enterprise is located in Ostrów Wielkopolski. Since its establishment, the company has specialized in the production of high-quality herring products, using raw materials obtained exclusively from certified suppliers. Its product range includes over 400 items distributed to both domestic and foreign markets. The plant has a production area of 2,733.14 $m^2$ and warehouse facilities covering 5,711.46 $m^2$, which ensures efficient organization of technological and logistical processes.

High production quality is supported by a food safety management system compliant with international standards, including the IFS (International Featured Standards) certification. The enterprise operates a comprehensive quality control system encompassing continuous process monitoring, microbiological and physicochemical testing, and the archiving of storage samples. The latter plays a key role in ensuring the durability and safety of products throughout their declared shelf life.

## 3.3 Archiving of Storage Samples

The archiving of storage samples is a systematic process designed to monitor the stability and quality of finished products during storage. Samples are collected from current production in a representative manner – covering products from the beginning, middle, and end of each production batch. This approach allows for a comprehensive assessment of product quality throughout the production cycle.

The archiving process in the company is highly structured and includes the following stages:

1. *Preparation of the production plan and determination of samples*

The laboratory technician prints the daily production schedule, marking the number and type of samples to be archived.

2. *Sampling and transport*

Line supervisors set aside appropriate product quantities in the cold storage area, from which the laboratory technician transfers them using a transport trolley. Cold chain integrity and full sample traceability are maintained.

3. *Data registration and electronic documentation*

The technician records each sample in a Microsoft Excel sheet, assigning an article number, production date, expiry date, production shift code, and machine identifier. The program enables product searches by name or customer, improving data entry efficiency.

4. *Verification and labeling*

Each sample is manually labeled with the article number, archiving date, expiry date, and sample purpose.

5. *Sample storage and monitoring*

After registration, samples are placed in designated storage zones. Depending on the product type, periodic quality assessments are performed, including sensory, microbiological, and physicochemical analyses.

6. *Documentation and result analysis*

Results are archived in Excel sheets and paper-based product cards.

This procedure constitutes an integral element of the company's quality assurance system, confirming that the product maintains its declared quality throughout storage. The process also provides valuable data for trend analysis and external certification audits.

## 3.4 Analysis of Process Time and Structure

In order to determine the actual time required for individual activities, work was observed during two shifts. A comparison of the distribution of tasks in both shifts showed that archiving samples is the dominant element of work regardless of the time of day, although in the first shift the share of this process in the total working time is slightly higher. In practice, this means that other tasks, such as physical and chemical analyses or administrative activities, are performed to a limited extent, often under time pressure. In some cases, it is also necessary to work outside standard hours, because the total working time exceeds the standard 8 h (as can be seen in Table 1), which indicates organizational overload.

**Table 1.** List of duties for shift I and II employees.

| Tasks | Working time [h] – shift I | Percentage share – shift I | Working time [h] – shift I | Percentage share – shift II |
|---|---|---|---|---|
| Physicochemical analysis | 3,0 | 24% | 0,45 | 5% |
| Archiving of finished product samples | 7,5 | 60% | 6,5 | 78% |
| Work assigned by the manager | 0,5 | 4% | 0,0 | 0% |
| Preparation of a tasting of the finished product along with a tasting report | 0,5 | 4% | 0,0 | 0% |
| Working in the system | 1,0 | 8% | 0,4 | 5% |
| Checking for bubbles at a temperature of 22 °C | 0,0 | 0% | 1,0 | 12% |
| **Total** | **12,5 h** | | **8,35 h** | |

The percentage breakdown of shift I and II shows that recording storage samples takes up the most time. Due to the need to optimize this process, the recording of storage samples has been subjected to detailed analysis. Table 2 presents the basic tasks involved in the sample archiving process, along with measurements of their duration.

The average total number of archived samples during two work shifts was 346.67, and the standard deviation was 25.39. Based on the recorded data concerning the work of laboratory employees in a given period, the average time for archiving storage samples was calculated. In addition, the amount of total working time spent on sample archiving was calculated. The results are presented in Table 3.

**Table 3.** Presentation of the average sample archiving time in hours and percentages.

| Shift | Average sample archiving time [h] | Time spent on archiving samples [%] | Time spent on other tasks [%] |
|---|---|---|---|
| I | 7,51 | 93,67 | 6,33 |

(*continued*)

**Table 3.** (*continued*)

| Shift | Average sample archiving time [h] | Time spent on archiving samples [%] | Time spent on other tasks [%] |
|---|---|---|---|
| II | 6,54 | 81,67 | 18,33 |

### 3.5 Identification of Inefficiencies

Identifying irregularities in the storage sample archiving process was one of the key stages of the analysis. Particular attention was paid to the method of record keeping in Excel, which currently serves as the main tool for managing archiving data. Based on a review of the documentation used and observations of the work, a number of irregularities and limitations were identified:

- Time-consuming work in Excel spreadsheets – database is complex, contains numerous references and hyperlinks, which causes slowdowns, program crashes, and difficulties in searching for information,
- Lack of adequate document protection – spreadsheet is used by a wide range of users who can both view and modify data what increases the risk of accidental changes, deletion of some information, or unauthorized modification,
- Manual updating of information and susceptibility to errors – all changes are made manually, which, with a large number of entries, leads to errors and inconsistencies, duplication or omission of data,
- No systematic backup system – using a single main file without a formal backup procedure poses a real risk of losing all data in the event of hardware failure or file corruption,
- Limited performance with large data sets – as the number of records and formulas increases, opening, saving, and editing files becomes increasingly difficult, hindering efficient information processing,
- No possibility of effective parallel work – although the file can be opened on several workstations, the practical ability to enter and save changes is limited, which reduces workflow efficiency and forces employees to wait for access,
- Manual report preparation and lack of standardization – creating reports based on spreadsheet data is time-consuming and prone to errors, and the lack of automation makes it difficult to maintain a consistent structure and comparability of reports.

The identified irregularities indicate that the current method of recording and archiving data in Excel is not keeping pace with growing organizational and quantitative requirements. Maintaining this solution in the long term may lead to a number of negative consequences: a decline in work efficiency, a reduction in data quality and consistency, an increased risk of information loss, and difficulties in responding quickly to changes in the market environment.

**Table 2.** Summary of data on the registration times of storage trials for shifts I and II (1 – Recording trials and labeling [h], 2 – Placing samples in designated locations [h], 3 – Total sample archiving time [h], 4 – Number of items [pcs.], 5 – Number of archived samples [pcs.], 6 – Total number of articles [pcs.], 7 – Total number of archived samples [pcs.]).

| No. | Shift | 1 | 2 | 3 | 4 | 5 | 6 | 7 |
|---|---|---|---|---|---|---|---|---|
| **1** | I | 5,1 | 2,05 | 7,15 | 55 | 170 | **97** | **319** |
| | II | 4,2 | 2,2 | 6,4 | 42 | 149 | | |
| **2** | I | 5,50 | 1,55 | 7,05 | 71 | 181 | **118** | **328** |
| | II | 5,05 | 1,35 | 6,4 | 47 | 147 | | |
| **3** | I | 5,5 | 2,05 | 7,55 | 64 | 200 | **119** | **356** |
| | II | 4,4 | 2,0 | 6,4 | 55 | 156 | | |
| **4** | I | 5,45 | 2,55 | 8,0 | 61 | 189 | **103** | **359** |
| | II | 5,1 | 2,34 | 7,44 | 42 | 170 | | |
| **5** | I | 5,55 | 2,47 | 8,02 | 73 | 223 | **120** | **387** |
| | II | 4,55 | 1,55 | 6,1 | 47 | 164 | | |
| **6** | I | 5,17 | 2,13 | 7,3 | 56 | 187 | **102** | **331** |
| | II | 4,38 | 2,1 | 6,48 | 46 | 144 | | |

## 4 Archiving of Storage Process Optimization

### 4.1 The Concept of Optimization

In practice, Excel spreadsheets still play an important role in quality departments, but their capabilities quickly become exhausted when the amount of data increases, greater automation is required, or several people need to work simultaneously. In this context, the implementation of dedicated quality management software becomes a natural direction for development. It is an investment that allows you to streamline processes, reduce costs, and increase the transparency of operations, which in turn strengthens customer confidence and the company's market position. However, before a company decides on a full-scale IT system, it is possible to consider intermediate or alternative solutions. These include, among others, individual spreadsheets, simple task management programs, partial outsourcing of processes to external companies, and the use of a label printer in the sample archiving process. Each of these solutions has its advantages, but also clear limitations. A SWOT analysis (Table 4) of each concept was performed to decide which solution would bring the most benefits.

**Table 4.** SWOT analysis for the concept of a new sample archiving process.

| Concept | Strengths | Weaknesses | Opportunities | Threats |
|---|---|---|---|---|
| Individual spreadsheets | • Flexibility<br>• Low initial costs<br>• Familiarity with the tool | • Lack of centralization<br>• Risk of errors<br>• Lack of integration | • Possibility of automation<br>• Development of support systems | • Increased complexity<br>• Lack of standardization |
| Simple task management programs | • Ease of use<br>• Accessibility<br>• Increased organization | • Limited functionality<br>• Lack of integration with other systems<br>• Lack of personalization | • Increased popularity of online tools<br>• Possibility of development | • Integration issues<br>• Scalability limitations |
| Outsourcing parts of processes to external companies | • Focus on key competencies<br>• Reduction of operating costs<br>• Access to specialist knowledge | • Lack of quality control<br>• Security risks<br>• Dependence on external suppliers | • Possibility to focus on innovation<br>• Possibility of scaling | • Loss of internal knowledge<br>• Quality and timeliness issues |
| Label printers | • Increased efficiency<br>• Precision<br>• Personal control over labels | • Initial cost<br>• Space required | • Integration with ERP systems<br>• Development of printing technology | • Compatibility issues<br>• Equipment damage |

A comparison of these concepts in a SWOT analysis shows that each of them can be useful in specific conditions, but their common weakness is limited scalability and difficulty in ensuring full consistency and transparency of processes. In the long term, the most promising direction remains the implementation of an IT system for quality management, or at least a gradual transition to semi-automated solutions (such as printer labeling), which may be a transitional stage before full digitization.

### 4.2 Automation of the Storage Sample Archiving Process

Due to the company's budget constraints, it is not possible to implement an IT quality management system at this time. For this reason, it was decided to implement intermediate solutions that will enable rapid improvement of the process with low financial outlay.

Firstly, a label printer was purchased prior to the implementation of full automation. This tool had already been implemented earlier, while preparatory work for the implementation of automation was underway. Even just implementing a label printer contributed to increased efficiency by automating labeling processes and ensuring high precision in sample marking. Its use enabled improved process organization and error reduction, which was a significant benefit in the short-term transition period. After the introduction of the label printer, a time analysis of the process (Table 5) of archiving storage samples was carried out to assess the impact of this optimization on the quality and efficiency of work.

**Table 5.** Average time spent archiving storage samples before and after implementing the improvement: label printers.

| Shift | Average sample archiving time [h] | | Time spent on archiving samples [%] | |
|---|---|---|---|---|
| | Before | After | Before | After |
| I | 7,51 | 5,64 | 93,67 | 70,54 |
| II | 6,54 | 4,37 | 81,67 | 54,67 |

The average total number of archived samples during two work shifts was 363, and the standard deviation was 47.79 but the results of implementing this solution showed a significant acceleration in the sample archiving process. Working time was reduced by nearly two hours on both shifts, which indicates the high effectiveness of this solution, making the label printer the most effective tool in the pre-implementation stage, pending the full introduction of a more advanced quality management system.

In the meantime, an analysis was carried out and preparations were made to automate the storage sample archiving process. The main goal of automation was to reduce the time needed to register samples by limiting manual activities. The plans included reducing manual searches for article numbers, reducing the amount of data entered manually into finished product cards, and eliminating manual label printing. In the first stage, a process diagram was developed in Bizagi Modeler, which allowed us to identify areas particularly susceptible to improvement. A fragment of the storage sample archiving process that has been automated is presented in Fig. 2. Figure 3 presents the planned changes in this area.

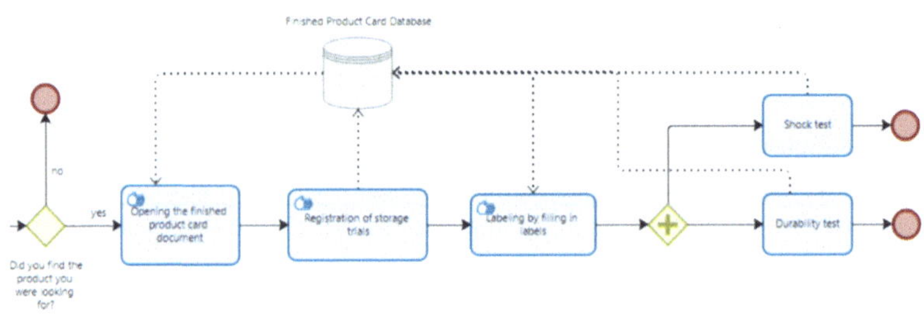


**Fig. 2.** Diagram of the storage sample archiving process (initial process).

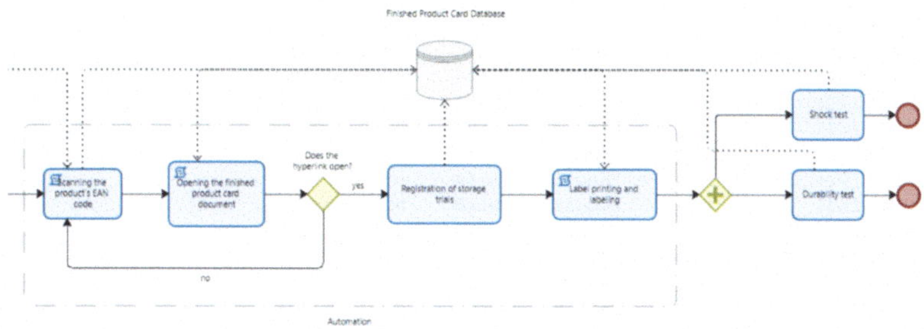


**Fig. 3.** Diagram of the storage sample archiving process after automation implementation.

In accordance with the adopted concept, automation included, among other things, scanning EAN codes using a handheld scanner. This replaced manual searching for products by name or item number and significantly reduced identification time. Another element was the automatic opening of the finished product card via a hyperlink assigned to a given item, which reduced the risk of errors. The assumptions also provided for automatic filling in of the current date after opening the document in order to reduce errors in determining expiry dates, which is particularly important in the case of complaint analyses. The last key improvement was the replacement of handwritten labels with labels printed from a label printer.

After defining the functional requirements, a list of necessary equipment was drawn up. First, a label printer and self-adhesive labels were purchased and then installed at the computer station in the laboratory. Previous experience with its use in a semi-automatic solution (described at the beginning of this chapter) confirmed the validity of the investment. The next step was to equip the workstation with an HDWR HD10 handheld barcode scanner. The use of the Plug&Play standard enabled its seamless integration with Excel – scanned codes are transferred directly to the active cell of the spreadsheet, which speeds up work and reduces the risk of errors.

The sheet intended for archiving was expanded to include a column with assigned EAN codes and a field for scanning the code in real time. The codes were entered using a scanner, using the quality control department's label catalog, which ensured data consistency.

The implementation was carried out in the Visual Basic for Applications (VBA) environment. It was assumed that the main task of the macro would be to search for the correct item in the EAN code column based on the scanned code, go to the appropriate row, and launch a hyperlink to the finished product card. For security reasons, the assignment of fixed file access paths was abandoned in order to avoid problems when transferring documents. The following columns were adopted as functional columns: column A – with EAN codes and column H – for scanning the code (Fig. 4).

| | A | B | C | D | E | F | G | H |
|---|---|---|---|---|---|---|---|---|
| 1 | EAN Code | Article Number | Article Name | Client | Packaging | Weight | | Searching for EAN codes |
| 2 | 5901510175286 | 278945 | Product 1 | Client A | box | 0,15 | | |
| 3 | 5901510175694 | 278982 | Product 2 | Client B | box | 0,15 | | |
| 4 | 5901510175245 | 279019 | Product 3 | Client c | box | 0,15 | | |
| 5 | 5901510175138 | 279019 | Product 4 | Client c | box | 0,15 | | |
| 6 | 5901510175164 | 279056 | Product 5 | Client A | box | 0,15 | | |
| 7 | 5901510175214 | 279093 | Product 6 | Client B | box | 0,15 | | |
| 8 | 5901510175467 | 279130 | Product 7 | Client B | box | 1,2 | | |
| 9 | 5901510175262 | 279167 | Product 8 | Client C | tray | 1,2 | | |
| 10 | 5901510175349 | 279204 | Product 9 | Client A | tray | 1,2 | | |
| 11 | 5901510175569 | 279241 | Product 10 | Client C | tray | 0,1 | | |
| 12 | 5901510175627 | 279278 | Product 11 | Client D | box | 0,1 | | |

**Fig. 4.** Excel program for archiving storage samples, extended with a table containing EAN codes and a field enabling their search using a scanner.

Several technical problems were encountered during implementation. The most significant one concerned opening hyperlinks – despite correct VBA code, some product cards did not launch. Analysis showed that the cause was the way file names were written: files containing an underscore (_) in their name were interpreted incorrectly. After changing the naming convention (e.g., from "278945_Product 1" to "278945 Product 1"), the problem was resolved. Another error was related to saving the file in .xlsx format, which does not support macros – this resulted in the loss of the VBA code. After changing the format to.xlsm, the problem did not occur. However, it was not possible to fully resolve the issue of automatically saving the current date in product cards opened by scanning the code – the functionality only worked when opening the hyperlink manually.

After completion of the implementation stage, the VBA project was password protected. The solution used is basic in nature – it prevents unauthorized persons from viewing and editing the code, but does not provide complete protection against advanced interference. A standard project protection procedure was used in the VBA editor, which consists of blocking access to the code and requiring a password to be entered when attempting to view it.

Functional tests were conducted in laboratory conditions, on company equipment, using actual data from the label catalog. The codes of the most frequently manufactured products were entered into the program, and their correctness was verified by quality control department employees. Four employees participated in the tests, performing their daily duties at the same time. This allowed the program's performance to be evaluated under typical load conditions.

The difficulties observed mainly concerned getting used to previous manual working methods. Suggestions were also made regarding further development of the solution, including integration with automatic label generation and printing, and automatic date filling in product cards. Due to budget constraints, the need to maintain compatibility with the existing printer infrastructure, and the additional training burden, these extensions were not included in the current implementation phase. Despite these limitations, the tests were evaluated positively and the program was considered stable enough for practical use.

A program for archiving storage samples was implemented in the laboratory of a food production plant. The solution, based on Excel and VBA macros, enables the recording, filtering, and searching of information about samples (including production date, batch number, product name, customer, and storage location). The implementation included installing the program in the production environment, preparing and migrating data from previously used catalogs, and training quality department employees.

The preparation and implementation of automation was treated as a project, so in accordance with project management methodology, a project and implementation team was established, comprising representatives of key departments: quality control, production, technical, and IT. To ensure transparency and control over the course of the work, project approval stages and a schedule were developed, covering concept preparation, program configuration, data migration, testing, and phased implementation and training. The project was implemented according to schedule, without the need for organizational changes in the laboratory's work. Key activities – configuring the spreadsheet to handle EAN codes, testing under conditions similar to real-life ones, and data validation – did not reveal any critical irregularities. As a result, the implementation was successful, and the developed solution met the functional assumptions and provides a basis for further expansion of the system in the future.

After the implementation of the automated storage sample archiving program was completed, its observation began in the conditions of everyday laboratory work. In the initial period of use, typical difficulties arose related to the staff getting used to the existing working methods. The most frequently reported issues were: continuing to archive in the traditional manner, incorrect connection of the EAN code scanner, scanning codes in the wrong field of the sheet, and maintaining paper documentation in parallel despite the new procedure being in force. The adaptation period to the automated solution lasted approximately 2–3 weeks. After this period, systematic collection of data on sample registration times and their placement in designated locations began. The average total number of archived samples during two work shifts was 382, and the standard deviation was 42.73. A summary of detailed values for shifts I and II is presented in Table 6.

**Table 6.** Summary of data on the registration times of storage trials for shift I and II after the implementation of automation (1 – Recording trials and labeling [h], 2 – Placing samples in designated locations [h], 3 – Total sample archiving time [h], 4 – Number of items [pcs.], 5 – Number of archived samples [pcs.], 6 – Total number of articles [pcs.], 7 – Total number of archived samples [pcs.]).

| No. | Shift | 1 | 2 | 3 | 4 | 5 | 6 | 7 |
|---|---|---|---|---|---|---|---|---|
| 1 | I | 1,55 | 1,46 | 3,01 | 50 | 167 | **102** | **340** |
| | II | 1,59 | 1,46 | 3,05 | 52 | 173 | | |
| 2 | I | 1,47 | 1,53 | 3,0 | 55 | 183 | **99** | **330** |
| | II | 1,25 | 1,10 | 2,35 | 44 | 147 | | |
| 3 | I | 2,06 | 2,05 | 4,11 | 57 | 190 | **124** | **413** |
| | II | 2,2 | 2,19 | 4,39 | 67 | 223 | | |

(*continued*)

**Table 6.** *(continued)*

| No. | Shift | 1 | 2 | 3 | 4 | 5 | 6 | 7 |
|---|---|---|---|---|---|---|---|---|
| 4 | I | 2,15 | 1,55 | 3,7 | 56 | 187 | **110** | **367** |
| | II | 1,49 | 1,51 | 3,0 | 54 | 180 | | |
| 5 | I | 2,02 | 2,05 | 4,07 | 59 | 197 | **121** | **403** |
| | II | 2,05 | 2,2 | 4,25 | 62 | 207 | | |
| 6 | I | 2,2 | 2,12 | 4,32 | 78 | 260 | **131** | **437** |
| | II | 1,57 | 1,44 | 3,01 | 53 | 177 | | |

## 5 Discussion

The analysis of the distribution of working time between individual tasks for both shifts, in three consecutive variants of the process organization (standard procedure, optimized process using a label printer, and fully automated sample archiving), allows for an assessment of the actual impact of the changes introduced on work efficiency.

The use of a label printer can be considered the first significant step towards streamlining the process. Already at this stage, a reduction in the time spent on archiving storage samples was noted, which translated into more time available for other tasks. However, it was only the implementation of automation, including the use of an EAN code scanner and automated sheet handling, that led to a clear balancing of responsibilities during the work shift. After the implementation of automation solutions, the amount of time spent on archiving decreased significantly, which allowed for better allocation of human resources and increased flexibility in work organization. This effect was particularly noticeable during the first shift, where the greatest reduction in the time spent on archiving was recorded. At the second shift, the improvement was primarily in the form of a more even distribution of time between archiving and other tasks, which indicates an overall improvement in work organization.

A comparison of the standard procedure, the optimized process, and the automated process (Fig. 5) also shows that optimization alone, while beneficial, is not a sufficient solution when the initial process is highly time-consuming. Reducing the total archiving time by about a quarter on the first shift and a third on the second shift after optimization alone was a clear improvement, but it still meant a significant burden on employees with repetitive tasks. It was only automation that brought about a reduction of approximately 50% in archiving time compared to the standard procedure, which significantly changed the structure of the working day (Fig. 6).

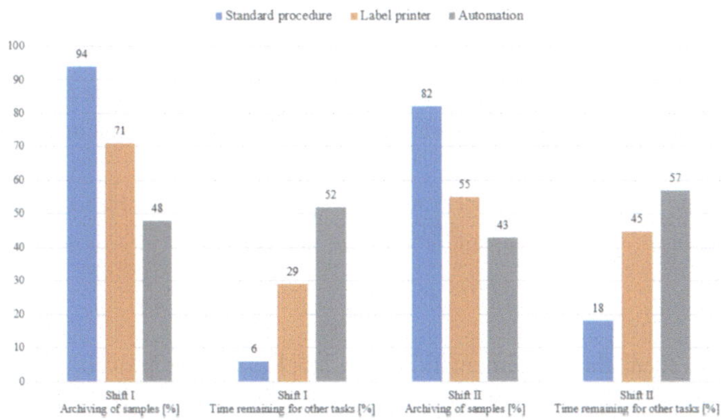


**Fig. 5.** Chart showing the total sample archiving time for shifts I and II: comparison of the standard procedure, the optimized process, and the automation of the sample archiving process.

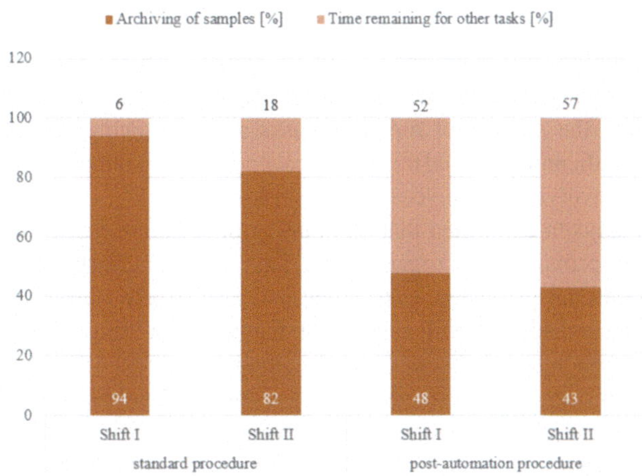


**Fig. 6.** Percentage distribution chart for shifts I and II: comparison of the standard procedure and the optimized sample archiving process.

It is noteworthy that after the label printer was introduced, the second shift achieved a higher share of time spent on other tasks compared to the first shift. This may be due to both better adaptation to new solutions and different working conditions (e.g., lighter workload, different task profile during the day). After the implementation of automation, these disparities decreased, and both shifts achieved a much more favorable distribution of time, with the effect being more pronounced in the morning shift, which is usually burdened with a greater number of operational tasks.

Accelerating work through partial automation led to changes in productivity. For each measurement, the number of archived samples was divided by the total sample archiving time, i.e., productivity was calculated. Table 7 shows the average productivity rates for both changes in the initial process, after the implementation of the label printer and after the implementation of the automated sheet.

**Table 7.** Average productivity for both shifts before and after individual improvements.

| Shift | Initial process [pcs/h] | Label Printer [pcs/h] | VBA sheets and label printer [pcs/h] |
|---|---|---|---|
| I | 25.49 | 35.46 | 53.61 |
| II | 23.76 | 37.67 | 56.24 |

As can be seen, the implementation of a label printer improved productivity by approximately 40% (for the I shift) and approximately 29% (for the II shift) compared to the initial process. The implementation of two improvements, i.e., a label printer and automated sheets, improved productivity compared to the initial process by approx. 110% (for the I shift) and approx. 136% (for the II shift).

## 6 Conclusions

As a result of implementing automation of the storage sample archiving process, a significant increase in work efficiency was noted on both shifts. The completed project, aimed at reducing the number of repetitive tasks and shortening the time needed to perform them, significantly reduced the time spent on archiving samples, which increased the possibility of involving employees in other tasks.

The measures taken confirmed that eliminating or reducing routine operations is a key element in improving work organization. This analysis clearly shows that modifications to operational processes, even when using relatively simple tools such as a label printer and a spreadsheet with automation features, can bring measurable benefits in terms of work efficiency and better use of human resources. A systematic, step-by-step approach to process improvement, ranging from simple optimization measures to automation, can be an effective strategy for improving the efficiency of laboratories in manufacturing companies. Importantly, the implemented solutions did not require the use of advanced IT systems, which increases the possibility of adapting similar measures in other organizations with limited financial and infrastructure resources.

The solutions implemented proved effective, but the results obtained also indicate further potential for optimization. The introduction of a more integrated IT system dedicated to archiving and data management processes could further improve cooperation between departments, improve task coordination, and make even better use of staff working time.

In summary, the automation of the storage sample archiving process has contributed to increased operational efficiency and more rational use of human resources. The results obtained justify continuing efforts towards further digitization and automation of auxiliary processes in the company.

**Acknowledgements.** The work presented in the paper has been co-financed under grant no. **0613/SBAD/4940.**

## References

1. George, M.L.: Lean Six Sigma: combining Six Sigma quality with lean speed. McGraw-Hill Education Ltd., New York (2002)
2. Castillo, E.: Process optimization. Springer, New York (2007). https://doi.org/10.1007/978-0-387-71435-6
3. Dumas, M., Mendling, J., Reijers, H.A., La Rosa, M.: Business Process Management (in Polish). Polish Scientific Publishers PWN, Warsaw (2022)
4. Duczmal, M.: Optimization Methods in Management (in Polish). Higher School of Management and Administration in Opole, Opole (2014)
5. Brzozowska, J., Gola, A.: Computer aided assembly planning using MS Excel software—A case study. Appl. Comput. Sci. **17**, 70–89 (2021). https://doi.org/10.23743/ACS-2021-14
6. Kupriyanov, O., Lamnauer, N.: Improvement of the Assembling Technology for Precision Joints Using Dimensional Information. Lecture Notes in Mechanical Engineering, pp. 52–60. Springer (2021). https://doi.org/10.1007/978-3-030-68014-5_6
7. Kosny, M.: Basics of Programming in MS Excel (in Polish) (2014).
8. Bissett, B.: Automated Data Analysis Using Excel. Wiley, Hoboken (2007)
9. Alexander, M., Kusleika, R.: Excel 2016 PL: Programming in VBA (in Polish). Helion, Gliwice (2016)
10. Clough, D.E.: Improve your Spreadsheet Productivity with VBA. Apress, Berkeley (2017)
11. Kralewski, D.: Creating a data model using Power Pivot. In: Buchnowska, D. (ed.) Management Dashboards in Business Data Analysis and Presentation Using MS Excel 2016 (in Polish), pp. 105–118. Faculty of Management, University of Gdańsk, Gdańsk (2016)
12. Opara, U.J.: Automating Forecasting Using Excel VBA (2023).
13. Blayney, P.J., Sun, Z.: Using Excel and Excel VBA for preliminary analysis in big data research. In: Advances in Data Analytics, pp. 110–136 (2019). https://doi.org/10.4018/978-1-5225-7277-0.CH007
14. Zabroń, M., Wołoszyn, J.: Business intelligence tools dedicated to big data analytics. Teach. Comput. Sci. **18**, 185–193 (2023). https://doi.org/10.15584/DI.2023.18.15

# Balancing Costs, Productivity and Well-Being: A Multi-dimensional Assessment of Maintaining Safety Measures in Organizations

Marcin Butlewski, Fryderyk Wachowiak(✉), and Krzysztof Zawodny

Poznan Univeristy of Technology, Rychlewskiego 2, 60-965 Poznan, Poland
Fryderyk.wachowiak@put.poznan.pl

**Abstract.** The COVID-19 pandemic compelled organizations to implement a wide range of safety and organizational measures to protect employees and maintain operational continuity.

**Objective:** This paper evaluates the long-term usefulness of these solutions in the context of CSR, with emphasis on employee well-being, the financial burden, and their impact on employee.

**Methods:** A multi-criteria evaluation method was used to assess 27 organizational safety measures implemented during pandemic. The measures were evaluated across three dimensions: costs, employee performance, and user comfort. Data sources included a literature review, and interviews with OHS professionals and workers.

**Findings:** The analysis revealed a variation in the perceived effectiveness and sustainability of the measures. These solutions were cost-effective, well accepted by employees, and supported stable performance. In contrast, measures such as mandatory PPE usage or the closure of social areas received low scores due to user discomfort or minimal perceived impact on employee. Significant findings are the discrepancy between cost-effectiveness and employee acceptance.

**Recommendations:** Presented model offers an evidence-based approach to evaluating which measures are worth retaining, refining, or discarding in the long term.

**Contribution:** Organizations that integrate such assessments into their management practices will be better positioned to balance economic realities with employee well-being.

**Keywords:** COVID-19 · Workplace Safety · Corporate Social Responsibility · Change Management

## 1 Introduction

Worker involvement significantly impacts the success rate of sustainable change initiatives in the workplace by fostering employee engagement, enhancing trust, and promoting adaptive and proactive behaviors. Employee engagement is crucial for successful change management, as it ensures that employees are committed to and actively participate in transformation efforts, thereby improving change outcomes and durability [1].


A. Hamrol et al. (Eds.): MANUFACTURING 2026, LNME, pp. 76–91, 2026.
https://doi.org/10.1007/978-3-032-21701-1_6

High involvement management strategies, which integrate employee relations into organizational planning, have been shown to enhance productivity and efficiency, leading to improved organizational effectiveness [2]. Trust, a critical factor for successful change, is bolstered through employee involvement, which supports innovation and co-creation, enabling organizations to leverage collective knowledge for effective change implementation [3]. Furthermore, employee involvement is essential in creating a healthy workplace, as it allows organizations to identify high-leverage practices that benefit both employees and the organization [4]. In the context of organizational adaptability, employee involvement through empowerment, participation, and teamwork has been shown to facilitate transformation, especially when combined with humble leadership [5]. Additionally, job-crafting and work engagement promote change-oriented behaviors, which are vital for sustainable organizational change [6]. Despite these benefits, challenges such as managerial reluctance and insufficient emphasis on worker involvement in policy frameworks can hinder the diffusion of these practices [7, 8]. However, integrating green human resource management with change management can align HR policies with sustainability goals, increasing employee engagement and fostering a culture of eco-consciousness, thereby driving positive environmental and organizational outcomes [9]. Overall, worker involvement is a pivotal element in the success of sustainable change initiatives, as it enhances engagement, trust, and adaptability, ultimately leading to more effective and enduring organizational transformations. Corporate Social Responsibility strongly relates to employee safety [10], including mental safety [11] and wellbeing [12]. During the SARS-CoV-2 pandemic, many companies have attempted to implement various solutions to protect employees while ensuring business continuity. In the post-pandemic era, some solutions have become a permanent part of the organization's existing practices. Some are still used within organizations as part of their more human oriented practices, and thus understood as CSR strategies, especially these which were based on digitalization of processes [13]. Developed methods, models, and strategies are being stored within organizations in case the pandemic situation repeats, but the SARS-CoV-2 pandemic, starting from 2020, served as a global turning point that put organizational change processes under unprecedented pressure. It created a real-time laboratory for testing how organizations of different types and sizes respond to rapid, large-scale disruptions. The crisis did not only challenge existing safety protocols - it fundamentally questioned how agile, responsible, and people-centered organizational change could (or should) be implemented under extreme conditions. In the face of this health emergency, organizations were compelled to rethink the very foundations of how they function. Change was no longer a strategic choice or gradual process - it became a necessity for survival. As such, the pandemic revealed much about the capacity of organizations to absorb shock, to act responsibly under uncertainty, and to implement change in a way that balances operational needs with employee well-being.

One of the most visible domains of organizational adaptation was the management of workplace safety, particularly through the introduction of bio-safety measures. Practices that were once considered optional or marginal became central to organizational functioning. These included reorganizing work models - such as implementing remote work widely into industry [14] and flexible workplace solutions [15], as well as more technical responses: mandatory mask usage, hand disinfection, use of gloves, temperature checks, mass testing, and the regular sanitization of shared spaces and equipment [16]. These interventions were not isolated technical decisions; they formed part of a broader organizational response to change - one that involved coordination, communication, and often, a renegotiation of the relationship between employer and employee. New spatial arrangements and revised workflows contributed to what has been widely referred to as the "new normal" [17], marking not just a temporary adjustment but a shift in expectations and norms. The question that arises from these events is not simply whether specific safety measures should or could be maintained, but how organizations approach change more broadly with workers consent [18] in the post-pandemic and CSR context. What lessons have been learned about readiness, responsiveness, and responsibility in managing change? Has the experience led to more inclusive, participatory, and flexible change processes - or have decisions remained largely top - down and reactive? Another interesting question is to what extent employees will accept various types of solutions in the future, given their experience with the implementation of previous solutions, and thus also the level of acceptance of restrictions on certain types of employee freedoms.

The pandemic thus provides a valuable lens for analysing organizational change: it highlights the tension between short - term crisis response and long - term strategy, between technical solutions and human factors, and between compliance - driven actions and values - based responsibility. As organizations face future challenges – whether related to health, climate, technology, or social expectations – their ability to draw from these experiences will shape not only their resilience, but also their legitimacy and trustworthiness in the eyes of employees, as a responsible employer who cares for employee mental health and well-being [19]. In this article, the authors present the results of research assessing the validity of the continuous (after the pandemic) use of biosafety measures in the integrated safety management of the enterprise in the context of employee safety as an element of Corporate Social Responsibility.

Building on this background, the present study addresses the following research questions: (1) Which COVID-19 biosafety measures implemented in organizations are perceived as the most beneficial when evaluated jointly in terms of financial cost, employee performance, and user comfort? (2) Which of these measures are worth maintaining as part of long-term CSR and occupational safety strategies in a post-pandemic context? To answer these questions, we developed and applied a multi-criteria assessment model that integrates financial, organizational, and human-centred dimensions.

## 2 Material and Methods

During the COVID-19 pandemic, a multi-criteria evaluation method was employed. This approach integrated three key dimensions: user perception, financial cost, and impact on employee performance. As part of the preliminary research – which consisted of interviews with representatives of occupational health and safety departments, a literature review, and an analysis of Internet sources (blogs and specialist OHS forums) – 26 solutions used by enterprises to counteract the SARS-CoV-2 pandemic were selected. These solutions were consistent with management staff perceptions of risk reduction in the work environment [16]. Each solution was rated based on (1) comfort and convenience as reported by employees, (2) implementation and maintenance costs calculated by organizations, and (3) perceived influence on operational performance. This composite method enabled the development of a practical hierarchy of solutions, balancing user experience with economic and organizational effectiveness, and provided a foundation for strategic decision-making in post-pandemic Corporate Social Responsibility (CSR) practices. The selected solutions are:

1. Establishment of an "anti-crisis team".
2. Additional personal protective equipment: masks, gloves, face shields.
3. Washing and disinfecting rooms and surfaces.
4. Disinfectant dispensers.
5. Disinfectant atomizers.
6. Information on the principles of healthy living.
7. Maintaining social distance.
8. Creating partition between positions.
9. Videoconferencing.
10. Limiting the number of people staying in one place.
11. Stretching work shifts.
12. Turning off the coffee, drinks and snacks machine.
13. Closing of social rooms in the part concerning the preparation of drinks and meals.
14. Giving up on external food delivery.
15. Cancellation or reduction of business trips and delegations.
16. Transition to e - learning training.
17. Employee training to strengthen the sense of social responsibility.
18. Reducing employee accumulation during a shift.
19. Remote work.
20. Introducing a break between shifts.
21. Removing paper towel from toilet.
22. Ban smoking or limit the number of people in smoking rooms.
23. Encouraging employees to self - monitor their health and stay home if they experience any signs of illness.
24. Open door policy wherever possible.
25. Strict epidemiological rigor of external visitors to the company.
26. Limiting the number of outside visitors to the company.

In order to create a hierarchy of solutions used by companies, it was decided to relate them to three criteria – financial, impact on employee performance and convenience of use for users.

The financial criterion was based on a self - calculation of the costs of purchase, operation, maintenance, withdrawal and disposal of assets. 30 working days were assumed as the cost calculated per month. One-time cost - to simplify the calculation - was assumed as a cost incurred in one month.

Particular attention should be paid to solutions that are cost-free, as they are often relatively easy to implement in a company. By contrast, solutions associated with fixed costs – such as the use of masks and gloves or hiring additional staff due to increased cleaning and disinfection – can be more problematic. The cost analysis presented in Table 1 is based on the costs incurred by a furniture manufacturer.

**Table 1.** Cost analysis of used solutions. Example from furniture company. Source: own study.

| Remedy | Purchase cost | Implementation cost | Cost of maintaining | Disposal/Removal/Withdrawal Cost |
|---|---|---|---|---|
| "Anti-cristis" team | --- | the cost of updating the scope of employee duties - salary supplement | working time of 5 employees 2 h per week 1300 $\frac{PLN}{month}$ working time of 5 employees 4 h per week 2600 *PLN* | --- |
| Gloves | 0.30PLN 5 pairs per day monthly 45 $\frac{PLN}{employee}$ | --- | --- | 0.032PLN / month 4.8 $\frac{PLN}{employee}$ |
| Masks | 0.20PLN 2 pcs per day monthly 60 $\frac{PLN}{employee}$ | --- | --- | 0.004PLN /month 1,2 $\frac{PLN}{employee}$ |
| Face shields | 13,0PLN 2 pcs/month 26 $\frac{PLN}{employee}$ | --- | --- | 0.035PLN / month 0.07 $\frac{PLN}{em.ployee}$ |
| Increased frequency of cleaning and disinfection | disinfection fogger 1000 $\frac{PLN}{pcs}$ | Employee recruitment process 500PLN | additional employee salary 1700PLN/month disinfectant liquid 60PLN/month | --- |
| Disinfectant dispensers | automatic station 200 $\frac{PLN}{pcs}$ | one - time refill with disinfectant 6 $\frac{PLN}{pcs}$ | disinfectant monthly 15 $\frac{PLN}{pcs}$ | --- |
| Disinfectant atomizers at work stations | atomizer 2.30 $\frac{PLN}{pcs}$ | one - time refill with disinfectant 1.20 $\frac{PLN}{pcs}$ | disinfectant monthly 1,20 $\frac{PLN}{pcs}$ | --- |
| Poster campaign on health during the epidemic | poster printing 1.50 $\frac{PLN}{pcs}$ | --- | --- | --- |

(*continued*)

**Table 1.** (*continued*)

| Remedy | Purchase cost | Implementation cost | Cost of maintaining | Disposal/Removal/Withdrawal Cost |
|---|---|---|---|---|
| Maintaining social distance | --- | spreading out the positions – 16 h of work of two Maintenance employees – 660PLN | --- | return to the pre - epidemic situation - 16 h of work for two Maintenance employees – 660PLN |
| Creating partitions between positions | plexiglass plate 200 $\frac{PLN}{worskateg}$ | 20 min of work by two Maintenance employees - 13.75 $\frac{PLN}{workstage}$ | --- | 10 min of work by two Maintenance employees - 13.75 $\frac{PLN}{workstage}$ |
| Videoconferencing | --- | --- | --- | --- |
| Limiting the number of people in one place | --- | --- | --- | --- |
| Stretching work shifts | --- | --- | increase in electricity consumption | --- |
| Turning off the drinks/snacks machine | --- | --- | reduced cost related to electricity consumption | --- |
| Closure of social rooms | --- | --- | --- | --- |
| No outside meals | --- | --- | --- | --- |
| No/limited business trips and delegations | --- | --- | reduced expenses related to the organization of business trips and delegations (no per diem allowances) | --- |
| E - learning training | --- | --- | cost of periodic e-learning training 35 $\frac{PLN}{workstage}$ vs. standard training 45 $\frac{PLN}{workstage}$ | --- |

(*continued*)

**Table 1.** *(continued)*

| Remedy | Purchase cost | Implementation cost | Cost of maintaining | Disposal/Removal/Withdrawal Cost |
|---|---|---|---|---|
| Employee training of CSR | --- | --- | production downtime (15 min) $3300 \frac{\left\{PLN\right.}{month}$ | --- |
| The rule is that an employee arriving earlier replaces an employee on an earlier shift | --- | --- | overtime - +50% vs. standard hourly rate | --- |
| Remote work | --- | training on, among others, safety 500PLN | --- | --- |
| Introducing a break between shifts | --- | --- | in the case of a continuous operation system – the production time needs to be shortened by 2 h | --- |
| Removing paper towels from toilets | purchase of automatic dryers $150\ \frac{PLN}{pcs}$ | 10 min of work by two DUR employees $-\ 6.90\ \frac{PLN}{pcs}$ | increase in electric energy consumption purchase of paper towels 25 $\frac{PLN}{day}$ | --- |
| No smoking | --- | removal of the smoking room $200\ \frac{PLN}{pcs}$ | --- | --- |
| Employees who have symptoms will stay at home | --- | --- | overtime replacing +50% vs. standard hourly rate | --- |
| Open Door Policy | --- | --- | heating costs (winter) costs related to running air conditioning (summer) | --- |

*(continued)*

**Table 1.** (*continued*)

| Remedy | Purchase cost | Implementation cost | Cost of maintaining | Disposal/Removal/Withdrawal Cost |
|---|---|---|---|---|
| Strict epidemiological rigor for outsiders | non - contact thermometer 50PLN | --- | --- | --- |
| Limiting the number of people from outside | --- | --- | --- | --- |

The developed survey allowed the users invited to the study to evaluate the solutions in terms of their impact on work comfort and the convenience of using the solutions. However, it is important to emphasize the complexity of the phenomenon and the connections between the individual criteria. Many times, despite a good assessment in one of the criteria, the solution also has a low assessment in another criterion. Examples include cost - free solutions that negatively affect efficiency or solutions that positively affect efficiency but are associated with high costs. Therefore, the assessment should be approached in a holistic manner, covering all the criteria.

## 3 Results

To assess user comfort, a survey was conducted. In total, 205 employees evaluated the solutions in terms of comfort and ease of use. The following scale was proposed to evaluate comfort: 1 - definitely uncomfortable (very much interferes with the performance of duties, negatively affects work comfort); 2 - rather uncomfortable; 3 - indifferent; 4 - rather comfortable; 5 - definitely comfortable (very helpful in the performance of duties, positively affects work comfort). Additionally, if the employee had no contact with the assessed protective equipment, it was possible to select the "not applicable" option (Table 2).

After conducting a survey among the respondents and collecting and processing the data for each of the analyzed solutions, the average of the collected responses was calculated. Figure 1 presents the five solutions that received the highest average scores in terms of user comfort according to the respondents.

In order to create a model of classification of the best solutions that bring the most benefits, it was decided to create a hierarchy. In each of the assessed categories, a solution could receive from 1 to 5 points. The final score for each solution was calculated as the product of the three criteria (financial, performance, comfort).

Rules for allocating points for the financial criterion:

In the case of the financial criterion, points were allocated depending on the costs defined for each solution. The point system for solutions is as follows:

- no costs – 5 points,
- low one - time costs (up to PLN 5,000) – 4 points,
- medium one - time costs (PLN 5,000 - 10,000) or very low fixed costs (less than PLN 1,000 per month) - 3 points,

**Table 2.** Comfort of use. Source: own study.

| No. | Solution | Rate | | | | | | Average |
|---|---|---|---|---|---|---|---|---|
| | | 1 | 2 | 3 | 4 | 5 | n/a | |
| 1. | Establishing of an "Anti-crisis team" | | | | | | | |
| 2. | Mask | 101 | 73 | 14 | 0 | 0 | 17 | 1,54 |
| | Gloves | 23 | 52 | 34 | 0 | 0 | 96 | 2,10 |
| | Face-shields | 74 | 84 | 44 | 3 | 0 | 0 | 1,88 |
| 3. | Washing and disinfecting rooms and surfaces | 1 | 4 | 74 | 82 | 44 | 0 | 3,80 |
| 4. | Disinfection dispensers | 0 | 4 | 57 | 74 | 60 | 10 | 3,97 |
| 5. | Equipping workstations with disinfection atomizers | 0 | 3 | 69 | 78 | 55 | 0 | 3,90 |
| 6. | Information on the principles of healthy and safe | 0 | 12 | 100 | 59 | 24 | 10 | 3,49 |
| 7. | Maintain social distance | 51 | 77 | 67 | 9 | 1 | 0 | 2,18 |
| 8. | Use of shields/partitions between stations | 62 | 53 | 30 | 1 | 1 | 58 | 1,82 |
| 9. | Videoconferencing | 0 | 0 | 3 | 10 | 17 | 175 | 4,47 |
| 10. | Limiting the number of people staying in one place, e.g. in the canteen, cloakroom | 52 | 68 | 55 | 28 | 2 | 0 | 2,32 |
| 11. | Stretching work shifts | 0 | 7 | 45 | 58 | 53 | 42 | 3,96 |
| 12. | Turning off the coffee, drinks and snacks machine | 61 | 60 | 42 | 0 | 0 | 42 | 1,88 |
| 13. | Closing of social rooms in the part concerning the preparation of drinks and meals (kitchens) | 19 | 6 | 7 | 0 | 0 | 173 | 1,63 |
| 14. | Giving up on external food delivery | 28 | 13 | 1 | 0 | 0 | 163 | 1,36 |
| 15. | Cancellation or reduction of the number of business trips and delegations | 7 | 6 | 9 | 1 | 0 | 182 | 2,17 |
| 16. | Transition to e-learning training | 0 | 1 | 14 | 25 | 11 | 154 | 3,90 |
| 17. | Employee training to strengthen the sense of social responsibility | 15 | 29 | 66 | 46 | 17 | 32 | 3,12 |
| 18. | Reducing employee accumulation during a shift | 12 | 13 | 15 | 10 | 1 | 154 | 2,51 |
| 19. | Remote work (for people not directly involved in employee service) | 42 | 80 | 56 | 15 | 2 | 10 | 2,26 |
| | Remote work (for people directly involved in employee service) | 8 | 20 | 118 | 32 | 17 | 10 | 3,15 |
| 20. | Introducing a break between shifts | 16 | 5 | 4 | 0 | 0 | 180 | 1,52 |
| 21. | Removing paper towels from toilets | 48 | 64 | 45 | 6 | 0 | 42 | 2,06 |
| 22. | Ban or limiting the number of people in smoking rooms | 87 | 24 | 60 | 14 | 10 | 10 | 2,16 |

*(continued)*

**Table 2.** (*continued*)

| No. | Solution | Rate | | | | | | Average |
|---|---|---|---|---|---|---|---|---|
| | | 1 | 2 | 3 | 4 | 5 | n/a | |
| 23. | The need to stay at home if any signs of infection occur | 6 | 20 | 60 | 71 | 48 | 0 | 3,66 |
| 24. | Open door policy wherever possible | 28 | 61 | 71 | 28 | 7 | 10 | 2,62 |
| 25. | Strict epidemiological rigor of external visitors to the company | | | | | | | |
| 26. | Limiting the number of outside visitors | 3 | 16 | 124 | 39 | 13 | 10 | 3,22 |

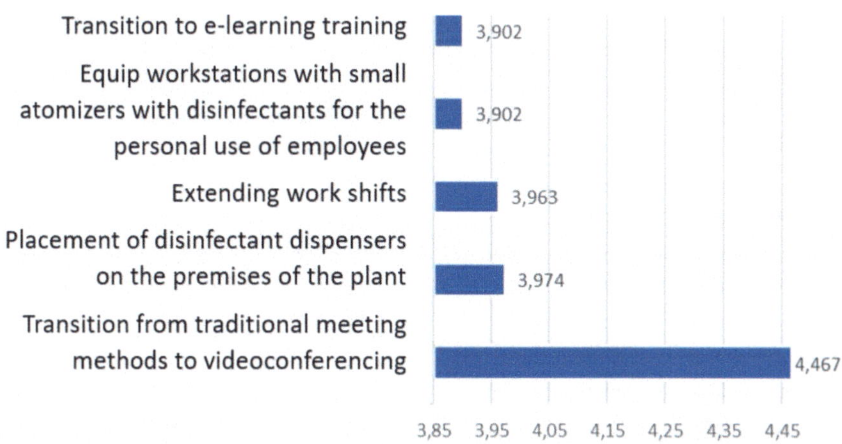


**Fig. 1.** Best solutions acc. to the respondents. Source own study.

- high one - off costs (over PLN 10,000) or low fixed costs (PLN 1,000–5,000 per month) – 2 points,
- medium and high fixed costs (over PLN 10,000 per month) – 1 point.

Principles for allocating points for the criterion of impact on employee performance:

In the case of the criterion of impact on employee performance, points were allocated depending on the rating assigned to each solution. The point system for solutions is as follows:

- positive impact on employee performance – 5 points,
- no impact on employee performance – 3 points,
- negative impact on employee productivity – 1 point.

Principles for allocating points for the convenience and impact on users' criteria:

In the case of the convenience and impact on user comfort criteria, points were allocated depending on the average ratings assigned to each solution in the survey.

Formula (1) shows the method of calculating the average points.

$$D = \frac{\sum_{i=1}^{5} A_i * O}{n} \quad (1)$$

where:

D - number of points allocated

Ai - number of ratings issued

O - rating of respondents (from 1 to 5)

n - number of respondents

Table 3 presents a summary of the scores for all solutions and their assigned place in the classification of the best safety measures implemented in enterprises during the epidemiological threat.

**Table 3.** Hierarchy of implemented solutions. Source: own study.

| Solution | Financial factor | Productivity factor | Comfort factor | Score |
|---|---|---|---|---|
| Switching from traditional meeting methods to video conferencing | 5 | 5 | 4.4 | 111.6 |
| Remote working (of people not directly involved in employee services) | 4 | 5 | 3.1 | 63.0 |
| Switching to e-learning training | 5 | 3 | 3.9 | 58.5 |
| Reducing the number of people in one place i.e. canteen, cloakroom | 5 | 5 | 2.3 | 57.9 |
| Remote working (of persons directly related to employee services) | 4 | 5 | 2.2 | 45.1 |
| Introduction of an absolute ban on smoking related mainly to the accumulation of employees in the smoking room | 4 | 5 | 2.1 | 43.1 |
| Promoting information on healthy living (more frequent hand washing, disinfection, distance, information on coronavirus) | 4 | 3 | 3.4 | 41.8 |
| Creation of a company "crisis unit" in charge of implementing solutions related to the epidemiological situation | 2 | 5 | 4.0 | 40.0 |
| Strict epidemiological rigor for external visitors to the company | 4 | 3 | 3.2 | 38.6 |
| Deployment of disinfectant dispensers in the workplace | 3 | 3 | 3.9 | 35.7 |

*(continued)*

**Table 3.** (*continued*)

| Solution | Financial factor | Productivity factor | Comfort factor | Score |
|---|---|---|---|---|
| Equipping workplaces with small atomisers with disinfectants for the personal use of employees at the workplace | 3 | 3 | 3.9 | 35.1 |
| Switching off the coffee and snack vending machines | 5 | 3 | 1.8 | 28.2 |
| Closing the social areas in the area of beverage and meal preparation (kitchens) | 5 | 3 | 1.6 | 24.3 |
| Staggering of working shifts (i.e. one department 6:00–14:00, another 6:05–14:05, etc.) | 2 | 3 | 3.9 | 23.7 |
| Introducing a break between shifts, i.e. 6 am to 2 pm, 3 pm to 11 pm (for non - continuous businesses) | 5 | 3 | 1.5 | 22.8 |
| More frequent washing/disinfecting of premises and surfaces (especially used by more than one employee), introduction of disinfection registers | 2 | 3 | 3.8 | 22.8 |
| No more provision of meals from outside | 5 | 3 | 1.3 | 20.3 |
| Training and talks to employees to introduce them to the purpose of the measures implemented in the establishment and to foster a sense of social responsibility | 2 | 3 | 3.1 | 18.7 |
| Removal of paper towels from toilets, leaving only automatic hand dryers for drying hands | 3 | 3 | 2.0 | 18.5 |
| Open door principle wherever possible | 2 | 3 | 2.6 | 15.6 |
| Introduction of a rule that an employee who arrives early for his/her shift exchanges an employee who can leave early to go home | 2 | 3 | 2.5 | 15.0 |
| Abandoning or reducing the number of business trips and business missions | 5 | 1 | 2.1 | 10.8 |
| Limiting the number of outside visitors to the enterprise | 5 | 1 | 2.0 | 10.0 |

(*continued*)

**Table 3.** (*continued*)

| Solution | Financial factor | Productivity factor | Comfort factor | Score |
|---|---|---|---|---|
| Use of guards/partitions between posts | 1 | 5 | 1.8 | 9.0 |
| Ensuring that employees maintain a social distance (i.e. by having more distance between workstations) | 4 | 1 | 2.1 | 8.7 |
| Encouraging employees to self - monitor their health and to stay at home if any signs are present | 2 | 1 | 3.6 | 7.3 |
| Equipping workers with additional protective equipment (i.e. gloves) | 1 | 1 | 2.1 | 2.1 |
| Equipping workers with shields | 1 | 1 | 1.8 | 1.8 |
| Equipping workers with masks | 1 | 1 | 1.5 | 1.5 |
| Introducing a break between shifts i.e. 6 am to 2 pm, 3 pm to 11 pm (for continuous work enterprises) | 1 | 1 | 1.2 | 1.5 |

## 4 Conclusion

The COVID-19 epidemic, caused by the SARS-CoV-2 virus, has significantly affected the perception of workplace safety, which is a key element of Corporate Social Responsibility. In many cases, employees no longer feel fully safe in the work environment, which directly affects their mental health, especially when their work involves frequent exposure to pathogens. Daily exposure to pathogens transmitted by both co - workers and external visitors to the company forced management to take action to reduce risk. The most frequently used tool for this purpose was various types of personal and collective protective equipment. Various prohibitions, orders, and restrictions implemented through procedures and instructions were also frequently implemented. Thanks to these actions, many companies achieved their intended goal of maintaining the highest possible employee performance. This was possible thanks to intermediate goals, which were primarily ensuring employee safety and minimizing infections within the company to the greatest extent possible. Over time, it became clear that some of the solutions were very well received in the workplace, others so much so that discussions began about the need for their permanent implementation within companies. Supporters of new solutions often emphasized the benefits of some of them, including those related to well-being i.e. flexible working time or home - office work, hoping they would continue after the pandemic ends. However, employees themselves often emphasized that some of the implemented measures negatively impacted the work process and, consequently, the company.

An analysis based on three criteria revealed significant differences in the perception of the implemented solutions. This is best illustrated by the difference in scores, with the best solution receiving 111,6 points and the worst receiving 1,5 points. Seven safety measures were considered viable solutions to be maintained after the pandemic ends:

- establishing a "crisis team" within the company responsible for implementing solutions related to the epidemiological situation.
- switching from traditional meeting methods to videoconferencing.
- limiting the number of people present in one location, e.g., in the cafeteria or locker room for all situations.
- switching to e-learning trainings.
- remote work (for those directly involved in employee service).
- remote work (for those not directly involved in employee service).
- introducing a strict smoking ban, primarily due to employee gathering in the smoking area.

In light of the above, it can be concluded that the presented model for evaluating adaptive safety practices and change management proved to be accurate. The example of costs for furniture manufacturer of maintaining solutions implemented during the pandemic do not outweigh the benefits of their continued use, therefore, it is justified to maintain some employee health and safety measures after the pandemic ends, as an element of the CSR strategy in the context of employee safety and health.

## 5 Discussion

The conducted research demonstrates that the COVID-19 pandemic has not only disrupted standard organizational practices but also created a unique testing ground for assessing the adaptability and sustainability of different organizational change practices like for example implementing workplace safety measures. The comprehensive evaluation model proposed in this study based on financial cost, user comfort, and employee performance, has proven to be an effective tool in identifying which interventions offer long-term value and align with Corporate Social Responsibility (CSR) goals and participation of users in assessing solutions. Looking ahead, further waves of organizational change will be driven not only by future health crises but also by the rapid diffusion of artificial intelligence and profound demographic shifts, which will force enterprises to redesign work systems - contexts in which the proposed computational evaluation model could again support evidence-based decision-making [20].

One of the most significant findings is the discrepancy between cost-effectiveness and employee acceptance. For instance, while some of the cheapest solutions (e.g., mandatory mask-wearing) scored poorly in terms of user comfort, others with minimal financial investment (e.g., videoconferencing or remote work) were highly rated across all criteria. This suggests that financial affordability alone does not determine the long-term viability of a safety measure; employee experience and operational efficiency are equally crucial for successful integration into post-pandemic organizational strategies.

This study nevertheless has certain limitations. The research was based on a specific sample of employees and interviews with occupational health and safety specialists from the furniture industry, which may limit the generalizability of the findings to other sectors and organizational contexts. The evaluation model considered only three dimensions, which were, financial costs, user comfort, and employee performance, while other relevant factors such as long-term psychological effects, organizational culture, or environmental sustainability were not included. Potential future losses combined with risk factor could be also an interesting aspect of research [21]. Moreover, the data were collected during and shortly after the COVID-19 pandemic, which may have influenced respondents' perceptions and acceptance of specific measures. Finally, the cost analysis was obtained from the furniture industry, restricting the transferability of results to sectors with different cost structures. Future studies should therefore broaden the scope, include multiple industries, adopt a longitudinal perspective, and integrate additional criteria to strengthen the robustness and applicability of the proposed model.

Despite these limitations the results underscore the importance of involving employees in the process of change implementation. Measures that were better received tended to be those that respected worker autonomy, supported task performance, or enhanced flexibility such as remote work, staggered shifts, or reducing interpersonal density in shared spaces. These findings support broader CSR narratives emphasizing empathy, inclusion, and long-term well-being rather than short-term compliance. The analysis also reveals that certain reactive solutions such as closing social areas or banning external food delivery were often poorly rated despite being easy to implement. This points to the need for a more thoughtful, human-centered approach to safety interventions, where social dynamics and workplace culture are considered alongside epidemiological rigor.

From a practical CSR perspective, the findings suggest that organizations should prioritise measures that simultaneously support health protection, maintain or enhance performance, and respect employee comfort. In particular, flexible work arrangements and digital solutions such as remote work, videoconferencing, and e-learning appear to offer a favourable balance between costs and benefits. By contrast, purely restrictive measures that undermine social interaction without clear performance gains should be applied cautiously and, where possible, replaced with more human-centred alternatives. Finally, the developed ranking system allows organizations to prioritize actions not only based on immediate health concerns, but also with a view toward resilience, strategic continuity, and social responsibility. It also provides a framework for responding to future crises, enabling decision-makers to select interventions that are both sustainable and aligned with organizational values.

## References

1. Vaja, J.R., Bhatt, J.S.: Employee engagement and organizational change: A study on the role of employee engagement in successful change management initiatives. Int. J. Manag. Dev. Stud. **7**(3), 52–61 (2018). https://doi.org/10.53983/ijmds.v7i3.007
2. Gollan, P.J., Davis, E.M.: Employee involvement and organisational change: The diffusion of high involvement management in Australian workplaces. In: Models of Employee Participation in a Changing Global Environment: Diversity and Interaction. Routledge (2018). https://doi.org/10.4324/9781315189284-4
3. Berggren, R.: Capabilities for change: Trust and involvement in organization development. Master's thesis,. University of Gothenburg (2019). https://doi.org/10.13140/RG.2.2.15202.17608
4. Grawitch, M.J., Ledford, G.E., Ballard, D.W., Barber, L.K.: Leading the healthy workforce: the integral role of employee involvement. Consult. Psychol. J.: Pract. Res. **61**(2), 122–135 (2009). https://doi.org/10.1037/a0015288
5. Bah, M.O.P., Sun, Z., Hange, U., Edjoukou, A.J.R.: Effectiveness of organizational change through employee involvement: evidence from telecommunications and refinery companies. Sustainability. **16**(6), 2524 (2024). https://doi.org/10.3390/su16062524

6. Baik, S.J., Song, H.D., Hong, A.J.: Craft your job and get engaged: sustainable change-oriented behaviour at work. Sustainability. **10**(12), 4404 (2018). https://doi.org/10.3390/su10124404
7. Cremers, J.: Management and worker involvement: Cat and mouse or win–win? Transfer. Eur. Rev. Labour Res. **17**(2), 199–210 (2011). https://doi.org/10.1177/1024258911401379
8. Bartkowiak, G., Krugiełka, A., Dachowski, R., Gałek, K., Kostrzewa-Demczuk, P.: Attitudes of polish entrepreneurs towards knowledge workers aged 65 plus in the context of their good employment practices. J. Clean. Prod. **280**, 124366 (2021). https://doi.org/10.1016/j.jclepro.2020.124366
9. Viterouli, M., Belias, D., Koustelios, A., Kakkos, N., Sdrolias, L.: Fostering sustainability through the integration of green human resource management and change management. In: Belias, D., Koustelios, A. (eds.) Advances in Human Resources Management and Organizational Development, pp. 199–221. IGI Global (2023). https://doi.org/10.4018/979-8-3693-0235-4.ch011
10. Apanel, A.P.: The importance of corporate social responsibility (CSR) for MSP companies in Poland: Benefits, barriers and good practice examples. Econ. Reg. Stud. **18**(1), 108–119 (2025)
11. Glavas, A., Radic, M.: Corporate social responsibility: an overview from an organizational and psychological perspective. In: Oxford Research Encyclopedia of Psychology. Oxford University Press (2019). https://doi.org/10.1093/acrefore/9780190236557.013.90
12. Widawska-Stanisz, A.: Corporate social responsibility, work-life balance and well-being—The employee perspective, vol. 209, pp. 473–487. Scientific Papers of Silesian University of Technology. Organization and Management Series (2024)
13. Mazur-Wierzbicka, E.: E-communication and CSR—A new look at organizations' relations with stakeholders in the time of digitalization. Procedia Computer Science. **192**, 240 (2021). https://doi.org/10.1016/j.procs.2021.09.240
14. Kohn, V., Frank, M., Holten, R.: How sociotechnical realignment and sentiments concerning remote work are related – insights from the COVID-19 pandemic. Bus. Inf. Syst. Eng. **65**(3), 369–384 (2023). https://doi.org/10.1007/s12599-023-00798-8
15. Rowley, C.: Back to the future: Post-pandemic work and management. Pers. Rev. **52**(2), 0770 (2023). https://doi.org/10.1108/PR-11-2022-0770
16. Ewertowski, T., Butlewski, M.: Managerial perception of risk in an organization in a post-COVID-19 work environment. Int. J. Environ. Res. Public Health. **19**(22), 14978 (2022). https://doi.org/10.3390/ijerph192214978
17. Vyas, L.: "New normal" at work in a post-COVID world: Work–life balance and labor markets. Polic. Soc. **41**(1), 155–167 (2022). https://doi.org/10.1093/polsoc/puab011
18. Church, A.H., Margiloff, A., Coruzzi, C.: Using surveys for change: An applied example in a pharmaceuticals organization. Leadersh. Organ. Dev. J. **16**(4), 89049 (1995). https://doi.org/10.1108/01437739510089049
19. Mościcka-Teske, A., Sadłowska-Wrzesińska, J., Butlewski, M., Misztal, A., Jacukowicz, A.: Stressful work characteristics, health indicators and work behavior: The case of machine operators. Int. J. Occup. Saf. Ergon. **23**(4), 1197577 (2017). https://doi.org/10.1080/10803548.2016.1197577
20. Revtiuk, Y., Pawlowska-Nowak, M.: Skills shortages as a driver of AI adoption: evidence from developed countries. In: Proceedings of the 33rd International Conference on Information Systems Development (2025). https://doi.org/10.62036/isd.2025.10
21. Revtiuk, Y., Bielawski, M.: Estimation of losses from the use of volume units to account for natural gas for households in Ukraine. In: Proceedings of MDSMES (2019). https://doi.org/10.2991/mdsmes-19.2019.9

# Current Challenges in Failure Mode and Effect Analysis and General Recommendations

Lennart Grüger[1,2](✉), Tim Sebastian Tübbicke[1,2], Conrad Uwe Lehmann[1], and Ralf Woll[1,3]

[1] QECB Quality Engineering Cottbus GmbH, Virchowstraße 43, 03044 Cottbus, Deutschland
lennart.grueger@b-tu.de

[2] Chair of Hybrid Manufacturing, Brandenburgische Technische Universität Cottbus-Senftenberg, Konrad-Wachsmann-Allee 17, 03046 Cottbus, Deutschland

[3] Chair of Quality Management, Brandenburgische Technische Universität Cottbus-Senftenberg, Siemens-Halske-Ring 14, 03046 Cottbus, Deutschland

**Abstract.** Failure Mode and Effects Analysis (FMEA) is a methodology for risk and failure prevention which has been known for decades. This method is used, among other things, in the automotive sector and in aerospace. Despite numerous existing research projects, there are still persistent and frequently recurring challenges. In this paper, over 60 FMEA reports created in accordance with the VDA/AIAG manual were examined for deviations from the standard and the most frequent anomalies were analyzed. It became clear that the main challenges lie in the transfer of documents from design to process FMEA, the functional and failure network analysis, and the formulation of functions and causes of failure, as well as the actions to be taken. For this purpose, clusters were created from over more than 300 deviations found and standard challenges were identified. This database is intended to serve as an overview to raise awareness of FMEA to be processed and as input for an failure database to support FMEA.

**Keywords:** Failure Modes and Effect Analysis · Challenges · Recommendations

## 1 Introduction

The automotive industry is an industry with high quality requirements. One of the key quality management methods for maintaining high standards is Failure Modes and Effect Analysis (FMEA). This is a team-oriented, qualitative analysis method for risk prevention and documentation of lessons learned. For the purpose of standardizing the procedure, the German Association of the Automotive Industry (VDA) and the Automotive Industry Action Group (AIAG) [1] which was founded in North America, published a joint handbook on FMEA processing for the first time in 2019 [2].

Despite this standardization, six years later the automotive industry still faces various challenges in processing FMEA. In addition to team composition and the high amount of time required, the increasing complexity of variants and increasing globalization confront the responsible parties with challenges. Furthermore, the ever-shorter development cycles increase the pressure on the capacities of automobile manufacturers [3].


A. Hamrol et al. (Eds.): MANUFACTURING 2026, LNME, pp. 92–106, 2026.
https://doi.org/10.1007/978-3-032-21701-1_7

Therefore, the study aims to analyze existing field data in the form of existing FMEA's, cluster possible deviations, and derive recommendations for action. Consequently, FMEA teams should be supported and empowered to carry out FMEA processing in a targeted and time-efficient manner.

## 2 Method

### 2.1 Methodological Principles and FMEA Procedure

The FMEA is a team-oriented, systematic, and qualitative analysis method that focuses on minimizing technical risks. This approach has been around since 1949 and is still important, especially in the automotive industry [4]. The main objective is to evaluate potential technical risks within a product or process. FMEA examines the causes and consequences of failures, documents the respective prevention and detection actions, and then recommends further actions to reduce risk. In addition, the methodology supports quality improvement, increased customer satisfaction, targeted communication within the corporate structure, and the development of a knowledge base [2, 5].

The limitations of FMEA are that the analysis is exclusively qualitative, not measurable, and always team-dependent. Furthermore, only single failures can be considered. For quantitative analyses and multiple failure considerations, other methods such as fault tree analysis or FMEDA (Failure Modes, Effects, and Diagnostic Analysis) must supplement FMEA [2].

There are basically two different types of FMEA: Design FMEA (D-FMEA) and Process FMEA (P-FMEA). D-FMEA is primarily used to analyze the general product design. Accordingly, it is used to check that potential failure modes and their causes or mechanisms have been considered and addressed before a component is approved. P-FMEA focuses on the general product development process and considers the possible failures that can occur during manufacturing, assembly, and logistics processes. The main objective of P-FMEA is to analyze the process and describe the implementation of actions before the actual start of the process so that undesirable failures do not occur in the first place [2, 5].

In addition to the methodological aspects, the FMEA team is also of essential importance in FMEA processing. The core team for FMEA processing always includes the FMEA moderator and a quality engineer, who contribute their methodological expertise to the team. The other team members differ depending on whether D-FMEA or P-FMEA processing is involved. In D-FMEA, at least one design engineer, one system developer, one component developer, and one test engineer are also involved. For P-FMEA, the moderator and quality engineer are supported by at least one process, system, and process validation engineer. For both types of FMEA, the FMEA team can be supplemented with additional responsibilities as needed, such as technical experts, project managers, or assembly line workers. It is fundamentally important to coordinate all specialist areas within the FMEA process, as they are the main sources of input and thus have a significant influence on the outcome of the FMEA.

Despite the differences in definition between D-FMEA and P-FMEA, both types are structured in the following seven steps when created: 1. Planning and preparation, 2. Structural analysis, 3. Functional analysis, 4. Failure analysis, 5. Risk analysis, 6.

Optimization, and 7. Documentation of results. The first three steps form the system analysis group, steps four to six form the failure analysis and risk reduction group, and the seventh step forms the risk communication group [2, 5].

In the first step of planning and preparation, the purpose and schedule of the FMEA are defined. The most important point is to describe the limits of the analysis and thus what should be included in the FMEA. Furthermore, existing basic FMEA's are determined, including lessons learned (LeLe) [2].

LeLe refers to all knowledge gathered from previous projects or current quality issues. This knowledge is incorporated into product and process development and thus also supports current FMEA processing. The aim is to increase customer satisfaction and quality while reducing the time required. Process-related quality management (PQM) can help to feed the knowledge gained back into the design and process creation process. Part of PQM can be standardized 8D reports, which record results within an eight-step problem-solving process [6–8].

The main objective in the first step of FMEA creation is to establish an optimal starting point for the subsequent steps, in particular structural analysis. This forms the basis for successful FMEA processing [2, 5].

**Structural analysis** is used to identify design interfaces (D-FMEA) or process flows (P-FMEA). This identification is based on a structure tree or similar (boundary diagram (D-FMEA)/process flow diagram (P-FMEA)). Particular attention is also paid to the cooperation between the customer's development teams and the supplier. Here, it is necessary to clarify who is responsible for which interfaces. The structural analysis is complete when the definition of the individual interfaces or process steps has been fully completed [2, 5]. Figure 1 illustrates the structural analysis of D-FMEA and P-FMEA.

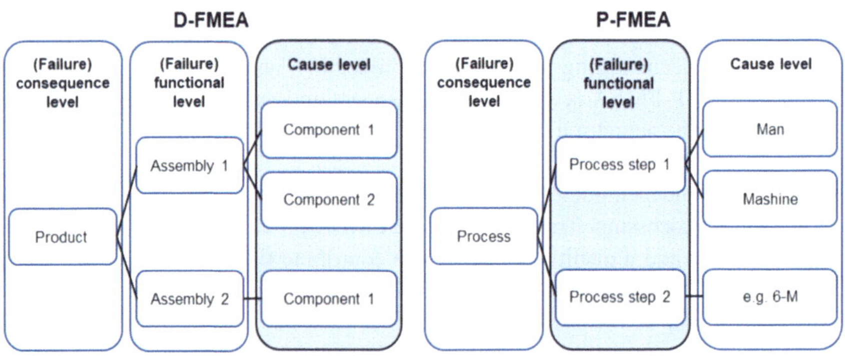


**Fig. 1.** Visualization of Step 2: Structural analysis by type of FMEA.

The final step in system analysis is **functional analysis**. This describes the tasks that a product or process is intended to perform and how the individual components or sequences interact with each other. The aim is to clearly structure the functions so that failures and their consequences can then be systematically identified. In this context, product/component characteristics must essentially be defined in the D-FMEA on the last level, which are then transferred to the middle level of the P-FMEA.

The P-FMEA must contain process parameters that can be adjusted to achieve the corresponding characteristics [2, 5].

**Failure analysis**, as the first step in the higher-level group of failure analysis and risk assessment, describes the possible consequences, types, and causes of failures. This description is carried out separately for each function defined in the functional analysis. The aim is to identify all potential risks or weak points of the product or process [2, 5]. Figure 2 shows the links between functional analysis and failure analysis in D-FMEA and P-FMEA.

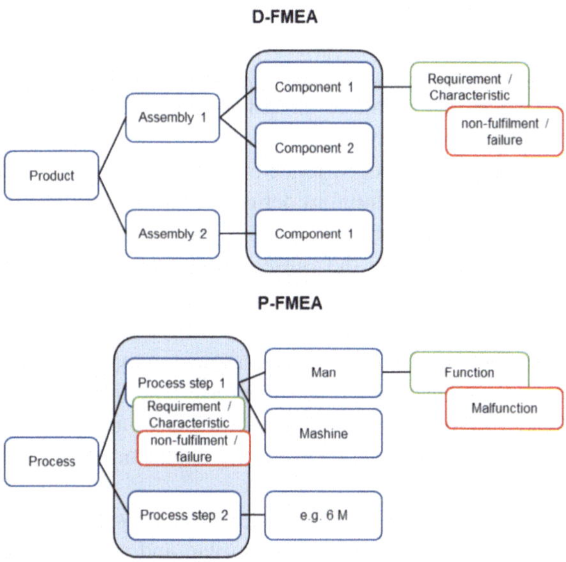


**Fig. 2.** Connection between functional and failure analysis and visualization of the level link (blue) from D-FMEA to P-FMEA.

In the **risk analysis**, the failures described are evaluated in three categories: significance (S), occurrence (O), and detection probability (D). Significance refers to the most serious consequence of the failure. The letter O indicates the expected frequency of occurrence of the cause of the failure. The E rating provides information about how the capability or maturity of the detection method is assessed. The rating within the three categories is always between one and ten, with one representing low significance and probability of occurrence as well as high probability of detection. Accordingly, a rating of one in each category is the optimum. Ratings of ten, on the other hand, indicate that the potential risk in the area under consideration is very high. The S, O, and D ratings are factors in the formation of the risk priority number (RPN), which ultimately determines the task priority for risk reduction. The RPN is calculated by multiplying the three individual ratings and can therefore take values between one and 1,000. In addition to considering the RPN in isolation, it is also possible to represent the risk assessment of malfunctions using an action priority matrix (AP-matrix). This matrix describes three areas that determine whether further risk minimization actions need to be taken or whether the current status is methodologically sufficient. Within the green area, the potential risk assessment is low and therefore no immediate adjustment of actions is necessary. In the yellow area of the AP matrix, it is up to the FMEA team to identify further actions that reduce the A and/or E rating of the malfunction. In this case, a cost/benefit analysis or other predefined key figures must be considered. Malfunctions that fall within the red area must be described with new prevention and/or detection actions. High risks can only be maintained as sufficient in absolutely exceptional cases and with detailed justification and documentation. In principle, all functions in the high-risk area should be reworked in the optimization phase. The goal should always be to reduce the risk to a minimum [2, 5].

The **optimization** phase is the final step in failure analysis and risk reduction. In this phase, conclusions are drawn for the implementation of risk reduction actions. The goal is to reduce the risk to an acceptable level (at least the yellow area of the AP matrix). It is important that the effectiveness of a new action is tested and proven. Once this has been done, the assessment of occurrence and/or detection can be lowered. Within the optimization process, it is essential to define responsibilities and deadlines for the implementation of actions. Without this definition, the optimization process can quickly miss its planned target. Throughout the entire optimization step, it is essential that the FMEA team, management, customer, and supplier work together [2, 5].

The final step in creating an FMEA is the **documentation** of results. Within this step, the results must be compiled and all assessments, improvements, and proofs of effectiveness must be stored in a comprehensible form. The aim is to create transparency and provide evidence of risk control. Furthermore, the documentation also serves as a basis for future projects, in which the knowledge gained from the processing should be directly incorporated [2, 5].

The result, which is produced after completing the steps described, is a product or process that has minimal risks according to the plan and is created on the basis of a proven risk prevention method. At this point, it is important to mention that an FMEA is essentially a living document that should be reviewed at defined intervals to ensure that it is up to date and that any changes are continuously reflected in the FMEA [2, 5].

## 2.2 Systematic Data Analysis of the Collected Data

The identification of clusters of repetitive standard deviations from the data collected in the FMEA reviews is based on a descriptive secondary analysis of existing FMEA files, which were analyzed by FMEA moderators from QECB – Quality Engineering Cottbus GmbH in the operational area. The analysis is evaluated using an evaluation tool with a traffic light system. Deviations are documented accordingly with comments that explain them in more detail. The analysis began with the individual examination of more than 60 FMEA. For each deviation identified, the project name, FMEA type, rating, and respective comment were derived and transferred to an Excel document. The data resulted in a total of 301 individual data records. Subsequently, 16 clusters were formed step by step using keywords from the evaluation tool, which summarize similar deviations in conjunction with the stored comments.

Due to the selected keywords, it is possible to assign a deviation to several clusters, so that more deviations are represented in the clusters than data records are analyzed. A total of 90.7% of all comments belong to at least one cluster. The data sets that cannot be assigned to a cluster due to the individuality of the comment are not considered in further evaluations.

# 3 Results–Challenges of FMEA and Recommendations

## 3.1 Major Challenges in Failure Mode and Effect Analysis

The systematic analysis of the collected data reveals 16 key challenges. Figure 3 shows the frequency of these challenges in a Pareto chart.

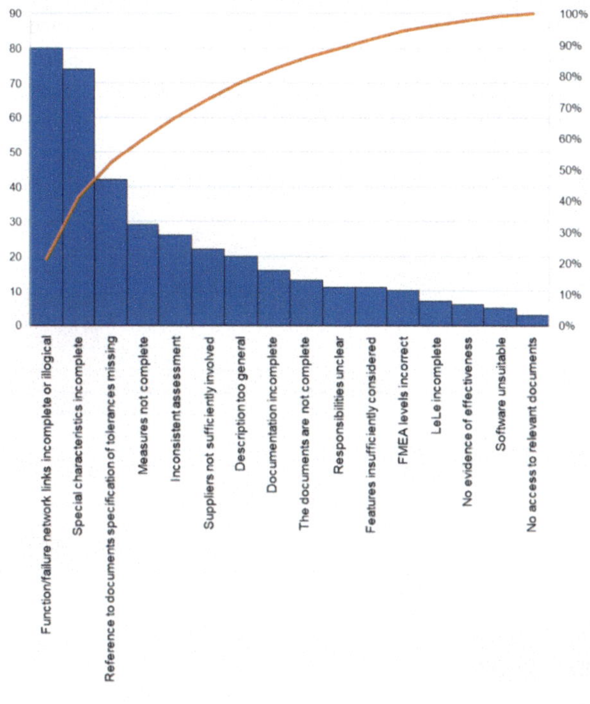


**Fig. 3.** Pareto chart of the grouped challenges resulting from the analysis.

This shows that more than 50% of the 375 deviations relate to the first three challenges:

1. "Link between function and failure network incomplete" (21.3%),
2. "Special features incomplete" (19.7%),
3. "Reference to relevant documents or specification of tolerances missing" (11.2%).

In the automotive industry, special characteristics are understood to be quantifiable attributes that require special consideration based on increased safety or functional requirements or regulatory actions. Such characteristics are filtered out during product or process development and are always marked in a special way [2, 9].

Very few deviations, on the other hand, are caused by a lack of access to relevant documents (0.8%) and the use of unsuitable software (1.3%). A software application is classified as unsuitable if it does not allow the respective visualization of function and failure networks or does not allow complete derivation of documentation. Accordingly, a significant improvement in the quality of FMEA would be possible if the focus were placed on avoiding the first three to four challenges. For a more detailed analysis and overview of when failures can occur in the FMEA creation process, these have been assigned to the respective steps of FMEA processing and visualized in Fig. 4. The green cells in the tables represent the earliest possible time of a potential non-conformity to a requirement according to the FMEA manual. The yellow and red cells visualize the ongoing and final potential for the occurrence or the last opportunity to correct a deviation. For example, special characteristics could be transferred incompletely from the D-FMEA to the P-FMEA (non-conformity in step 1), then forgotten in the functional analysis (step 3) and finally in the failure analysis (step 4).

**Fig. 4.** Breakdown of non-conformity according to FMEA processing steps and their initial (green), ongoing (orange), and final possible cause/workaround (red).

It becomes clear that two of the three major deviations – the lack of "special features" and the lack of document referencing – can already arise in the **planning phase**. Like the other deviations that arise during planning, these are due to insufficient transfer of information.

Deviations in the **structural analysis** mainly relate to a lack of consideration of interfaces, including those with (sub)suppliers, which also points to insufficient communication and delineation of responsibilities. In addition, the FMEA level structure could be incorrectly constructed, i.e., it could not correspond to the logic of failure modes – failures – failure causes.

This type of inadequate structural analysis can have serious consequences for the further course of the FMEA. As a result, the link between the function network and the failure network is flawed or illogical.

The **function and failure analysis** guarantee a significant potential for deviation. In addition to identifying the special characteristics, this also includes the actual linking of the function and failure networks. Furthermore, the tolerances for the respective characteristics examined must be specified or reference must be made to corresponding documents that contain these tolerances. Furthermore, all three of the greatest potentials for deviation are contained in these sub-steps. Added to this is the well-known challenge of clearly formulating the respective work steps or characteristics.

Referencing also plays a role in **risk analysis** and **optimization**. However, it is less the tolerances of the characteristics that are particularly relevant, but rather the specification of norms and standards or document numbers. These are required, to clearly assign work instructions or test specifications for prevention or detection actions. In addition, evidence of effectiveness, e.g., in the form of measurement system analyses or similar, must also be provided for the actions. Other deviations that have been identified are inconsistent assessments of occurrence or detection probability that do not comply with the VDA standard. Furthermore, unclear responsibilities are also an issue. In this case, entire (work) groups are frequently specified instead of clear responsibilities for tracking actions. Finally, insufficiently documented lessons learned from, for example, PQM or 8D reports should be mentioned. If available, these must be traced back to the respective FMEA for documentation purposes.

In the final step of the documentation, missing evidence of effectiveness was also identified. In addition, incomplete or patchy documentation was also noted, in which, for example, the significant risks or evidence of effectiveness were not communicated transparently.

Since almost 40% of the deviations (148 out of 375) arise in the first step, these were additionally analyzed separately for D-FMEA and P-FMEA. The result is shown in Table 1.

**Table 1.** Non-conformities that may arise in step 1 divided in D-FMEA and P-FMEA.

| Deviation | Total number | Total number D-FMEA | Total number P-FMEA |
|---|---|---|---|
| Special characteristics | 74 | 32 | 42 |
| Incomplete Referencing of relevant documents or specification of tolerances missing | 42 | 16 | 26 |
| Documents incomplete | 13 | 7 | 6 |
| Responsibilities unclear | 11 | 1 | 10 |
| Software unsuitable | 5 | 0 | 5 |
| No access to relevant | 3 | 1 | 2 |
| Total | 148 | 57 | 91 |

This makes clear that incorrect or incomplete consideration of special characteristics is almost always transferred from D-FMEA team to colleagues working on P-FMEA. Incomplete documents, on the other hand, tend to play a greater role in D-FMEA. This is probably due to the respective starting point of D-FMEA and possibly missing information.

However, the gap is most evident in the clarification of responsibilities (D-FMEA = 1 /P-FMEA = 10) and the use of the required software solution (D-FMEA = 0 /P-FMEA = 5). It is important to note that D-FMEA is often created centrally in the manufacturers' competence centers, whereas P-FMEA is the responsibility of the plants. Accordingly, responsibilities and software solutions for D-FMEA have often been clearly defined for years and the relevant know-how is available. With P-FMEA, on the other hand, plants may face new challenges, and communication with plants often quickly becomes very complex as their number increases and globalization progresses.

Apart from the pure numbers, the authors consider missing special characteristics and incorrect or incompletely linked failure networks to be the most critical issues. In the worst case, these issues could result in safety-related special characteristics being missed or risks being underestimated. Accordingly, this could result in incorrect prioritization of tasks.

## 3.2 Derived Recommendations

**Recommendations for Improving the First Step – Planning**

The following recommendations for action can be derived from the preceding analysis of deviations. In the first step, planning, the scope of the analysis, including the analysis boundaries, the general project plan, and all necessary documents, should be fully defined and available. In addition to drawings, special characteristics lists, and boundary diagrams (D-FMEA) or flow charts (P-FMEA), the most important documents are possible basic FMEAs, which can serve as templates and contain lessons learned from the past. Furthermore, it is essential to agree on clear communication within the topic and FMEA team, as well as a clear definition of responsibilities. Finally, the software solution used must be checked for suitability and up-to-datedness. If these points are fully agreed upon and are also implemented during the further creation steps, then the prerequisites for an FMEA that complies with the manual are in place.

**Recommendations for Improving the Second Step – Structural Analysis**

In structural analysis, according to the boundary diagram, an interface analysis and an analysis of the components provided by suppliers must also be considered. This applies in particular if these components have special characteristics. A frequent criticism of this approach in practice is the possible violation of the assumption that components are delivered in good condition. However, this can only be determined along the process flow, as the respective delivery condition was considered in the previous process step. To ensure that this is also the case for externally supplied components, it is recommended that standard actions such as initial sampling or supplier certificates be documented as preventive actions, and that requalification audits or incoming goods inspections be documented as possible detection actions, and that supplier quality be considered accordingly in the FMEA.

With regard to the **wording**, it should be noted that it should be chosen in such a way that it can also be clearly understood by uninvolved third parties in the future. Furthermore, the wording should be chosen in such a way that it refers to the actual activities and characteristics that are being examined. To this end, particular care must be taken to ensure that the wording provides added value in the level structure. This is easier to achieve in an automated process than in a manual process carried out by an employee. In this case, the wording refers more to activity steps that should correspond to the work instructions given to production personnel.

In terms of the level structure, the recommendation is to build the middle level based on the individual components (D-FMEA) or the process steps or workstations (P-FMEA). In this regard, the high relevance of the planning phase and the careful creation of the boundary diagram or process flowchart becomes clear once again. A special case is (offline) rework. Here, the authors recommend building the middle level based on the components to be reworked, as these are often offline stations in the style of workshop production. Accordingly, only one station would be created on the middle level, which in turn would quickly lead to confusion.

**Recommendations for Improving the Third Step – Functional Analysis**

As described in Section 2.1, the functional analysis in the D-FMEA should refer to the actual functions of the components. Based on a careful D-FMEA, the functional analysis of the P-FMEA essentially consists of transferring the third or characteristic level to the second or functional level. This approach is also intended to avoid an incomplete consideration of the characteristics. In particular, the "special characteristics" should be stored in the description or designation with unique ID numbers and descriptions. These are usually specified by the FMEA requester (usually the OEM) and must be considered consistently in D-FMEA and P-FMEA. Given the high number of product variants that are common today, it is particularly important to address the relevance of all "special characteristics" transferred to the respective variant and to clarify an exact understanding of these. If a designation is unclear, it is recommended to clarify it in the comments and not in the actual designation in order to ensure consistency across all levels (OEM-supplier-(sub)supplier).

For this purpose, the description must be transferred to all relevant controlled documents such as drawings, test plans, production control plans, etc. This procedure in turn leads to a lower risk of insufficient supplier involvement.

With regard to linking, it should be noted that links should only be made within a linear structure and not across different levels. Figure 5 illustrates this.

This leads to a particular challenge in the wording and may mean that handling processes must be considered both at the end of one station and at the beginning of the next station.

**Recommendations for Improving the Fourth Step – Failure Analysis**

The recommendations for failure analysis are similar to those for functional analysis. However, special attention should be paid to the linking of failure networks, as these define the significance of the respective failures. This in turn leads to the criticality of individual topics. In addition, it is possible to recheck the completeness of the "special features" and the interface analysis. Furthermore, linking error networks allows the

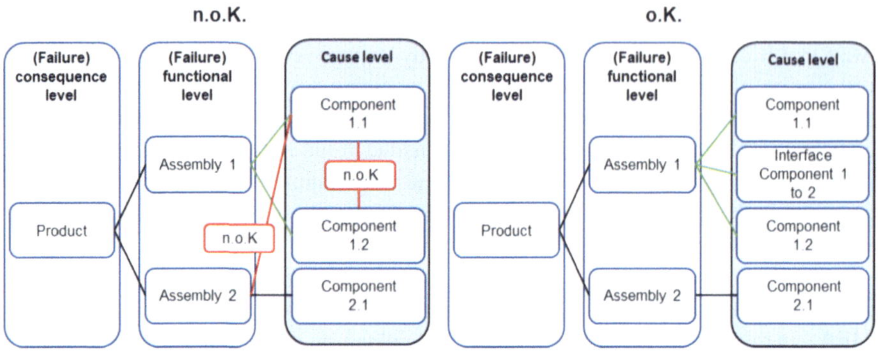


**Fig. 5.** Example of an incorrect cross-link (left) and a correct link along the line structure (right).

plausibility of error causes and errors to be checked. This must lead to a clear cause-and-effect relationship. Figure 6 shows a negative example of linking causes (right) to the supposed consequence (left) as observed in practice in the upper section, and the authors' corresponding suggestion of what a proper cause-effect link might look like in the lower section.

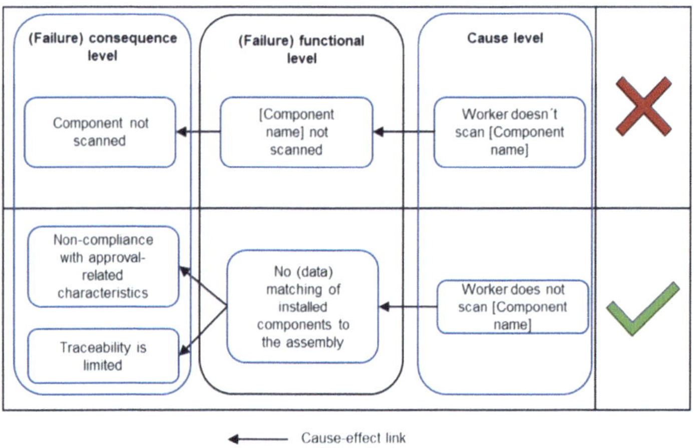


**Fig. 6.** Incorrect or illogical link (top) between analyzed levels and authors' suggestion (bottom).

Another major challenge in this process step is the fact that the assessment must be carried out as if no risk prevention actions were in place. This is particularly challenging when revising existing FMEA's, as various **prevention and detection actions** are usually already defined. However, these actions only reduce the probability of occurrence or detection and **should not be considered** when assessing the significance.

**Recommendations for Improving the Fifth Step – Risk Analysis**

Risk analysis includes the evaluation of existing risk prevention strategies. This involves assessing the probability of occurrence (O) and detection (D) based on these actions.

Particular attention must be paid to the consistency of the actions. A probability of occurrence of one should only be assigned if the actions are actually Poka-Yoke solutions. Accordingly, an assessment of the occurrence probability of $O = 1$ always results in a detection probability of $D = 1$, because it is assumed that the failure will definitely be avoided by design.

A visual inspection by an employee is to be assessed with $D = 8$. A documented dual control principle justifies a reduction to an assessment of $D = 7$.

Automatic detection of the failure in the subsequent process is rated $D = 4$, while automatic detection in the same process step is rated $D = 3$. It should be noted here that if the detection occurs in the subsequent process, this should still fall within the area of responsibility. A discovery by the next (internal) customer is not to be considered as long as there are no interface agreements that guarantee an inspection of the relevant characteristics for the corresponding quality. An automatic discovery of the cause of the mistake before it occurs is to be rated $D = 2$.

Ratings of $D = 5/D = 6$ refer to short/long-term inspection intervals for the measured sample, or small/large sample sizes. These relative designations must be adjusted depending on the scope or size of production. For example, a sample test at the start of a shift in two-shift operation would be a short interval, whereas a test once a month would be more equivalent to a longer test interval and would be rated $D = 6$. Another way of reducing the probability of detection from $D = 7$ or $D = 8$ is if a possible defect can be detected in the subsequent process using camera technology. In this case, the expertise of the FMEA team is particularly important in order to assess the respective probability of detection. For example, in some cases, it is possible to detect mistakes from previous steps using camera technology in subsequent steps, provided that they are still visible to the camera technology, scanner, or (contour) sensor technology.

With regard to the **wording of actions**, those relating to employees pose a particular challenge. These are mostly work instructions, which in turn require the reference of unique document numbers. Instruction in the work instructions should be usable in accordance with common practice and should not require any additional actions. Experiments or tests must also be referenced with document numbers or the standards according to which the tests are carried out. Corresponding test results or the results of measurement system analyses do not constitute actions in themselves, but are **evidence for effectiveness** of the actions. For example, a machine capability study (MFU) is proof of the effectiveness of teaching a robot, but not a detection action for the series process. A corresponding detection action would be (usually) integrated overload protection in the robot control system itself.

**Recommendations for Improving the Sixth Step – Optimization**

During the optimization phase, the FMEA team faces similar challenges to those encountered in the initial risk analysis. It is supposedly easier to work on the topics at this stage, as more extensive information and documented lessons learned are already available, depending on the processing status. Feedback on these lessons can be a great advantage and is therefore recommended, as information on root cause analysis, e.g., from 8D reports, can further improve the quality of the FMEA. On the one hand, the FMEA team must be made aware of this fact, and on the other hand, the organization must provide a system that supports the moderator and FMEA team in integrating the lessons learned.

Furthermore, the optimization phase offers the last chance for troubleshooting for seven of the 16 main challenges. This phase must therefore be assessed critically.

One operational recommendation made by the authors is to use several action statuses to document the different degrees of completion. An example of implementation in the APIS software is shown in Fig. 7 using an automatic welding process.

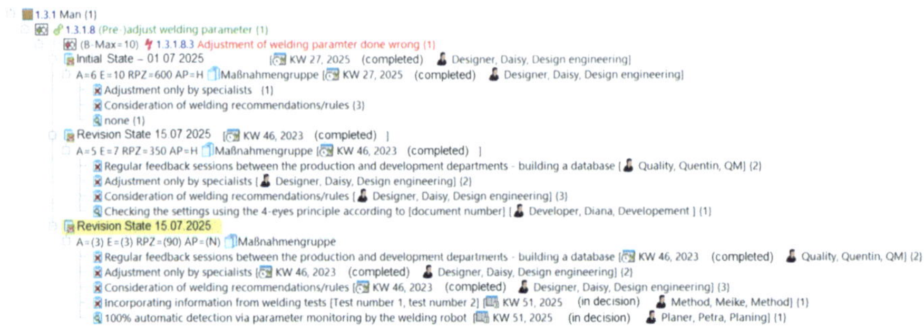


**Fig. 7.** Example illustrating possible actions and their impact on action priority.

This illustrates that different implementation options for the discovery were identified for the same level of actions. A dual control principle would improve the probability of discovery, but would still entail a high task priority. A discovery about automated temperature control in the welding process using camera technology or similar would result in a low task priority, but would require a correspondingly high investment. The FMEA team has therefore identified the risk and a management decision is now required as to whether the investment will be made or the high risk accepted. The result must be documented accordingly.

**Recommendations for Improving the Seventh Step – Documentation**
Documentation is the final step and is essential for communication with management, suppliers, and customers. Therefore, the authors recommend coordinating the type and scope of documentation with all parties involved/interested at the start of the project. Furthermore, the current standard of documentation and project communication should be reviewed/coordinated in regular meetings.

## 4 Discussion and Conclusion

### 4.1 Discussion

Despite the extensive basic data analysis and the uniform evaluation scheme, the analyses, like any audit, represent random samples. These are therefore based on snapshots. Furthermore, despite attempts at objectivity, the findings are subject to the reviewer's human subjectivity. A more far-reaching cross-company analysis could further reduce this influence. However, due to the high levels of confidentiality, this is only feasible with a great deal of effort, if at all.

In the future, machine learning and AI approaches in combination with a standardized vocabulary could improve the quality of the FEMAs as well as the reviews. Initial approaches in this regard are already being researched within the company.

## 4.2 Conclusion

In this study, more than 60 FMEAs were examined for deviations that were created in accordance with the current version of the VDA/AIAG handbook. In total, 375 deviations were identified and summarized in 16 main challenges. The three key issues, which make up more than 50% of the deviations, and the recommendations for addressing them are listed in Table 2.

**Table 2.** Summary of recommendations to address the key issues.

| Deviation | Recommendation |
| --- | --- |
| 1)Incomplete/unlogic links between the function/failure network (21.3%) | **D-FMEA:** Detailed structural analysis divided into assembly (failure level) and component (cause level)<br>**P-FEMA:** Transfer all D-FMEA documents/information and transfer the cause level of the D-FMEA to the failure level of the P-FMEA. |
| 2) Incompletely considered "special characteristics" (19.7%) | **D-FMEA:** Creation of databases/lists from previous projects and lessons learned for the respective components.<br>Check the relevance of corresponding generic special features at the start of FMEA creation and regular monitoring of associated documents (e.g., drawings) with a corresponding feedback process across all detail levels.<br>**P-FEMA:** Transfer of all D-FMEA documents/information as well as regular monitoring of associated documents. |
| 3) Missing references to relevant documents or missing tolerance specifications (11.2%). | **D-FMEA/P-FMEA:** Insert the document numbers directly into the respective FMEA deadline. If documents are not available in the initial phase, unique placeholders should be provided and assigned with the corresponding responsibility and deadline. |

Overall, maintaining the interdisciplinary nature of the team is relevant for avoiding all of the issues mentioned. Accordingly, at least the core team members must be present for every meeting.

A further breakdown according to the seven steps of FMEA into initial, ongoing, and final creation/workaround steps suggests that the majority of deviations arise during the planning phase. A separate investigation of deviations relevant to planning in D-FMEA and P-FMEA confirms this assumption. The communication of requirements for documentation, software, and corresponding responsibilities is particularly important. The reason for this is considered to be the fact that D-FMEA is usually carried out in the manufacturer's development competence centers and P-FMEA in the respective plants. Accordingly, increasing variant diversity and globalization are increasing the complexity of project communication between locations.

In addition to the standard recommendation to carefully assemble the FMEA team, the authors consider meaningful and detailed boundary diagrams or process flowcharts to be of great importance for communication and further processing. Furthermore, detailed consideration of special characteristics and their consistent consideration and clear referencing throughout the entire supply chain are important. In addition, it is not the sheer number of special characteristics that is relevant to the quality of the FMEA, but their integration. Regular communication between the parties involved with the aim of integrating lessons learned is also important.

Ultimately, the comprehensive analysis of each of the process steps enabled specific recommendations to be made for addressing the challenges. However, these recommendations are not exhaustive and must be continuously expanded and refined in the spirit of continuous improvement.

## References

1. Belcaro, P.L.: 9-object-model FMEA: A conceptual model for creating technical risk analyses (in German). Springer Vieweg, Wiesbaden (2025)
2. Automotive Industry Action Group, German Association of the Automotive Industry: Failure Mode and Effects Analysis – FMEA Handbook: Design FMEA, Process FMEA, Supplemental FMEA for Monitoring & System Response. Automotive Industry Action Group, Southfield, Michigan (2022)
3. Nganso Yanghu, G.D.: Influence-based screening of technical risks in the early phase of product development (in German). Academic Dissertation, Brandenburg University of Technology Cottbus-Senftenberg (BTU) (2025).
4. Jakoby, W.: Quality management for engineers: A practice-oriented textbook for planning and controlling quality processes (in German). Springer Vieweg, Wiesbaden (2019)
5. Beuth Verlag: DIN EN 60812:2015–08 – Failure mode and effects analysis (FMEA). (in German). https://www.dinmedia.de/de/-/-/236280158
6. Rhodes, L., Dawson, R.: Lessons learned from lessons learned. Knowl. Process. Manag. **20**, 154–160 (2013). https://doi.org/10.1002/kpm.1415
7. Wagner, K.W., Käfer, R.: Process-oriented quality management (PQM): Practical implementation of ISO 9001 (in German). Hanser, Munich (2023)
8. Elangovan, S., Jusoh, M.S., Muhd Yusuf, D.H., Ismail, M.S., Hj Din, M.S.: 8D problem solving methodology: Continuous improvement in automation organization. J. Phys. Conf. Ser. **2129**, 012017 (2021). https://doi.org/10.1088/1742-6596/2129/1/012017
9. Haefele, S., Westkämper, E.: Identification of product safety-relevant tasks for global automotive manufacturers. Procedia CIRP. **17**, 326–331 (2014). https://doi.org/10.1016/j.procir.2014.02.052.2

# Big Data Analytics as a Tool Supporting the Sustainable Development of Manufacturing Enterprises

Katarzyna Ragin-Skorecka(✉), Daria Motała, and Agnieszka Krugiełka

Poznan University of Technology, Faculty of Engineering Management, Poznan, Poland
katarzyna.ragin-skorecka@put.poznan.pl

**Abstract.** This article examines the impact of big data analytics (BDA) on the sustainable development of manufacturing enterprises in Poland. The study was conducted in the fourth quarter of 2024 among medium-sized and large companies, with responses collected from directors and managers evaluating the level of BDA use and its effects across seven operational areas: efficiency, natural environment, working conditions, social environment, strategic resource security, social responsibility, and interpersonal relations. Grey System Theory and Grey Relational Analysis were applied to calculate Grey Relational Grade scores at both the variable level and across triple bottom line dimensions. The findings reveal a positive and consistent relationship between BDA adoption and performance in all assessed areas, with the strongest effects observed in the socio-organizational dimension. Large enterprises exhibited stronger relationships than medium-sized firms, particularly in the environmental and economic dimensions. The results also indicate complementarity between efficiency improvements and positive environmental and social responsibility outcomes. The study highlights the importance of aligning BDA initiatives with integrated economic, environmental, and social objectives. Key limitations include the cross-sectional, self-reported nature of the data and the use of GRA, suggesting avenues for future research using objective performance indicators and advanced modeling approaches.

**Keywords:** Big Data Analytics · Sustainability · Manufacturing industry

## 1 Introduction

In the era of widespread industrial digitization and the implementation of Industry 4.0 solutions, big data analytics (BDA) is becoming a key tool for improving the competitiveness of manufacturing enterprises. Analysing large, heterogeneous data streams allows for process streamlining, waste reduction, and faster response to disruptions [1–4]. At the same time, regulatory and societal pressure is growing to achieve economic results in a sustainable manner, i.e., while respecting the environment and the well-being of employees and local communities [5]. In this context, BDA is perceived not only as a technology supporting efficiency, but also as a factor accelerating the transformation towards sustainable development [6–8].


A. Hamrol et al. (Eds.): MANUFACTURING 2026, LNME, pp. 107–121, 2026.
https://doi.org/10.1007/978-3-032-21701-1_8

The literature provides research results regarding the partial impact of BDA on operational efficiency and selected environmental indicators. In the environmental dimension, the literature points to promising mechanisms, such as reducing energy consumption, emissions, and waste, also indirectly through green innovations [3, 7]. In the economic dimension, the relationship between BDA and productivity and financial performance is examined; however, these findings are less frequently integrated with softer aspects, such as analytical culture, organizational learning, or security of strategic resources [9, 10]. In the socio-organizational dimension, research focuses primarily on economics and the environment, while working conditions, interpersonal relations and social responsibility are omitted [5, 11, 12].

Based on the above considerations, there is a lack of comprehensive, empirical analyses that simultaneously assess the impact of BDA in three dimensions of sustainable development: environmental, economic and socio-organizational, which is confirmed by various researchers [3, 5].

An additional gap observed is the perspective of enterprise size. Small and medium-sized enterprises (SMEs) face specific barriers to BDA implementation – limited resources, a shortage of technical competencies, and less management support, which contributes to varying outcomes associated with BDA implementation compared to large organizations [5, 13]. Furthermore, comparative analysis of medium-sized and large manufacturing enterprises is rarely found in research, and conclusions from other geographical contexts may not necessarily translate to the realities of Poland and the European regulatory environment [6, 14].

This article presents the results of research conducted in medium-sized and large manufacturing enterprises. The analysis was conducted using Grey System Theory (GST) and the Grey Relational Analysis (GRA) model. The conclusions drawn from this study focus on the perceived impact of BDA on seven areas of enterprise performance: enterprise efficiency (EE), natural environment (NE), working conditions (WC), social environment (SE), security of strategic resources (SS), social responsibility (SR), and interpersonal relationships (IR), enabling a coherent assessment of the three dimensions of sustainable development. This article fills an empirical gap and offers practical recommendations for managers planning to use BDA as a tool to support the implementation of sustainable development goals in industry.

## 2 Literature Review

Sustainable development of manufacturing enterprises is a key direction of contemporary management, integrating environmental, social, and economic goals within a single operational strategy [9, 15–17]. Companies operate under increasing regulatory pressure, social expectations, and the need to maintain global competitiveness, which leads to the need to implement modern technologies, including automation, robotics, the Internet of Things (IoT), artificial intelligence (AI), and big data analytics (BDA) [18–22]. These technologies support both reducing the environmental impact of industrial activities and improving work quality and economic efficiency [23–26]. At the same time, the intensive use of BDA allows enterprises to more comprehensively analyse operational processes, combining economic, environmental, and social aspects in a coherent manner [12, 27–29].

In the environmental area, automation and robotics enable precise management of energy and material consumption, leading to reduced losses and emissions [30, 31]. Furthermore, the integration of sensors and IoT systems allows for real-time process monitoring, identification of inefficiencies, and taking preventive measures [19, 32]. Big data analytics support the prediction of environmental load patterns, enabling the optimization of production process parameters [8, 27, 33]. In many enterprises, BDA is also integrated with energy management systems and renewable energy sources, which increases energy efficiency and system stability [12, 16, 28]. In this context, the growing scale of data use in large enterprises enables a more comprehensive assessment of the environmental impact of production activities, supporting strategic decision-making [5, 34].

Implementing circular economy (CE) principles combined with BDA allows for waste minimization, reuse of raw materials, and optimization of product life cycles [5, 11, 35]. These activities bring both tangible economic and ecological benefits, increasing the company's resilience to the volatility of the raw materials market. The use of BDA in circular economy processes further facilitates the assessment of the impact of these activities on all dimensions of sustainable development, allowing companies to better integrate environmental and economic strategies [6, 28, 36].

At the same time, the digitization of processes significantly enhances the social aspects of sustainable development. Automation reduces monotonous and dangerous tasks performed by employees, increasing safety and work comfort [37]. Virtual reality (VR) and augmented reality (AR) technologies enable advanced training in a simulation environment, while BDA supports risk analysis and preventive measures [10, 38, 39]. Integrating technology with corporate social responsibility (CSR) policies contributes to increased transparency, social responsibility, and a positive corporate image [40]. Research indicates that in large enterprises with broader access to data, the social effects of digitization and data analytics are particularly noticeable, supporting the development of a culture of safety and employee engagement.

Economically, modern technologies and data analytics contribute to increased operational efficiency, process stability, and market competitiveness. Automation and digital twins of processes increase production repeatability, reduce failures and downtime, and limit raw material losses [18, 41, 42]. Data analytics enables more effective risk management in supply chains, logistics optimization, and increased operational resilience of enterprises [2, 43, 44]. The benefits of using modern technologies vary depending on company size – large companies achieve an advantage primarily in operational stability, while smaller ones benefit primarily from product innovation and market flexibility [45, 46]. In this context, BDA allows for assessing the extent to which improving economic efficiency can support other dimensions of sustainable development, including environmental impact and social responsibility [9].

The integration of digital technologies and big data analytics allows manufacturing companies to pursue environmental, social, and economic goals in a coherent and systemic manner. The condition for fully utilizing their potential is the appropriate organizational and competency preparation of employees, as well as the adaptation of implemented technologies to the specifics of the industry and region [9, 36].

## 3 Materials and Methods

The aim of the study was to identify and assess the impact of Big Data Analytics on environmental, economic, and socio-organizational aspects in medium-sized and large manufacturing enterprises. The following research questions were posed:

- RQ1: What relationships between BDA and sustainable development are described in the literature?
- RQ2: How is the impact of BDA perceived as a tool supporting sustainable development? This question refers to the following dependent variables: EE – enterprise efficiency, NE – natural environment, WC – working conditions, SE – social environment, SS – security of own strategic resources, SR – social responsibility; IR – interpersonal relations in the enterprise.

The chosen research method – a survey – was used due to the research objective and research questions, as it allows for capturing participants' opinions and experiences. The questionnaire used closed-ended questions. Responses to the questions were constructed based on a Likert scale. The following scale was used to assess the impact of BDA on individual variables (RQ2): 1 – very negative, 2 – negative, 3 – neutral, 4 – positive, 5 – very positive. Limitations of the conducted research include: the lack of additional explanation of the question's intention, diverse interpretations of questions and answers by respondents, lack of control over external factors accompanying the research, and the lack of dynamic interaction between the researcher and respondents [47].

The geographic scope of the study is related to the selection of enterprises – medium-sized and large organizations conducting production-related activities in Poland. Respondents were selected purposively based on the size of the enterprise they represent (174 medium-sized, 126 large) and their position as director (131) or manager (169). This selection of respondents increases the credibility of the results by obtaining answers from individuals knowledgeable about the enterprise's operations and the solutions used. Among the respondents were 124 women and 176 men. The study is representative, assuming a 95% confidence level and an estimated error of 5.6% [48]. The population of medium-sized and large enterprises was assumed to be 18,240, according to the Central Statistical Office of Poland [49]. The study was conducted in the fourth quarter of 2024.

The data analysis utilized an approach based on Grey System Theory (GST), proposed by Deng, which is used to analyse systems characterized by incomplete, imprecise, or partially available information [50]. GST is particularly useful in studies based on qualitative data, small samples, and ordinal scales, such as Likert scales, because it allows for assessing the relative strength of relationships between variables without having to meet restrictive assumptions regarding data distribution [51–55]. The level of BDA use in operations (BDA_use) was measured with a single-item indicator on a 1–5 scale (where 1 means definitely not and 5 means definitely yes), referring to the current use of Big Data analytics in operational processes. The index's construction is based on the consistency of the content with the question's purpose and is used for perceptual assessments with a narrow range of meanings. In the analyses, it was treated as an observable indicator of the intensity of BDA use. The dependent variables used in the study are marked with abbreviations and are understood as follows:

- EE – enterprise efficiency – perceived operational efficiency, e.g. productivity, quality, timeliness,
- NE – natural environment – perceived environmental impact, e.g. energy/material consumption, waste/emissions,
- WC – working conditions – work safety and organization, ergonomics, downtime affecting health and safety,
- SE – social environment – impact on the local community and external stakeholders,
- SS – security of strategic resources – protection of data, know-how, critical infrastructure,
- SR – social responsibility – CSR practices, transparency, compliance with social expectations,
- IR – interpersonal relations – communication, cooperation, organizational climate.

In this study, one of the key methods, Grey Relational Analysis (GRA), was used to assess the relationship between the level of big data analytics use and dimensions of corporate sustainability. The selected method enables comparison of relations between variables with different measurement scales after appropriate normalization, copes well with data limited in number and of a perceptual nature, and allows for obtaining synthetic assessments of Grey Relational Grade (GRG) both at the level of variables and TBL dimensions.

The following research hypotheses were formulated:

- H1. The level of BDA use in operational activities demonstrates a strong positive relationship with assessments of BDA's impact on individual aspects of enterprise performance (EE, NE, WC, SE, SS, SR, IR).
- H2. The relationship between the level of BDA use and the economic dimension of sustainable development (EE, SS) is stronger than the relationship between the level of BDA use and the environmental (NE) and social dimensions (WC, SE, SR, IR).
- H3. The relationship between the level of BDA use and the three dimensions of sustainable development is stronger in large enterprises than in medium-sized enterprises.
- H4. The assessment of BDA's impact on enterprise performance (EE) demonstrates a positive relationship with assessments of environmental impact (NE) and social responsibility (SR).

To test the hypotheses, a two-stage Grey Relational Analysis procedure [56] was used. In the first stage (model 1), the level of BDA utilization in operations was treated as the reference variable, while the assessments of the impact of BDA on EE, NE, WC, SE, SS, SR, and IR were treated as comparative sequences. The following notations were adopted: $x_0(k)$ – values of the reference sequence (level of BDA utilization) and $x_i(k)$ – values of the comparative sequence i (EE, NE, WC, SE, SS, SR, IR) for observation k. Because higher values on the Likert scales correspond to a higher level of BDA utilization or a more positive impact, all variables were treated as benefit indicators and normalized using the "larger-the-better" transformation in the interval [0,1]:

$$x_i'(k) = \frac{x_i(k) - \min_k x_i(k)}{\max_k x_i(k) - \min_k x_i(k)} \tag{1}$$

Then, the distances between the reference sequence and the comparison sequences were calculated:

$$\Delta_i(k) = | x'_0(k) - x'_i(k) | \quad (2)$$

and the minimum and maximum distance in the entire data set were determined

$$\Delta_{\min} = \min_{i,k} \Delta_i(k) \quad (3)$$

$$\Delta_{\max} = \max_{i,k} \Delta_i(k) \quad (4)$$

On this basis, the coefficients of the grey relation $\xi_i(k)$ for the i-th sequence and the k-th variable were calculated for each variable and observation $k$:

$$\xi_i(k) = \frac{\Delta_{\min} + \rho\ \Delta_{\max}}{\Delta_i(k) + \rho\ \Delta_{\max}} \quad (5)$$

where $\rho$ is the distinguishing coefficient, assumed as $\rho = 0.5$ by default. Grey relational grades (GRG) $\gamma_i$ for the i-th sequence, obtained by aggregating $\xi_i(k)$ across variables, were determined as the average value of the grey relational coefficients for each of the dependent variables:

$$\gamma_i = \frac{1}{n}\sum\nolimits_{k=1}^{n} \xi_i(k) \quad (6)$$

which was interpreted as a measure of the strength of the relationship between the level of BDA use and individual aspects of the company's sustainable development and was used to rank these aspects (H1).

In the second stage of the analysis (model 2), the relationship scores were aggregated into three dimensions of the Triple Bottom Line. The variables EE and SS were assigned to the economic dimension, NE to the environmental dimension, and WC, SE, SR, and IR to the social dimension. For each TBL dimension, the aggregate relationship score was calculated as the average of the scores of the corresponding component variables, i.e.:

$$\gamma_{ECO} = \frac{\gamma_{EE} + \gamma_{SS}}{2} \quad (7)$$

$$\gamma_{ENV} = \gamma_{NE} \quad (8)$$

$$\gamma_{SOC} = \frac{\gamma_{WC} + \gamma_{SE} + \gamma_{SR} + \gamma_{IR}}{4} \quad (9)$$

The aggregated relationship assessments defined in this way allowed us to estimate the relative strength of BDA's impact on the economic, environmental, and social dimensions of sustainable development (H2). Hypothesis H3 was verified by comparing the values of $\gamma_{ECO}$, $\gamma_{ENV}$, and $\gamma_{SOC}$ (both for detailed variables and TBL aggregates) in groups of medium-sized and large enterprises. Hypothesis H4 was tested based on the relationship between the BDA impact assessment on efficiency (EE) and the environmental impact assessments (NE) and social responsibility (SR), using a similar procedure of normalization and calculation of coefficients and gray's relationship assessments.

## 4 Results

In the first part of the analysis, using Model 1 described above, the data were normalized, followed by the calculation of distances and grey relational coefficients. An excerpt of the data is presented in Table 1.

**Table 1.** Fragment of calculations performed in model 1.

| Delta_SR | Delta_IR | Delta_min | Delta_max | rho | ξ_SR | ξ_IR |
|---|---|---|---|---|---|---|
| 0,4167 | 0,4167 | 0,0000 | 1,0000 | 0,5000 | 0,5455 | 0,5455 |
| 0,1667 | 0,1667 | | | | 0,7500 | 0,7500 |
| 0,0833 | 0,0833 | | | | 0,8571 | 0,8571 |
| 0,5000 | 0,5000 | | | | 0,5000 | 0,5000 |
| 0,0833 | 0,0833 | | | | 0,8571 | 0,8571 |
| 0,4167 | 0,4167 | | | | 0,5455 | 0,5455 |

GRG scores were then calculated for each dependent variable. All scores range from 0.6659 to 0.6905 (with a small difference of 0.0247 between extreme values) – Table 2.

**Table 2.** Heatmap of GRG for each variable. Scale normalized to [0,1].

| Variable | Overall GRG |
|---|---|
| EE | 0,6684 |
| NE | 0,6659 |
| WC | 0,6905 |
| SE | 0,6727 |
| SS | 0,6780 |
| SR | 0,6768 |
| IR | 0,6845 |

A higher level of Big Data Analytics use in operations is associated with a higher positive assessment of the impact of BDA on the indicated dependent variables. Analysis of the obtained GRG results shows that BDA is strongly related to all analysed aspects. This indicates a consistent, positive relationship between BDA use and assessments of BDA's impact across all analysed areas, confirming hypothesis H1.

In Model 2, aggregated GRG scores were calculated for each TBL area, in accordance with the adopted assumptions. The results are presented in Table 3 an Fig. 1.

**Table 3.** Aggregate GRG score for TBL.

| $\gamma_{ECO}$ | $\gamma_{ENV}$ | $\gamma_{SOC}$ |
|---|---|---|
| 0,6732 | 0,6659 | 0,6811 |

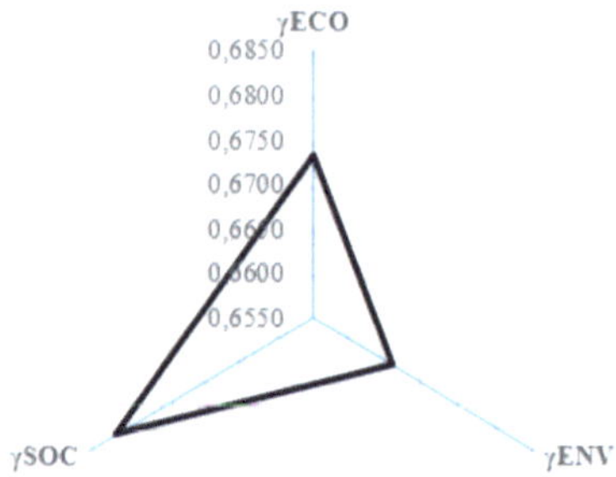


**Fig. 1.** GRG radar chart according to TBL dimensions.

Analysing the results indicates that hypothesis H2 is not supported. The components of the economic dimension do not exhibit higher GRG values than the socio-organizational dimension, for which the relationship strength is the highest. This suggests that respondents perceive BDA primarily as a tool that strengthens the socio-organizational sphere. Economic and environmental benefits of BDA are rated positively and at comparable levels.

The GRA results show that the relationships between the level of BDA use and the dimensions of sustainable development are stronger in large enterprises, which confirms hypothesis H3 (Table 4 and Fig. 2).

**Table 4.** Aggregated GRG rating for TBL by enterprise size.

| | $\gamma_{ECO}$ | $\gamma_{ENV}$ | $\gamma_{SOC}$ |
|---|---|---|---|
| Large | 0,6780 | 0,7268 | 0,6957 |
| Medium | 0,6683 | 0,6633 | 0,6887 |

The largest differences are observed in the environmental dimension, where the rating of the grey relationship in large enterprises is significantly higher than in medium-sized companies. The economic dimension also demonstrates a stronger correlation with BDA in large enterprises. In the social dimension, the differences are small at the aggregate level, but analysis of the component variables reveals a different profile of benefits. In large enterprises, BDA is more often associated with improved working conditions and interpersonal relations, while in medium-sized enterprises, the link between BDA and social responsibility is particularly strongly emphasized (Table 5).

The very high values of the grey correlation (Table 6) obtained between the assessment of the impact of BDA on enterprise efficiency (EE) and the assessment of the impact

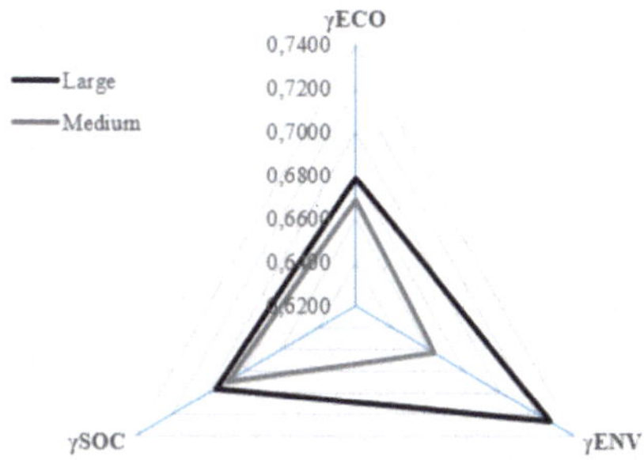


**Fig. 2.** GRG radar chart according to TBL dimensions, disaggregated by firm size.

**Table 5.** Heatmap of GRG for each variable by enterprise size.

| Variable | Large GRG | Medium GRG |
|---|---|---|
| EE | 0,6677 | 0,6674 |
| NE | 0,7268 | 0,6633 |
| WC | 0,7183 | 0,6692 |
| SE | 0,6719 | 0,6718 |
| SS | 0,6883 | 0,6692 |
| SR | 0,6929 | 0,7419 |
| IR | 0,6997 | 0,6721 |

on the natural environment (NE) and social responsibility (SR) indicate a strong coexistence of these assessments. Respondents who perceive BDA as a tool for improving efficiency also perceive its positive impact on the environment and CSR. This means that in the surveyed companies, economic, environmental, and social effects are perceived as complementary, not competitive, which confirms hypothesis H4.

**Table 6.** GRG coefficients between the effect of BDA on EE and the effect of NE and SR.

| GRG_EE_NE | GRG_EE_SR |
|---|---|
| 0,8023 | 0,8040 |

The conducted GRA analyses confirmed that a higher level of BDA use is associated with a strongly positive and largely complementary impact on all dimensions of sustainable development, with particularly strong relationships in the socio-organizational area and more visible effects in large enterprises.

## 5 Discussion

The observed relationships between BDA and TBL dimensions are consistent with studies presented in the literature [6, 9, 11], which demonstrate parallel improvements in operational and environmental indicators and emphasize the role of data-driven practices in CSR. This convergence particularly concerns the strengthening of the socio-organizational sphere through process transparency and improved decision quality. The results of our own research indicate a strong relationship between the level of BDA use and assessments in all studied areas of enterprise operation, which confirms hypothesis H1. The strongest relationships were observed in the socio-organizational dimension (working conditions, interpersonal relations, social responsibility), suggesting that BDA is perceived not only as a tool for improving efficiency and cost reduction, but also as a mechanism promoting process transparency, resource security, and the quality of the working environment. This is confirmed by other authors [7, 8, 25, 26], who indicate a positive impact of BDA on operational and strategic results in the analysed enterprises. However, the literature discusses the impact of BDA primarily on single dimensions of sustainable development. For example, the impact of BDA on the economic dimension is presented by Thanabalan et al. [3] and Popovič et al. [24]; on the environmental dimension: Belhadi et al. [9], Le et al. [12]; on the social dimension: Al Koliby et al. [6] and Khattab [11].

The lack of confirmation of hypothesis H2 in our own research indicates that social and environmental benefits are rated at least as high as economic ones, which weakens the traditional primacy of financial motivations in justifying investments in BDA. The literature indicates that the financial aspect is the main motivation for implementing BDA in an enterprise [3, 7, 11]. However, conclusions consistent with our results can be found regarding the positive impact of BDA on each of the dimensions of sustainable development [25, 26]. The relatively high rating for socio-organizational benefits may reflect the specific conditions of Polish industry: growing pressure for transparency and operational safety, skills shortages on production lines (emphasis on ergonomics and decision-making support), and data integration in the supply chain, which improves communication and work coordination [57, 58]. In these circumstances, BDA solutions are perceived as a tool that streamlines processes and reduces decision-making uncertainty, translating into better working conditions and relationships.

Comparing the relationship between BDA and sustainability dimensions by company size, it can be observed that it is stronger in large companies, especially in the environmental and economic dimensions, which confirms hypothesis H3. A likely explanation is the greater maturity of data and analytical infrastructure, the ability to integrate BDA with IT systems, and access to analytical competencies. In the social dimension, the differences at the aggregation level are small, but component analysis reveals a different profile of benefits: in large companies, BDA was more often associated with improved working conditions and relationships, while in medium-sized companies, the links with CSR activities were more strongly emphasized. There are few studies in the literature demonstrating the relationship between company size and the impact of BDA on sustainability [3, 7, 20, 26].

The positive correlation between BDA impact assessments on enterprise efficiency (EE) and natural environment impact (NE) and social responsibility (SR) confirms hypothesis H4 and indicates the complementarity of economic, environmental, and social goals. This result is consistent with approaches in which data analytics triggers intermediary mechanisms such as green innovation, process improvement, and organizational learning, which lead to simultaneous improvement of operational, environmental, and social indicators [11, 24–26, 59].

## 6 Conclusion

The study confirmed the existence of consistent, positive relationships between the level of Big Data Analytics use and seven areas of enterprise operation, with the strongest connections in the socio-organizational dimension. Analyses showed that social and environmental benefits are rated at least as highly by managers as economic ones, i.e., there is no dominance of the economic dimension. Furthermore, the analyses revealed a stronger relationship between BDA and TBL in large enterprises, particularly in the environmental and economic dimensions. The study confirmed that the use of BDA can simultaneously support economic, environmental, and social goals. These results fill an empirical gap, providing a holistic picture of the impact of BDA in the Polish manufacturing sector and provide a basis for multi-outcome program design. Practically, BDA initiatives should be designed to target the TBL simultaneously and be monitored with multi-dimensional KPIs.

Practical implications include prioritizing BDA implementations that integrate operational goals with ESG objectives (e.g., energy management, resource security, employee well-being), while adapting the scale and tools to organizational maturity. The study has the following limitations: the cross-sectional and self-reported nature of the data and the use of GRA. Further work should complement the results with triangulation with objective KPIs (e.g., energy and material consumption, emissions) and modelling of mechanisms and boundary conditions. It is worth examining, among other things, whether the impact of BDA is indirectly mediated through green innovation (BDA promotes green innovation, which in turn improves performance) and whether the strength of this impact depends on the quality of data governance, as well as the company's size and industry.

**Acknowledgments.** This research was funded by the Poznan University of Technology, Faculty of Engineering Management – project No. 0812/SBAD/4240 and 0813/SBAD/2986.

## References

1. Alorfi, A.S., Alsaadi, N.: Barriers to the adoption of big data analytics in Saudi Arabia's manufacturing sector: An interpretive structural modeling approach. Systems. **13**(4), 250 (2025). https://doi.org/10.3390/systems13040250
2. Raj, R., Kumar, V., Verma, P.: Big data analytics in mitigating challenges of sustainable manufacturing supply chain. Oper. Manag. Res. **16**(4), 1886–1900 (2023). https://doi.org/10.1007/s12063-023-00408-6
3. Thanabalan, P., Vafaei-Zadeh, A., Hanifah, H., Ramayah, T.: Big data analytics adoption in manufacturing companies: The contingent role of data-driven culture. Inf. Syst. Front. **27**(3), 1061–1087 (2025). https://doi.org/10.1007/s10796-024-10491-0

4. Ussembayeva, Z., Dautkanov, N., Tlevlessova, D., Mussayeva, S., Mateyeva, A.: Application of big data for the analysis and optimization of production lines. Period. Eng. Nat. Sci. **13**(3), 551–560 (2025)
5. Primadasa, R., Azzat, N.N., Kusrini, E., Mansur, A., Masudin, I.: Enhancing circular supply chain management (CSCM) in manufacturing SMEs: An integrated CODAS-ISM-MICMAC approach to big data analytics capability. Process Integr. Optim. Sustain. **9**(2), 695–715 (2025). https://doi.org/10.1007/s41660-025-00482-2
6. Al Koliby, I.S., Al-Swidi, A.K., Al-Hakimi, M.A., Mehat, N.A.B., Gelaidan, H.M., Al-Hattami, H.M.: From data to green innovation: The role of green organizational learning capability and total quality management. Sustainable. Futures. **10**, 101191 (2025). https://doi.org/10.1016/j.sftr.2025.101191
7. Haylemariam, L.G., Umar, R.M., Oduro, S.: Modelling the role of big data analytics capabilities in impacting corporate environmental performance: A serial mediation analysis. Bus. Strateg. Environ. **34**(6), 7791–7815 (2025). https://doi.org/10.1002/bse.4373
8. Waqas, M., Honggang, X., Ahmad, N., Khan, S.A.R., Iqbal, M.: Big data analytics as a roadmap towards green innovation, competitive advantage and environmental performance. J. Clean. Prod. **323**, 128998 (2021). https://doi.org/10.1016/j.jclepro.2021.128998
9. Belhadi, A., Kamble, S.S., Zkik, K., Cherrafi, A., Touriki, F.E.: The integrated effect of big data analytics, Lean Six Sigma and green manufacturing on the environmental performance of manufacturing companies: The case of North Africa. J. Clean. Prod. **252**, 119903 (2020). https://doi.org/10.1016/j.jclepro.2019.119903
10. Firdaus, R., Akbar, A., Jahan, S., Poulova, P., Yasmin, F.: Effects of analytics large data set on decision-making and organizational performance: A study on Chinese manufacture sector. J. Artif. Intell. Technol. **5**, 365–375 (2025). https://doi.org/10.37965/jait.2025.0788
11. Khattab, S.A.: The impact of big-data analytics on environmental performance: The mediating roles of Lean Six Sigma and green production. Jordan J. Bus. Adm. **21**(4), 539–562 (2025). https://doi.org/10.35516/jjba.v21i4.403
12. Le, H., Vu, K.C.: Big data analytics and environmental performance: The moderating role of internationalization. Financ. Res. Lett. **64**, 105484 (2024). https://doi.org/10.1016/j.frl.2024.105484
13. Justy, T., Pellegrin-Boucher, E., Lescop, D., Granata, J., Gupta, S.: On the edge of big data: Drivers and barriers to data analytics adoption in SMEs. Technovation. **127**, 102850 (2023). https://doi.org/10.1016/j.technovation.2023.102850
14. Willetts, M., Atkins, A.S.: Performance measurement to evaluate the implementation of big data analytics to SMEs using benchmarking and the balanced scorecard approach. J. Data Inf. Manag. **5**(1), 55–69 (2023). https://doi.org/10.1007/s42488-023-00088-8
15. Mani, M., Madan, J., Lee, J.H., Lyons, K.W., Gupta, S.K.: Sustainability characterisation for manufacturing processes. Int. J. Prod. Res. **52**(20), 5895–5912 (2014). https://doi.org/10.1080/00207543.2014.886788
16. Zhang, T., Ji, W., Qiu, Y.: A framework of energy-consumption driven discrete manufacturing system. Sustain. Energy Technol. Assess. **47**, 101336 (2021). https://doi.org/10.1016/j.seta.2021.101336
17. Awan, U., Braathen, P., Hannola, L.: When and how the implementation of green human resource management and data-driven culture to improve the firm sustainable environmental development? Sustain. Dev. **31**(4), 2726–2740 (2023). https://doi.org/10.1002/sd.2543
18. Bouyahrouzi, E.M., Benmimoun, R., El Kihel, Y., Bajjou, M.S.: Integrating Industry 4.0 technologies and maintenance 4.0 for sustainable manufacturing: A systematic literature review. Int. J. Adv. Manuf. Technol. **140**, 35–61 (2025). https://doi.org/10.1007/s00170-025-16194-3
19. Çiğdem, Ş., Yildiz, B., Meidute-Kavaliauskiene, I., Cincikaite, R.: The impact of IoT usage on big data analytics and supply chain integration. Stud. Bus. Econ. **20**(1), 19–42 (2025). https://doi.org/10.2478/sbe-2025-0002

20. Maroufkhani, P., Iranmanesh, M., Ghobakhloo, M.: Determinants of big data analytics adoption in small and medium-sized enterprises (SMEs). Ind. Manag. Data Syst. **123**(1), 278–301 (2023). https://doi.org/10.1108/IMDS-11-2021-0695
21. Kretzer, A.R., Benitti, F.B., Siqueira, F.: Challenges and opportunities in big data analytics for Industry 4.0: A systematic evaluation of current architectures. IEEE Access. **13**, 183419–183447 (2025). https://doi.org/10.1109/ACCESS.2025.3624558
22. Ragin-Skorecka, K., Motała, D.: Industry 4.0 technologies as a catalyst for lean transformation: Bridging the gap between declaration and practice. In: European Lean Educator Conference, pp. 46–62. Springer, Cham (2025). https://doi.org/10.1007/978-3-032-09806-1_5
23. Wu, W., Li, X., Ruan, G.: How big data analytics capability promotes green radical innovation? The effect of corporate environment ethics in digital era. Systems. **13**(5), 370 (2025). https://doi.org/10.3390/systems13050370
24. Popovič, A., Hackney, R., Tassabehji, R., Castelli, M.: The impact of big data analytics on firms' high value business performance. Inf. Syst. Front. **20**(2), 209–222 (2018). https://doi.org/10.1007/s10796-016-9720-4
25. Raut, R.D., Mangla, S.K., Narwane, V.S., Gardas, B.B., Priyadarshinee, P., Narkhede, B.E.: Linking big data analytics and operational sustainability practices for sustainable business management. J. Clean. Prod. **224**, 10–24 (2019). https://doi.org/10.1016/j.jclepro.2019.03.181
26. Er, C.H., Mosawi, T.A.: Effects of big data analytics on sustainable manufacturing: A comparative study analysis. Chin. J. Urban Environ. Stud. **10**(4), 2250022 (2022). https://doi.org/10.1142/S2345748122500221
27. Mehmood, K., Kautish, P., Rashid, M., Joshi, Y., Iftikhar, Y.: Digitalization in the circular economy: Synergistic impact of big data analytics, green internet of things, and ambidextrous green innovation. J. Clean. Prod. **509**, 145610 (2025). https://doi.org/10.1016/j.jclepro.2025.145610
28. Su, Y., Peng, D., Wang, Y., Tan, Z.: The impact of government open data on firms' energy efficiency: Analyse the mediating role of capacity utilization and biased technological progress. Energies. **18**(17), 4626 (2025). https://doi.org/10.3390/en18174626
29. Cheng, J., Singh, H.S.M., Zhang, Y.C., Wang, S.Y.: The impact of business intelligence, big data analytics capability, and green knowledge management on sustainability performance. J. Clean. Prod. **429**, 139410 (2023). https://doi.org/10.1016/j.jclepro.2023.139410
30. Chang, K.H., et al.: Big data analytics energy-saving strategies for air compressors in the semiconductor industry – An empirical study. Int. J. Prod. Res. **60**(6), 1782–1794 (2022). https://doi.org/10.1080/00207543.2020.1870015
31. Shokravi, S., Smith, A.J.R., Burvill, C.R., Maheswararajah, S.: Industrial environmental performance evaluation – Comparison and sensitivity analysis. In: International Congress on Environmental Modelling and Software (Paper 238) (2012)
32. Ikumapayi, O.M., Laseinde, O.T., Akinlabi, E.T.: Roles of IoT, big data analytics, and cyber-physical systems in a sustainable manufacturing. In: E3S Web of Conferences, vol. 552, p. 01046. EDP Sciences (2024). https://doi.org/10.1051/e3sconf/202455201046
33. Vacchi, M., Settembre-Blundo, D., Iattici, L., Ferrari, A.M., Rosa, R., Berselli, N.: From black box to analytical insight: A data-driven evaluation of technological sustainability in manufacturing supply chains. Supply Chain Analytics. **12**, 100171 (2025). https://doi.org/10.1016/j.sca.2025.100171
34. Munim, Z.H., Dushenko, M., Raaness, H., Westerberg, T.V.: The impact of big data analytics capabilities on the sustainability of maritime firms. In: Data Analytics for Supply Chain Networks, pp. 227–242. Springer, Cham (2023). https://doi.org/10.1007/978-3-031-29823-3_9

35. Mehmood, K., Kiani, A., Rashid, M.D.: Is data the key to sustainability? The roles of big data analytics, green innovation, and organizational identity in gaining green competitive advantage. Tech. Anal. Strat. Manag. **37**(5), 494–508 (2025). https://doi.org/10.1080/09537325.2023.2293867
36. Cheng, J., Sun, N., Zou, Y.N., Lu, K.B., Wang, D.D.: Leveraging big data analytics capability and green supplier integration for improved economic, social, and environmental sustainability in manufacturing companies. SAGE Open. **15**(3), 21582440251378683 (2025). https://doi.org/10.1177/21582440251378683
37. Hosseinzadeh, A., Chen, F.F., Shahin, M., Bouzary, H.: A predictive maintenance approach in manufacturing systems via AI-based early failure detection. Manuf. Lett. **35**, 1179–1186 (2023). https://doi.org/10.1016/j.mfglet.2023.08.125
38. Ndanguza, I.: A.P.: Managerial competence in integrating Industry 4.0 with corporate social responsibility for enhanced safety culture in manufacturing. Sustainability. **17**(10), 4678 (2025). https://doi.org/10.3390/su17104678
39. Dahiya, R., Le, S., Kroll, M.J.: Big data analytics and firm performance: The effects of human capital and mediating firm capabilities. J. Strat. Innov. **36**(1), 200535 (2025). https://doi.org/10.1016/j.jsinno.2025.200535
40. Li, L., Chai, Q., Lin, J., Luo, X.: Examining the interplay between blockchain capability and big data analytics capability in driving sustainable product innovation: A mixed-methods approach. Bus. Strateg. Environ. **35**(2), 2167–2191 (2025). https://doi.org/10.1002/bse.70278
41. Falconí, R., Chávez, J.: Predictive analysis of demand in manufacturing: Inventory optimization through big data. In: Proceedings of the 23rd LACCEI International Multi-Conference for Engineering, Education, and Technology (LACCEI 2025). LACCEI, Mexico City, Mexico (2025). https://doi.org/10.18687/LACCEI2025.1.1.877
42. Miel, A.: Reduction of scrap rate in automotive manufacturing by using big data. In: In: 2025 International Conference on Artificial Intelligence, Computer, Data Sciences and Applications (ACDSA), pp. 1–6. IEEE (2025). https://doi.org/10.1109/ACDSA65407.2025.11166205
43. Chen, H., Wu, J., Zhang, W., Guo, Q., Lu, H.: Study of the predictive mechanism with big data-driven lean manufacturing and Six Sigma methodology. In: IFIP International Conference on Advances in Production Management Systems, pp. 662–672. Springer, Cham (2021). https://doi.org/10.1007/978-3-030-85910-7_70
44. Kalaitzi, D., Tsolakis, N.: Supply chain analytics adoption: Determinants and impacts on organisational performance and competitive advantage. Int. J. Prod. Econ. **248**, 108466 (2022). https://doi.org/10.1016/j.ijpe.2022.108466
45. Raguseo, E., Vitari, C., Pigni, F.: Profiting from big data analytics: The moderating roles of industry concentration and firm size. Int. J. Prod. Econ. **229**, 107758 (2020). https://doi.org/10.1016/j.ijpe.2020.107758
46. Liu, N., Jian, Z., Tan, Y.: Exploring the effects of big data analytics capability on service innovation performance of manufacturing enterprises. Tech. Anal. Strat. Manag. **37**(13), 4168–4183 (2024). https://doi.org/10.1080/09537325.2024.2441807
47. Matejun, M.: Methodology of survey research in management sciences: A model-based approach. In: Lisiński, M., Ziębicki, B. (eds.) Contemporary Problems in the Development of Management Methodology, pp. 341–354. Cracow University of Economics Foundation (2016)
48. Cochran, W.G.: Sampling Techniques, 3rd edn. John Wiley & Sons (1977)
49. GUS–Statistics Poland: Enterprises Department: Activity of Non-Financial Enterprises in 2023, Statistics Poland (2024)
50. Deng, J.L.: Control problems of grey systems. Syst. Control Lett. **1**(5), 288–294 (1982)
51. Liu, S., Fang, Z., Yang, Y., Forrest, J.: General grey numbers and their operations. Grey Syst. Theory Appl. **2**(3), 341–349 (2012). https://doi.org/10.1108/20439371211273230

52. Mierzwiak, R., Więcek-Janka, E.: The analysis of successors' competencies in family enterprises with the use of grey system theory. Grey Syst. Theory Appl. **5**(3), 302–312 (2015). https://doi.org/10.1108/GS-03-2015-0007
53. Prakash, S., Agrawal, A., Singh, R., Singh, R.K., Zindani, D.: A decade of grey systems: Theory and application – Bibliometric overview and future research directions. Grey Syst. Theory Appl. **13**(1), 14–33 (2023). https://doi.org/10.1108/GS-03-2022-0030
54. Ragin-Skorecka, K.: Influence of the knowledge of selected concepts on the organization process maturity. In: Soliman, K.S. (ed.) Proceedings of the 37th International Business Information Management Association Conference (IBIMA), pp. 7972–7977. IBIMA Publishing, Cordoba, Spain (2021)
55. Więcek-Janka, E., Nowak, M., Borowiec, A.: Application of the GDM model in the diagnosis of crises in family businesses. Grey Syst. Theory Appl. **9**(1), 114–127 (2019). https://doi.org/10.1108/GS-06-2018-0030
56. Liu, S., Lin, Y.: Grey Information: Theory and Practical Applications. Springer (2006)
57. Loska, A., Paszkowski, W., Waszkowski, R.: The scenario approach to the concept of maintenance of technical systems of urban engineering. In: International Scientific-Technical Conference Manufacturing, pp. 84–97. Springer, Cham (2024). https://doi.org/10.1007/978-3-031-56444-4_7
58. Piechowski, M., Meller, A.: Use of KPI indicators to optimize production process performance – Case study. In: International Symposium on Distributed Computing and Artificial Intelligence, pp. 3–12. Springer, Cham (2023). https://doi.org/10.1007/978-3-031-38318-2_1
59. Sellitto, M.A., Borchardt, M., Pereira, G.M.: Modelling for environmental performance assessment in manufacturing operations. Manag. Prod. **17**, 95–109 (2010). https://doi.org/10.1590/S0104-530X2010000100008

# Smart Vision System in Quality Inspection of Medical Products – Case Study

Agnieszka Kujawińska[1](✉), Wenjie Dong[2], Magdalena Hryb[1], and Adrian Czernek[1]

[1] Faculty of Mechanical Engineering, Poznan University of Technology, Poznan, Poland
agnieszka.kujawinska@put.poznan.pl

[2] College of Economics and Management, Nanjing University of Aeronautics and Astronautics, Nanjing, PR China

**Abstract.** The article presents the concept, implementation, and effectiveness assessment of a vision system incorporating elements of artificial intelligence (AI) to support quality control in the packaging process of orthopedic bone wax. The research problem addressed the inherent limitations of human visual inspection -namely, variability in assessment outcomes, cognitive load imposed on operators, and difficulties in detecting defects on transparent or glossy surfaces - as well as the need to ensure repeatable and objective real-time inspection. A system was designed that leverages AI-based tools in the learning process of the inspection program. The effectiveness evaluation, conducted under real production conditions, demonstrated a reduction in the number of nonconformities identified during the final inspection (after the packaging operation), accompanied by a decrease in cycle time and improved process stability. The proposed approach constitutes a practical example of AI implementation in the medical device sector, emphasizing product compliance with regulatory requirements and system robustness against process disturbances.

**Keywords:** Vision System · Artificial Intelligence · Quality Control · Orthopedic Wax

## 1 An Introduction

Visual inspection is one of the oldest and most widely used methods of conformity assessment in manufacturing processes. The primary benefit of this approach is the immediate ability of the operator to observe and respond to any detected non-conformity. In a multitude of industries, particularly those that demand elevated standards of quality, such as the medical, electronic and automotive sectors, visual inspection constitutes a pivotal phase in the assurance of product quality and safety. Notwithstanding the advantages of traditional visual inspection, its application is constrained by numerous limitations arising from human physiological and cognitive capabilities. Prolonged observation has been shown to result in a state of fatigue, decreased concentration, and differences in perception between operators. This, in turn, leads to a reduction in the repeatability and


A. Hamrol et al. (Eds.): MANUFACTURING 2026, LNME, pp. 122–131, 2026.
https://doi.org/10.1007/978-3-032-21701-1_9

variability of assessments over time. Additionally, the detection of non-conformities on surfaces characterised by irregular structures, high gloss or transparency poses a significant challenge. In mass production settings, attaining the requisite levels of precision and expediency is often unfeasible [1–4].

The development of vision technologies and image processing tools has enabled inspection processes to be automated, eliminating subjectivity in assessment and increasing the consistency of results. Vision systems use algorithms for edge detection, object segmentation, and colour, texture, and shape analysis to allow image analysis. This not only makes it possible to detect non-conformities, but also to classify them and identify trends in their occurrence. In recent years, there has been a clear shift from manual inspection to automated vision systems that use AI tools. This shift is evidenced by the systematic incorporation of machine learning methods into AVI (Automated Visual Inspection) [5, 6].

The capabilities of vision systems used in industrial quality control have been expanded by artificial intelligence (AI) and machine learning (ML) methods. Rather than using classic algorithms based on fixed decision rules, these systems employ learning models that can adapt to changes in lighting, materials and geometry. Consequently, modern automatic visual inspection (AVI) systems are more flexible and resilient to environmental disturbances, rendering them an effective tool for ensuring the quality of production processes [7, 8].

Convolutional neural networks (CNNs) are among the most effective deep learning methods. Their ability to recognise patterns in visual data enables accurate image classification and defect detection, even when the differences between compliant and defective products are subtle and difficult for the human eye to discern. CNNs are used in supervised systems, which learn from labelled examples, as well as in semi-supervised and unsupervised systems, which can independently detect deviations from the pattern [9, 10].

From a practical application perspective, the accuracy of the algorithm is becoming increasingly important, as is its integration with control and management systems. Research shows that integrating a vision system with artificial intelligence elements into line control devices (e.g. PLC controllers) and production and quality management systems (MES and QMS) increases its efficiency. This enables online and in-line responses, such as the automatic rejection of defective products or their redirection for re-inspection. Such a combination also allows for full data traceability, i.e. the ability to retrieve information about when, during which shift and for which reason a given product was rejected [8].

However, the literature also points out that, despite significant progress in the development of vision systems using artificial intelligence, there are still several problems that limit their long-term reliability. The first of these is domain drift, whereby an AI model gradually loses accuracy when production conditions change. Examples of such changes include the use of a different material supplier or a change in colour, gloss or surface texture. This is known as overfitting. In such situations, the system may misclassify correct products as defective or vice versa [11, 12].

The second challenge is the inability of humans to understand the model's decisions. Operators often only see the result (*good* or *defective*), but do not know why the system evaluated the image in this way. This is why XAI (Explainable Artificial Intelligence) systems that can show which parts of the image influenced the model's decision are increasingly being developed [13].

Another issue is ensuring data security and compliance with data processing rules, particularly when images are stored, analyzed or transferred between different systems within a company. It is essential to ensure that data is securely stored and only used for quality control purposes [14, 15].

Another important issue is managing the uncertainty of results, i.e. determining how "confident" the model is in its decision. Modern solutions can recognize their own level of confidence and refer ambiguous results for re-analysis, for example. The literature suggests using adaptive thresholds, which vary depending on production conditions, and incremental learning. This allows the system to update its knowledge based on new examples confirmed by experts. This helps maintain system stability and minimise erroneous decisions in the long term [15, 16].

Based on the above, the research aimed to develop and implement a vision system incorporating artificial intelligence elements for the automatic detection and classification of products in the orthopedic wax packaging process. The system was intended to operate online, integrate with the production line and ensure repeatable, objective and rapid quality control. This article presents the system's conceptual assumptions, its evaluation in industrial conditions and an analysis of the impact of its implementation on key performance indicators in the process. Section 2 presents the materials and methods, including process description, quality requirements, system architecture, learning and testing procedures, and implementation and evaluation methodology. Section 3 presents the quantitative and qualitative results, along with the interpretation and analysis of their impact on process efficiency.

## 2 Materials and Methods

### 2.1 Orthopedic Wax

Orthopedic wax is used in surgery and belongs to the classic category of non-absorbable mechanical hemostats. Orthopedic wax is made from a mixture of purified beeswax and softening agents, most commonly pharmaceutical petroleum jelly. This composition gives the material plasticity and stability at body temperature, enabling the surgeon to mold the wax to the shape of the bleeding bone surface. The most important advantages of wax include ease of application, a hemostatic effect and stable physical properties. This chemically inert material does not react with blood or bone tissue; its action is purely mechanical. Hemostasis is achieved by physically closing the lumen of the vessels in the bone - known as tamponade - without the involvement of any biochemical processes. This makes wax an effective tool for controlling bleeding in situations where other methods are inadequate or impractical. Orthopedic wax is non-absorbable, meaning it can remain in tissue long-term. Nevertheless, it remains one of the most widely used and proven means of controlling intraoperative bone bleeding.

In clinical practice, orthopedic wax is widely used in orthopedics and traumatology, neurosurgery, thoracic surgery, and maxillofacial surgery, especially during procedures on long bones, ribs, the sternum, or the cranial vault. It is available in the form of flakes, blocks, or sticks, which facilitates its quick use and adaptation to various operating conditions (Fig. 1).

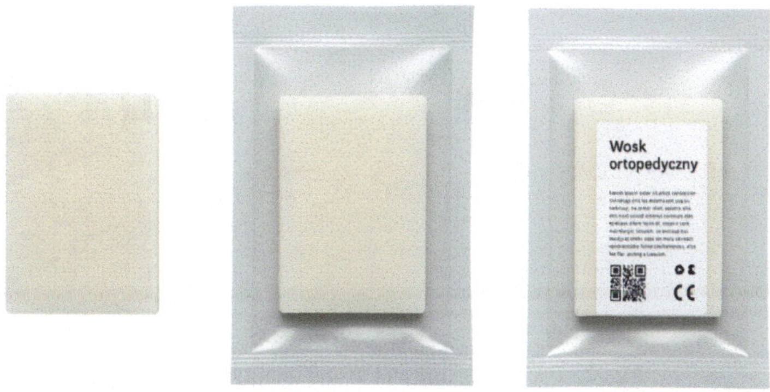


**Fig. 1.** Orthopedic wax in unpackaged and packaged form (own work).

The production process of orthopedic wax involves multiple stages and is subject to stringent quality requirements resulting from industry standards and internal Good Manufacturing Practice (GMP) rules. The finished product must have consistent physical properties, meet material purity requirements, and enable full traceability of each produced batch.

The production process begins with the selection of the raw materials: purified beeswax and pharmaceutical-grade petroleum jelly. Both ingredients are checked for chemical and microbiological purity. Next, the ingredients are measured precisely and homogenized under controlled conditions. This takes place in closed, sterile tanks to minimise the risk of contamination. The temperature, viscosity and homogeneity of the mass are constantly monitored to ensure a stable and repeatable mixture. The finished mixture is then divided into portions and formed into flakes. Each unit undergoes an additional visual inspection, during which the operator assesses its homogeneity, the absence of contamination and its consistency. Once approved, the product is secured with protective film. Next, the wax undergoes a sterilization process, typically using gamma radiation or pressurized steam. After sterilization, the product is packaged individually and marked with the batch number, expiry date, CE markings and data enabling full traceability throughout the product's life cycle. Final quality control includes both a visual inspection and an analysis of the production documentation. Only after positive verification is the product placed in collective packaging and sent for distribution.

The most important critical characteristics of orthopedic wax are sterility, uniform consistency, softness, plasticity, an absence of impurities, chemical stability and full traceability of each batch.

The orthopedic wax packaging process also includes a quality control operation, which is currently performed visually by a qualified operator. At this stage, each product unit is assessed for packaging integrity, correct film sealing, absence of contamination and compliance of the label with the production documentation. Among other things, the operator checks (see Fig. 2) that the packaging is undamaged, the batch number and expiry date are legible, and there are no foreign particles inside the sachet.

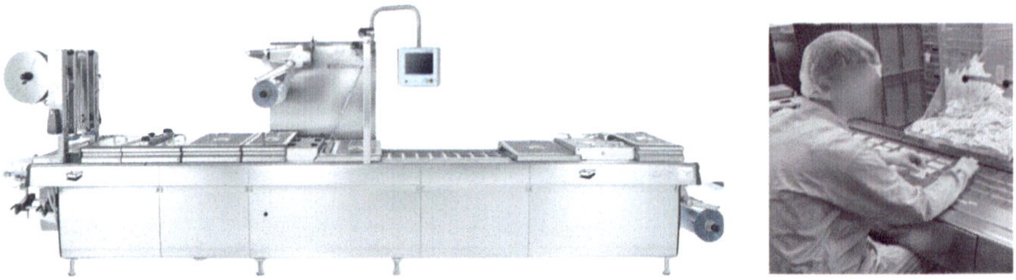

packaging line visual inspection

**Fig. 2.** Packaging line and visual quality control.

Visual inspection is a key element of the quality system. If contaminated wax reaches the recipient, it carries serious clinical and regulatory risks. Microbiological contamination or foreign particles can cause surgical site infections, foreign body reactions or chronic inflammation, and can interfere with bone healing. If a product does not meet sterility requirements, the entire batch must be recalled, the incident reported to the relevant authorities, and the manufacturer may face legal action. For this reason, final quality control, including a detailed assessment of visual cleanliness and packaging integrity, is essential for ensuring clinical safety and compliance.

Despite its simplicity, the current visual quality control method is limited in its effectiveness and very time-consuming. Studies have shown that operators can only detect between 60% and 80% of contaminants, and this figure drops to 50% for long-term work or work performed in suboptimal lighting conditions. Furthermore, visual assessment alone accounts for 20–30% of the entire packaging process, significantly reducing operational efficiency. The variability of human perception and the increasing complexity of regulatory requirements mean that manual visual inspection is an inadequate solution, making it necessary to implement more consistent, automated inspection methods.

## 2.2 Potential Product Defects

Five main classes of wax defect have been defined: D1: damage or deformation of the blister; D2: incomplete filling of the chamber with wax; D3: contamination/bubbles (solid inclusions or gas bubbles); D4: labelling errors (offset, rotated, creased or missing); D5: surface contamination (streaks or drops). Figure 3 shows two examples of non-conformity: contamination of the wax and a hair in the package.

Each non-conformance class was assigned geometric and photometric feature descriptors (e.g., contour, texture, and contrast metrics), which form the basis for preliminary segmentation and classification in the AI learning process. Acceptance/rejection criteria were defined based on product quality requirements, user risk, and operator experience.

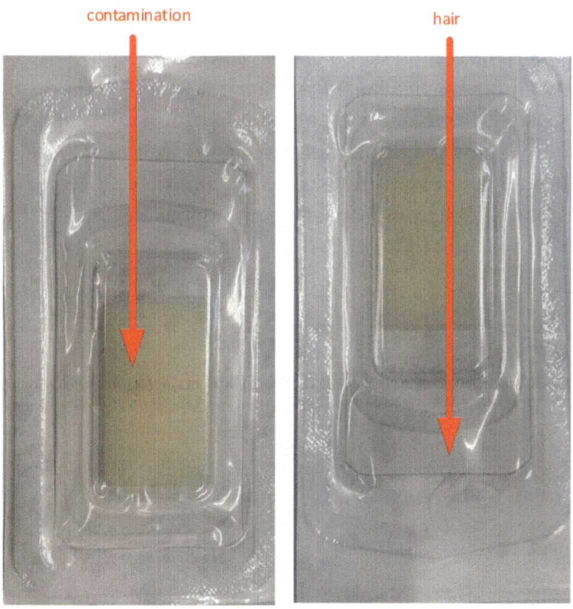


**Fig. 3.** Example of defects.

Complex defects (combinations of defects, e.g., incomplete filling + blisters) were also defined and treated as multi-labeled in the classification. For the purposes of root cause analysis, each detection is tagged with metadata for the batch, work change, operator, and process settings, which allows defects to be correlated with manufacturing parameters.

## 2.3 Vision System Architecture (Hardware and Software)

Figure 4 shows the general architecture of the system. It includes a lighting module to ensure stable illumination conditions, a camera to record the image of the product surface, a PLC controller to process signals and communicate with the production line, and a monitor to visualize the analysis results. The entire system is integrated with a conveyor belt, on which the product to be inspected is placed, and a rejection mechanism, which is responsible for separating non-compliant items. This diagram is for illustrative purposes only and shows the system's structure and the flow of information between its modules.

The optical module was designed as a system comprising a camera with a macro lens, a positioning mechanism and an optical filtration system. Diffuse ring lighting was employed to minimise reflections and local overexposure (hot spots) on transparent or shiny surfaces. Image acquisition was triggered by a signal from a linear encoder, while frame buffering ensured uninterrupted system operation. Preliminary data processing (luminance normalization, noise reduction and distortion correction) and artificial intelligence model inference were performed in the computing unit.

The proposed solution uses a Keyence camera in conjunction with VS Creator 1.5.4 software. This enables the configuration of image acquisition parameters and the implementation of artificial intelligence-supported analytical tools. The system was adapted to record images of the front and back of orthopaedic wax, enabling the assessment of surface uniformity and the detection of defects, as well as the analysis of label and imprint accuracy (including expiry dates and QR codes). VS Creator software enables object detection and the definition of inspection regions, as well as image filtering to correct shadows, compensate for uneven lighting and reduce noise.

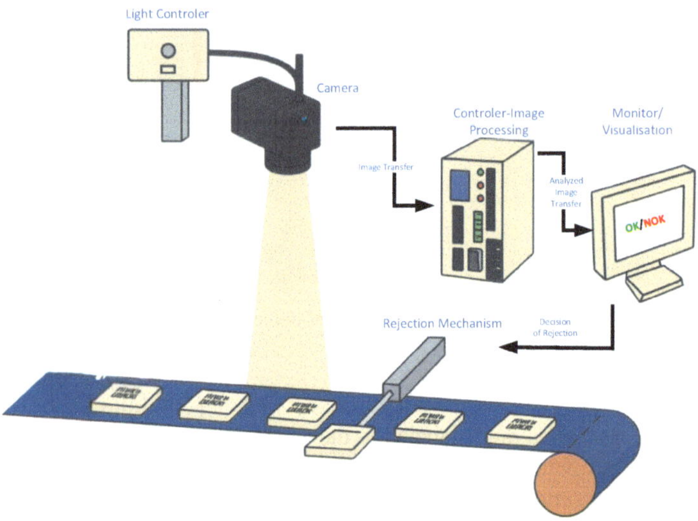


**Fig. 4.** General scheme of system.

The artificial intelligence module was a key element of the vision system's architecture. Not only did the software enable classic image processing, such as object detection, inspection region definition, lighting compensation and noise reduction, but it also integrated AI tools based on supervised learning. The model was trained by presenting examples of correct and incorrect products, enabling it to independently shape decision boundaries and increase the effectiveness of detecting subtle defects.

The AI OCR and AI QR modules enabled the automatic reading of text, numbers and two-dimensional codes while analysing print quality and detecting contaminants affecting reading accuracy. Consequently, the vision system could self-adjust the tolerance threshold, generalize data and reduce the number of false alarms, thereby significantly exceeding the capabilities of traditional visual inspection.

## 2.4 System Training and Testing

A collection of 480 products, obtained directly from the production line, was prepared to train the artificial intelligence model. This collection represented two versions of the product (good/bad) under constant lighting conditions. All images were manually labelled using a dedicated annotation tool and a double-checking process was employed to reduce labelling errors to below 0.5%.

The set was divided into: 70% for training, 15% for validation and 15% for testing. The data were separated by batch numbers and production changes. To increase data diversity, the following augmentations were applied: rotations of $\pm 5°$, shifts of up to 10 pixels. These measures improved the model's ability to generalise.

The base model was a medium-depth convolutional neural network, optimised for on-edge operation on a computing unit integrated with a Keynece camera. After implementation, incremental fine-tuning was activated using newly recorded non-compliant examples. Each model update required passing a regression test on a set of 20 reference samples.

To quantitatively assess the model's effectiveness, classic metrics used in automatic visual inspection (AVI) tasks were employed. Each metric describes a different aspect of the classifier's accuracy:

- *True Positive Rate* (TPR, or sensitivity) determines the proportion of actual defects that have been correctly classified as non-compliant. This metric is particularly important in safety- and quality-critical contexts, as it minimises false negatives.
- *Positive Predictive Value* (PPV) expresses the proportion of correct detections among all cases that the model considers defective. From a quality cost perspective, high precision is crucial, as it reduces the number of false positives and unnecessary rejections of compliant products.
- *True Negative Rate* (TNR, or specificity) describes the system's ability to confirm correct products. It determines how rarely the model incorrectly marks a compliant product as non-compliant.
- The *F1* measure is the harmonic mean of TPR and PPV, providing an overall measure of the quality of the classifier. It is particularly useful in systems where both failure to detect a defect and false positives are critical.

Table 1 summarizes the results of the effectiveness assessment of the developed model for the test data.

**Table 1.** Results of the AI model effectiveness assessment.

| The name of metrics | Acronim | Value |
|---|---|---|
| True Positive Rate | TPR | 0.88 |
| Positive Predictive Value | PPV | 0.92 |
| True Negative Rate | TNR | > 0.95 |
| The F1metric | F1 | 0.89 |

The obtained sensitivity, TPR = 0.88, indicates that the model correctly identifies 88% of actual defects in blisters containing orthopedic wax. This is significantly higher than the typical accuracy of manual inspection and translates into fewer undetected defects (false negatives), which is particularly important in the medical device sector. The precision (PPV) of 0.92 means that only a small proportion of cases marked as non-compliant were false alarms. In other words, over 92% of defect signals were justified, reducing the number of unnecessary rejections and improving process efficiency. TNR specificity of more than 0.95 confirms the stable and reliable identification of compliant products. The model does not tend to reject good products excessively, which is crucial for maintaining production flow and reducing losses. F1 score of 0.89 indicates a high level of overall effectiveness of the classifier, balancing sensitivity and precision. This confirms that the model performs well in both defect detection and minimizing misclassifications.

## 3 Summary and Results

The assessment of the effectiveness of the vision system incorporating artificial intelligence elements confirmed that the implemented solution far exceeds the capabilities of traditional visual inspection. The obtained metrics - high sensitivity (TPR = 0.88), precision (PPV = 0.92) and specificity exceeding 0.95 - clearly indicate that the system has effectively reduced both undetected defects and false rejections. The F1 value of 0.89 confirms the system's ability to balance both types of error, which is crucial in the inspection of medical products.

A significant effect of implementing the system on the production line was an increase in the speed and consistency of the inspection process. Thanks to automatic image acquisition and processing, the system can perform a full inspection of a blister in under 250–300 ms, enabling it to operate at a line speed of 100–110 pieces per minute without causing delays. Compared to manual inspection, where an operator would take several times longer to evaluate a single product, this represents a significant reduction in cycle time and an increase in throughput for the packaging operation.

Furthermore, automating the inspection process eliminated variability resulting from human factors such as eye fatigue, cognitive load and lighting conditions. Integrating the system with a rejection mechanism and a PLC controller enabled smooth transfer of OK/NOK status information and immediate responses to detected non-conformities. This streamlined material flow and reduced costs associated with handling non-conforming products in later stages of the process.

Using AI tools increased the system's resistance to optical interference and variability in production conditions even further. The learning module, which was based on examples of correct and incorrect products, enabled decision boundaries to be adapted to actual conditions and subtle defects to be detected that were invisible during manual inspection. Consequently, the developed system improved the quality of non-compliance detection and contributed to stabilizing the process, thereby increasing its operational reliability.

**Acknowledgments.** The presented results derive from a scientific statutory research conducted by Chair of Production Engineering, Faculty of Mechanical Engineering, Poznan University of Technology, Poland, supported by the Polish Ministry of Science and Higher Education from the financial means no 0613/SBAD/4940.

## References


1. Kujawińska, A., Diering, M.: The impact of the organization of the visual inspection process on its effectiveness. Int. J. Adv. Manuf. Technol. **112**, 1295–1306 (2021). https://doi.org/10.1007/s00170-020-06543-9
2. Kujawińska, A., Diering, M., Przybył, A.: Machine vision system for quality control of stents used in Angioplasty. In: Diering, M., Wieczorowski, M., Harugade, M. (eds.) Advances in Manufacturing IV. Manufacturing 2024 Lecture Notes in Mechanical Engineering. Springer, Cham (2024). https://doi.org/10.1007/978-3-031-56467-3_18
3. See, J.E.: Visual inspection: A review of the literature. Sandia Report 77 (2012).
4. Borkowski, S., Knop, K.: Challenges faced in modern quality inspection. Manag. Prod. Eng. Rev. **7**, 11–22 (2016). https://doi.org/10.1515/mper-2016-0022
5. Hütten, N., Alves Gomes, M., Hölken, F., Andricevic, K., Meyes, R., Meisen, T.: Deep learning for automated visual inspection in manufacturing and maintenance: A survey of open-access papers. Appl. Syst. Innov. **7**(1), 11 (2024). https://doi.org/10.3390/asi7010011
6. Zhou, J., Hartman, N.: A Model-Based Visual Inspection System (MBVIS) for critical plastic bottle dimensional measurements. Comput. Aided. Des. Appl. **21**, 270–280 (2023). https://doi.org/10.14733/cadaps.2024.270-280
7. Ameri, R., Hsu, C.-C., Band, S.S.: A systematic review of deep learning approaches for surface defect detection in industrial applications. Eng. Appl. Artif. Intell. **130**, 107717 (2024). https://doi.org/10.1016/j.engappai.2023.107717
8. Vazquez-Nava, A., Ahuett-Garza, H., Orta-Castanon, P., Saldana, C.: Fast inline inspection: Vision system for geometric deviation measurement and early defect detection in smart manufacturing. Int. J. Comput. Integr. Manuf. **38**, 1–24 (2025). https://doi.org/10.1080/0951192X.2025.2503316
9. Bhandari, B., Park, G.: Non-contact surface roughness evaluation of milling surface using CNN-deep learning models. Int. J. Comput. Integr. Manuf. **37**, 423–437 (2024). https://doi.org/10.1080/0951192X.2022.2126012
10. Papavasileiou, A., Michalos, G., Makris, S.: Quality control in manufacturing – Review and challenges on robotic applications. Int. J. Comput. Integr. Manuf. **38**, 79–115 (2025). https://doi.org/10.1080/0951192X.2024.2314789
11. Puttero, S., Verna, E., Genta, G., Galetto, M.: Automatic component recognition and defect detection in electronic board recycling process. Int. J. Comput. Integr. Manuf. **38**, 1–16 (2025). https://doi.org/10.1080/0951192X.2025.2515393
12. Maślanka, M., Jancarczyk, D., Rysinski, J.: Integration of machine vision and PLC-based control for scalable quality inspection in Industry 4.0. Sensors. **25**, 6383 (2025). https://doi.org/10.3390/s25206383
13. Ali, S., et al.: Explainable Artificial Intelligence (XAI): What we know and what is left to attain trustworthy Artificial Intelligence. Inf. Fusion. **99**, 101805 (2023). https://doi.org/10.1016/j.inffus.2023.101805
14. Fischer, S., Wenger, A.: Artificial intelligence, forward-looking governance and the future of security. Swiss Polit. Sci. Rev. **27**, 170–179 (2021). https://doi.org/10.1111/spsr.12439
15. Bécue, A., Praça, I., Gama, J.: Artificial intelligence, cyber-threats and Industry 4.0: Challenges and opportunities. Artif. Intell. Rev. **54**, 3849–3886 (2021). https://doi.org/10.1007/s10462-020-09942-2
16. Rudin, C., Chen, C., Chen, Z., Huang, H., Semenova, L., Zhong, C.: Interpretable machine learning: Fundamental principles and 10 grand challenges. Stat. Surv. **16**, 1–85 (2022). https://doi.org/10.1214/21-SS133

# Sustainable Development

# A Conceptual Framework for Assessing the Readiness and Suitability of Enterprises for Militarization and the Fulfillment of State Functions in the Context of Armed Conflict

Aleksander Jawor[2](✉) and Anna Borucka[1]

[1] Wrocław University of Technology, Wybrzeże Wyspiańskiego 28, 50-370 Wrocław, Poland
[2] Military University of Technology, Kaliskiego 2, 00-908 Warszawa, Poland
aleksander.jawor@pwr.edu.pl

**Abstract.** The geopolitical challenges of the 21[st] century necessitate closer cooperation between the government and enterprises in the implementation of tasks for the military and state services. This article presents a conceptual matrix designed to assess the suitability of a given enterprise in the context of an armed conflict. It also outlines the key factors influencing an enterprise's suitability and discusses how these factors can be incorporated into both corporate development strategies and governmental guidelines as well as national defense strategies. The implementation of militarization in the production process and its management requires meeting a set of criteria to ensure that this transformation is characterized by reliability. The enterprise assessment process constitutes an integral component of militarization.

**Keywords:** Production · Militarization · Production Management in War Time · Industry · Armed Conflicts

## 1 Introduction and Literature Review

The origin of this research is the essence and significance of manufacturing enterprise in national defense system. The national security is multi-meaning concept containing not only military sphere but also economic stability, social cohesion and the protection of national interests. It is dynamic category which is influence by internal and external factors. Its assurance is one of the most fundamental task of state. In Poland, which is analyzed in this research, the constitution [1] in art. 5 impose an obligation on the State duty to secure independence and inviolability of borders, to guarantee humans and citizens' rights and freedoms and its safeness. This constitutional obligation is a fundamental for all state activities in security area, especially for programs designed to prepare the economy and industry for militarization and all potentially crisis situation.

In the national security system the economy plays the crucial role as the fundament for state stability and the ability to respond for various threads. An effectively functioning and resilient economy is necessary to ensure all financial resources for defense


A. Hamrol et al. (Eds.): MANUFACTURING 2026, LNME, pp. 135–148, 2026.
https://doi.org/10.1007/978-3-032-21701-1_10

needs, to keep social cohesion and to enable state to work under crisis conditions. Among the crucial areas where the economy intersects with security should be listed especially financial, energy and food security as well as industrial defense potential [2]. The last one is defined as material resources within Poland's industry that secure state defense needs and it pose significant element of whole defense system. Facing modern civilizational threats characterized by increasing complexity and unpredictability, the national defense preparation becomes crucial. The evolution of threats including hybrid war, cyberattacks and geopolitical instability forces to use complex approach to security that extends traditional military abilities. In such complicated scenarios the economic resilience and the ability to quick adaptation the economic structure to military needs are as important as military readiness. Therefore, the responsibility for personal and material services concern both citizens and entrepreneurs. The document that regulates production and modernization tasks for entrepreneurs during wartime and peacetime is "Program Mobilizacji Gospodarki" (PMG – the Economic Mobilization Program) [3]. This document is secret but is fundamental for all state activities in security field, including economical mobilization.

The fundamental objective of the Economic Mobilization Program (PMG) is to prepare the national economy to fulfill tasks related to securing increased defense needs of the state, including the Armed Forces of the Republic of Poland, in conditions of national security threats and during wartime. This entails planning the potential conversion of part of the industry to war production, effectively allocating resources, and ensuring the continuity of supply of key products and services.

The preparation of the national economy for effective operation under conditions of threat and wartime, through securing essential supplies, services, production, and infrastructure, imposes an obligation on enterprises to prepare for these requirements. The mobilization of material resources embraces the utilization of raw materials, production capacities and infrastructure to secure the continuity of supply of strategic goods and services. The use of technological potential is significant to. It is necessary for the production of modern military equipment and the performance of specialized services. Entrepreneurs, especially those of significant economic-defense meaning, are also obligated to prepare mobilization plans within their establishments, which define the manner and principles of their operation under conditions of national security threats and during wartime.

Preparing the economy for militarization requires that each of its components be ready for such eventuality. It is crucial that in times of crisis all plans function are effectively and fully implemented. The process of production militarization comprises a set of principles and requirements for enterprises concerning the design, execution of manufacturing processes, and maintenance [4, 5]. In such a context, reliability is one of the key factors affecting system readiness [6]. Therefore, certain risks can be mitigated by pre-assessing the specified enterprises and excluding those that do not meet the criteria. The quality of preparation could be increased if all instructions and guidelines will be implemented.

There is a lot of examples in history where the human factor was the reason of lose. Management in army operates at multiple levels. The whole people, especially team leaders, should know the strategy course. Some researchers criticize that approach and suggest that human and organizational factors are not reasons of strategic errors. Nevertheless the strategy analysis on country level help to decrease that risk [7]. The assessment of a company in terms of its suitability for militarization should constitute one of the components of the national defense strategy.

The result of conducting such an assessment on a national scale will be the generation of a vast amount of data requiring processing. Therefore data handling becomes essential. Machine learning offers the capability to automate numerous tasks and to draw conclusions based on their outcomes. A significant increase in publications on this subject has been observed [8]. Formulating conclusions and based on them developing strategic guidelines for a country of significant size, such as Poland, may be complex without software support. Robust data analysis enables real-time responses and the implementation of strategies which translates into improved national preparedness for militarization.

It should also be noted that although companies of significant importance to the state have been identified there is currently no possibility to assess or compare them in terms of their readiness for militarization [9], as no evaluation model exists. Such a model would also enable companies to prepare themselves for assessment, while also serving as an effective tool for army decision-makers, allowing them to conduct systematic and comparable enterprise evaluation according to uniform criteria. This model would also make it possible to design clear criteria for judging enterprise militarization readiness and to determine or simulate [10] the level of readiness. The grading indicator would be measurable and could be used both for planning mobilization activities and for monitoring the evolution of economy's potential adaptation to the needs of national defense system. Achieving a certain evaluation score could contribute to the company's prestige and improve its perception in the market. Moreover, it would help ensure higher quality of military products. Production management involves monitoring the process and its key factors, such as human resources, identification of bottlenecks and material availability. It also encompasses a range of activities, including optimal utilization of equipment and machinery as well as assortment strategy. The most crucial measure of the process is its productivity [11]. This is particularly important in the era of Industry 4.0, where cost reduction and error elimination are of paramount importance [12]. Lean manufacturing tools such as 5S and TPM are already standard practice in many industrial facilities [13]. They should also be considered in the context of militarization, including in plants characterized by single-unit and small-batch production.

The decision-making process is also undergoing digitalization and is increasingly supported by artificial intelligence and machine learning systems [14]. While this phenomenon brings significant benefits in terms of data processing speed, operational optimization, and strategic scenario forecasting, in the context of the defense and armaments sector it simultaneously generates new challenges related to information security. The digitization and automation of processes increase the risk of interception, manipulation, or unauthorized access to strategic data, defense-related technologies, or classified information, which becomes particularly critical in situations of national threat. Moreover, the integration of decision-making systems with artificial intelligence–based solutions entails risks associated with the vulnerability of algorithms to cyberattacks, disinformation, and attempts to interfere with their functioning, which may lead to disruptions in the continuity of key defense processes. Therefore, it should be emphasized once again that business processes complement technological processes and should be considered together. In order to effectively monitor changes in both strategy and processes, it is essential to define and track key performance indicators (KPIs) [15]. In the case of militarization decisions have to be made very rapidly; therefor as part of the strategy, decision-making models should be developed to support achieving the required product quality at every stage of its development. These are mathematical models that provide concrete, comparable results [16]. They can also be applied to various processes, such as transportation [17], logistics and supply operations [18]. In the case of mass production, the use of tools such as Six Sigma or statistical analysis, including linear regression, can be highly beneficial. However these methods must serve as a complement to production planning, not a substitute for it [19]. The Kanban tool enables control over material flows within the production process and helps reduce production costs, primarily by minimizing waste. A similar approach is the Drum-Buffer-Rope method, which focuses on managing production capacity [20]. In a national-scale readiness assessment, these process-oriented methods can be operationalized through digitalized, standardized data collection and analysis; therefore, a minimal data architecture is required to ensure consistent inputs and to enable ML-based decision support. In this respect, each enterprise is represented by a stable identifier and a set of assessment records collected in recurring cycles (e.g., annual/biannual). A single assessment record includes, at a minimum:

- enterprise metadata (e.g., sector, ownership category, location),
- infrastructure and environmental factors (e.g., proximity to road/rail nodes, backup power availability, shelter/continuity arrangements),
- production factors (e.g., technology/material capability, capacity indicators, flexibility of assortment, quality and reliability KPIs),
- capital/business factors (e.g., logistics assets, lead times, contractual readiness),
- evidence references and timestamps to support traceability and longitudinal monitoring.

This minimal structure ensures comparability across assessment cycles and provides a consistent basis for analytics and ML-based decision support. Importantly, such outputs are intended to inform prioritization and verification activities rather than replace expert judgement.

## 2 Business Process Model and Notation

A lot of processes in enterprise including business and production processes, are very complex. Especially in big organizations. Sometimes it is necessary to present some processes in simple way to make it understandable for all discussion participants. There is a lot of tools to make it easier, including graphical methods like Business Process Model and Notation (BPMN).

It is a standardized method for describing various types of processes. These may include business processes, standardized work procedures, decision-making processes, or production processes. BPMN can be applied with three main objectives: description, simulation, and execution of processes [21].

In modern enterprises, processes are subject to continuous improvement in order to maximize efficiency. This tool enables production planning as well as a flexible approach to process management. It also allows for the input and analysis of specific data [22]. This does not preclude the use of other supporting tools, such as the Gantt chart.

The study also explored how process modeling can improve efficiency. This requires the collection of all necessary data as well as the verification of chosen key performance indicators (KPIs). BPMN was supported by additional tools, such as the Ishikawa diagram. The researchers concluded that BPMN is a highly useful tool in the context of process improvement because it allow to identify ineffective information flow and implement Lean rules. However, they also noted that many factors—such as data flow and the selection of appropriate KPIs—play a critical role, and that implementing changes is, of course, not an easy task [23].

An undeniable advantage of using process description tools is their accessibility to individuals without a background in mathematical modeling. This makes it possible to develop, for example, decision support systems for hierarchical production planning that can be used by anyone within the organization [24].

The above examples demonstrate that BPMN is useful not only for management and business model descriptions but can also be successfully applied in mechanical engineering, particularly in production management.

## 3 Scoring of the Enterprise

An enterprise can be evaluated from many different perspectives. In the case of economic analysis, the researcher will focus on financial aspects, a legal expert on legal issues, while in mechanical engineering the assessment will concern product quality and process indicators. This is due to the fact that the entire production facility can be evaluated, but it is also possible to assess individual processes, such as the evaluation of warehouse processes [25]. However, the production process itself can also be evaluated from operational, market, dynamic, technical perspectives, as well as according to profit criteria [26].

The enterprise evaluation concept proposed by the authors aims to create a database consisting of assessments of various companies across the entire country. Importantly the proposal includes both the description of the evaluation process and the evaluation criteria. The starting point is preparing the national economy for potential militarization or other crisis situations requiring a shift in production priorities, such as natural disasters.

The necessity for such measures is linked to the current geopolitical situation. One illustrative example is the production of large-calibre ammunition. The stated objective is currently an annual output of 200,000 rounds. However in 2024 the Russian armed forces employed by various estimates between 3 and 4.5 million rounds of such ammunition while Ukrainian forces used approximately 2 to 2.5 million rounds. In 2023 Poland produced about 5,000 rounds of this ammunition, by 2025 production had increased to 30,000 rounds. The planned target is to reach 200,000 rounds annually by 2028 [27].

It should be noted that these increases occur in the context of elevated spending on military production, but they do not amount to militarization. Certainly at the outset of an armed conflict parties typically maintain stockpiles of ammunition. Nevertheless there remains a pronounced quantitative disparity between even the 200,000-round objective and the volumes consumed in an actual armed conflict.

In such circumstances the demand for production of spare parts and components rises and ammunition is among the most deficit-prone items. Given the need to produce a variety of other components it is not feasible to retool production facilities to manufacture ammunition exclusively. What is possible, however is to mitigate the effects of a crisis and to prepare for it in advance. In formulating such a defense strategy and guidelines for industry enterprise assessment and awareness of available resources are invaluable to decision-makers enabling the efficient allocation of tasks to the most appropriate units.

In contemporary times few anticipated the outbreak of war in the middle of Europe. Nevertheless, the subsequent events have demonstrated the critical importance of maintaining a domestic defense industry for national security. According to some scholars, this sector is one of the most innovative branches of the economy, having undergone numerous transformations over the past three decades [28].

The conflict in Ukraine clearly shows that consumption of ammunition was unprecedented and it outgrow the European production capacity. It forced European countries to implement new instruments and investments (such as ASAP and then EDIP) to urgently increase production and rebuild the supplies [29]. The conflict also demonstrates that in long-term attrition war the advantage is gained by countries that have not only armament with advanced technology but above all fully developed and flexible industrial base that is able to rapid militarization and to scale the production.

Undoubtedly the product lifecycle in the defense sector is significantly longer than that of civilian solutions. However the technological competition is indeed at the highest level and necessitates the continuous pursuit of innovative solutions.

Knowledge management is also one of the key aspects of strategic enterprise management. This is particularly important for new technologies which initially are known by only a few individuals. For example additive manufacturing technologies, which until recently were relatively unknown, are now widely applied [30].

Other researchers have examined the maturity level of Industry 4.0 in strategic defense sector enterprises in Brazil. Strategic enterprises in Brazil are characterized, among other factors, by a predominance of domestic capital in their shareholding structure which the authors of this study also incorporated as an evaluation criterion. Maturity assessment models are tools used to analyze and continuously improve organizational performance. The areas most impacted by digital transformation were also identified. The evaluation covered the following domains: Strategy and Organization (capability to develop new business models), Smart Factory (interconnectivity of the value chain and product lifecycle), Smart Operations (horizontal and vertical integration of the enterprise), Smart Products (application of advanced technologies), Data-Driven Services (optimization of resource utilization) and Employees. The research methodology was provided and this questionnaire was sent to one chosen company. It is a structured survey with predefined answers and the method for calculating the evaluation score was also outlined [31].

Recently within the sphere of innovation greater emphasis has been placed on the human factor rather than the economic factor. With the increase in global defense sector expenditures technological innovations are playing an increasingly significant role which is also beneficial for civilian solutions. The authors also highlight the potential use of artificial intelligence for purposes such as predicting equipment failures. Overall AI tools can contribute to enhancing the efficiency of the defense sector and the military by fostering greater innovation [32].

## 4 A Conceptual Framework for Assessing the Readiness and Suitability of Enterprises for Militarization

The authors proposed a framework for the process of evaluating and compiling a database of assessed companies. On behalf of the government and the ministry, this task is carried out by the provincial governor in each voivodship who distributes the evaluation forms to selected manufacturing companies with a request for completion and sets a deadline for submission. While awaiting responses the governor simultaneously and continuously organizes meetings with experts. Infrastructure conduction inspections based on previously received guidelines are also necessary to ensure that preparations for potential militarization are of the highest standard.

Upon receiving the letter instructing them to perform the assessment the company forwards the questions regarding specific sections of the questionnaire to the responsible substantive departments whose task is to complete them with justifications and explanations. After gathering information from all units a final report is prepared and sent to the Provincial Governor.

The Provincial Governor's task is now to prepare a list of companies along with their evaluations for the ministry, as well as to conduct regular and ongoing monitoring of the listed companies regarding the implementation of improvements.

Based on data received from all Provincial Governors the Ministry incorporates the collected information into the country's defense strategy. Simultaneously using the evaluations it develops guidelines for companies and agencies responsible for infrastructure. Their task is to implement these guidelines in their activities and to report on this implementation back to the Ministry. Given the defense context, the assessment register requires explicit information security and data governance. First, each data element is labeled by sensitivity and handled under the "need-to-know" principle with role-based access control. Second, data are encrypted in transit and at rest; access is logged (audit trail) and periodically reviewed. Third, data minimization is applied-only fields necessary for readiness assessment and monitoring are collected and retained for a defined period, with versioning for accountability. Fourth, any ML development and inference are executed in a controlled environment separated from external networks, and model artifacts (training datasets, parameters, outputs) are governed as sensitive assets. Finally, model outputs are treated as recommendations; final decisions remain with authorized public-administration and expert bodies.

This process should be conducted regularly to continuously monitor changes and the state of the economy's preparedness as well as to refine the guidelines. This is particularly important because the situation of companies may change over time. Normally the BPMN is presented from left to right side, but for the needs of article format the authors present it from up to bottom. The process is shown in the Fig. 1 below.

The authors also proposed criteria for assessing the suitability of a company for militarization. These criteria form the basis of a questionnaire the completion of which serves as the foundation for evaluating the company. While the questionnaire includes responses that are assigned points contributing to the overall assessment. The company is required not only to complete the form but also to describe and justify each selected answer. The length of the justification varies depending on the specific criterion.

The criteria were divided into three parts: infrastructure and environment, production factors, and capital factors. The assessment is based on the percentage of points obtained for each criterion. For the combined scores from areas 1 and 2, achieving less than 60% corresponds to a rating of 1. Subsequent ratings are assigned in increments of 10%, as follows: 60%–70% — 2, 70%–80% — 3, 80%–90% — 4%, and above 90% — 5. If at least two-thirds of the points are obtained in area 3, an additional 0.5 is added to the score from areas 1 and 2, making the maximum possible rating 5.5.

It is within the ministry's authority to determine the tasks assigned to companies based on their respective ratings. However it should be taken into account that even a company with a low rating may have tasks to perform and be useful during the process of militarization. The individual criteria along with the corresponding point allocations are presented in the Figs. 2, 3 and 4 below.

When justifying the selection of criteria it is necessary to consider fundamental concepts related to national security such as critical infrastructure. This concept is relatively new in the field of security studies in Poland and it has not yet been clearly defined. Nevertheless it can be asserted that damage or failure of such infrastructure would significantly affect the security and operational efficiency of the country. The definition has been changed and has also found reflection in legislation after major incidents like terrorist attacks in US in 2001 and in Madrid in 2003.

It is crucial to classify facilities as critical because omission could provide to serious consequences. In the case of Poland the criteria for classification are not publicly available. Generally the most important factors are the impact of a facility on state and citizen security, as well as the impact of efficiency of administrative and state economy. Economic and defense-related actions are also important because they enable the rapid restoration of functionality to critical facilities [33].

According to Polish legislation critical infrastructure should be understood in a broad sense encompassing not only facilities directly related to the military but also systems such as emergency services, communications and healthcare. Threats to such infrastructure may arise from natural disasters, the absence of alternative solutions or routes or human actions of a terrorist nature. Attacks may be motivated by various objectives and can provide to diverse consequences. Therefore these facilities must be subject to special protection and associated risks should be systematically identified. Furthermore procedures related to protective and mitigation actions should be continuously refined and improved [34].

When planning a defense strategy and the protection of critical infrastructure it is essential to draw on the experiences of other countries including the most recent lessons from conventional warfare, such as those emerging from the war in Ukraine. It should be recognized that attacks may ạlso take a cyber form with potentially equally major consequences. Such actions often referred to as "hybrid warfare" can occur even before a crisis situation. They significantly affect citizens' sense of security and trust in public administration. During designing the new infrastructure threat factors related to conventional attacks should be taken into account integrating resilience measures into the

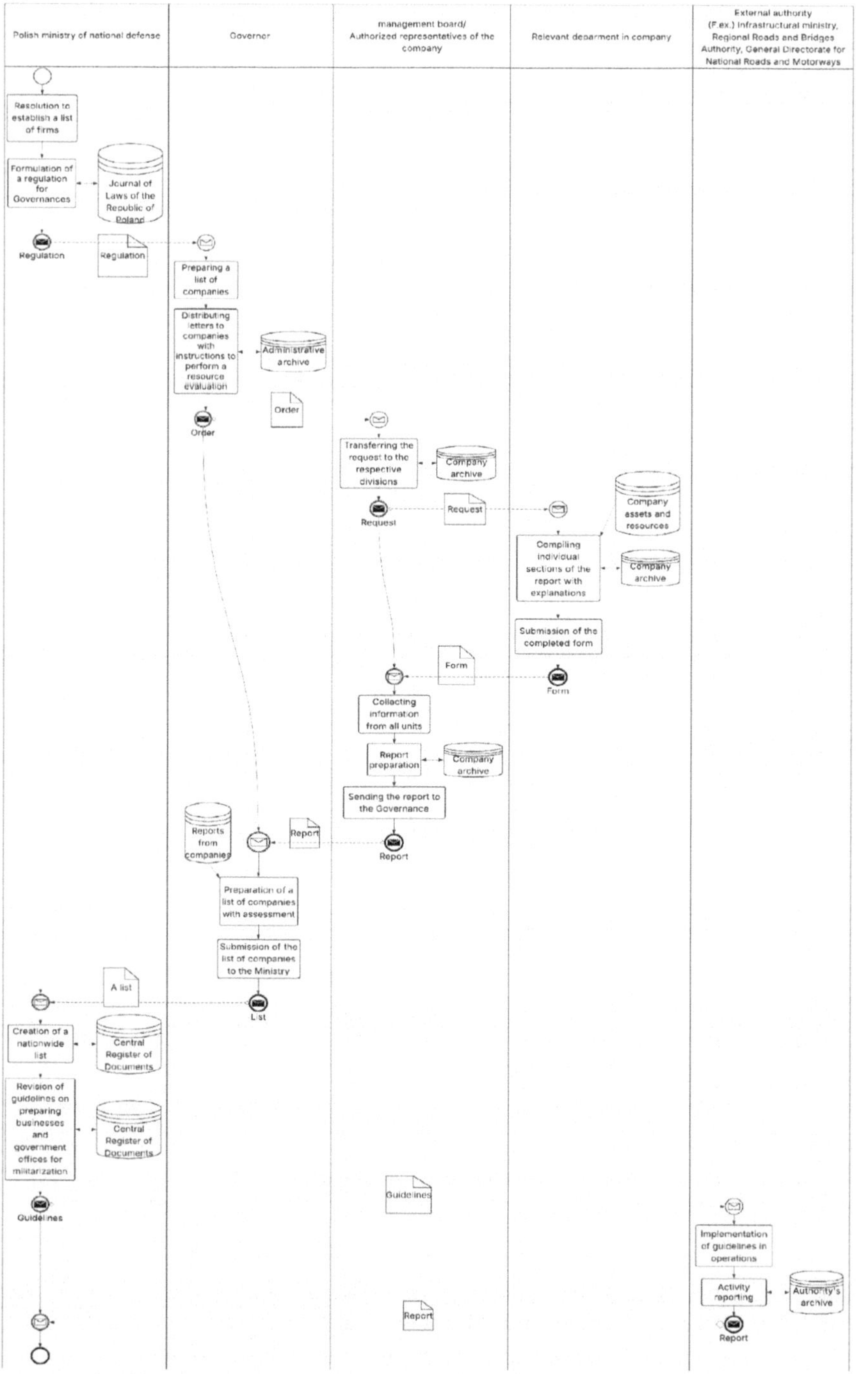


**Fig. 1.** The process of evaluating companies designated for militarization in BPMN.

| | *Infrastructure factors* | | | |
|---|---|---|---|---|
| 1 | Access to railways | | | |
| | railway siding on the plant area - 10 pkt | none - 0 pts. | | |
| 2 | Access to highway | | | |
| | <10 km - 10 pts. | 10-25 - 5 pts. | 25-50 - 3 pts. | >50 km - 0 pts. |
| 3 | helicopter landing pad | | | |
| | on plant area - 10 pts. | <10 km - 5 pts. | >10 km - 0 pts. | |
| 4 | Own energy sources (more than 1 answer possible) | | | |
| | diesel generators - 5 pts. | own photovoltaic farm - 3 pts. | none - 0 pts. | |
| 5 | Shelter on the plant area | | | |
| | yes, for 100% of workers - 10 pts. | yes, for not all workers - 5 pts. | No - 0 pts. | |
| 6 | military training area/test track for vehicles | | | |
| | yes - 5 pts. | No - 0 pts. | | |

**Fig. 2.** Evaluation of infrastructure factors.

| | *Production factors* | | | |
|---|---|---|---|---|
| 1 | Experience in procession of abrasion-resistant materials (hardox, armox) | | | |
| | Armox and hardox - 10 pts. | Armox or hardox - 5 pts. | other abrasion-resistant types - 3 pts. | none - 0 pts. |
| 2 | 3d Printers(more than 1 answer possible) | | | |
| | for metallic materials - 5 pts. | for polymers - 3 pts. | none - 0 pts. | |
| 3 | Own transportation sources (more than 1 answer possible) | | | |
| | trains - 5 pts. | trucks - 3 pts. | delivery vehicles - 1 pts. | none - 0 pts. |
| 4 | Experience in military production | | | |
| | yes and AQAP norm implemented - 10 pts. | Yes - 5 pts. | experience in production for heavy industry f.e. mining - 3 pts. | none - 0 pts. |
| 5 | Variety of assortment - possibility of adapting production lines | | | |
| | variety of products, freedom of arrangement - 5 pts. | easily adaptable line - 3 pts. | Lines adapted to one product - 1 pts. | |
| 6 | R&D department in company | | | |
| | yes, with experience in military projects - 10 pts. | yes, with experience in heavy industry- 6 pts. | Yes - 3 pts. | none - 0 pts. |

**Fig. 3.** Evaluation of production factors.

| | *Capital/business factors* | | |
|---|---|---|---|
| 1 | Transportation company in capital group (more than 1 answer possible) | | |
| | yes, train comp. - 3 pts. | yes, truck company - 1 pts. | none - 0 pts. |
| 2 | Steel producer in capital group | | |
| | Yes - 3 pts. | No - 0 pts. | |
| 3 | Company with the participation of the state treasury (or capital group owned by the company belongs to the State Treasury) | | |
| | 100% - 5 pts. | with the deciding vote - 4 pts. | without the deciding vote or without participation - 0 pts. |

**Fig. 4.** Evaluation of capital/business factors.

planning and construction of facilities. Resilient infrastructure in Ukraine decreased the impact of Russian attacks at the begin of the war. The use of dedicated diesel generators deployed across approximately 13,000 locations and helped society a lot [35].

Considering the above aspects proximity of infrastructure was identified as a priority criterion. It enables the transport of manufactured goods to end consumers. It is the reason why factors such as distance from highways and railways, as well as the possibility of helicopter landing within the facility are assigned high priority. Also having some transportation make this process easier.

Own energy sources is a alternative solution in the event of damage to high-voltage lines or power plants. A shelter located on the enterprise area is intended to provide employees with a sense of security and could be use as safe place for the population. Its presence aims to build also social trust. An military training area enables the testing of different solutions directly and allows the development department to acquire practical knowledge. Also 3d printers will help company to produce and test some prototypes.

Production for military purposes is characterized by certain specific requirements. Therefore experience in designing and manufacturing such products is very useful. Some of the materials are common used in heavy industry such as in the military sector. Consequently if a company has experience with such materials and technologies adaptation to the production of new items will be easier. It enable error reduction and faster decision-making. A facility that manufactures a wide range of products demonstrates operational flexibility, which is significant when adapting processes to new production requirements.

Capital-related factors are not the most critical. However depending on the situation they can either facilitate or hinder operations. In a crisis scenario time is a key constraint, so reducing delivery times and increasing access to transportation resources is advantageous. This is even easier when the company is state-owned as the administration can enforce the expected solutions through its management structures. The criteria along with their evaluations, are presented in the figures below.

The proposed assessment framework should be subjected to validation. For this purpose, the results of the study may be presented to public administration bodies for use in the evaluation of selected enterprises. In addition, a panel of experts should be appointed to carry out the assessment. The experts' profiles should cover diverse aspects of production management and national militarization. The panel should consist of ten experts, which would help to minimize randomness and errors. The evaluation criteria, however, are defined with sufficient precision to preclude subjectivity, as responses are provided on the basis of qualitative parameters or numerical values.

## 5 Conclusions

The model proposed in this article for assessing enterprises potentially destined for militarization represents a significant step toward strengthening the national security system and fills an existing research gap in this area. The proposed process engages both central and regional administrations, enabling the systemic collection and analysis of information concerning the state's industrial defense potential. It also allows for the identification of strategically important resources and the assessment of their readiness to meet national defense needs. The introduction of uniform valuation methodology additionally increase transparency and institutional accountability, providing the base for more deliberate national security planning.

The presented evaluation criteria, divided into relevant thematic areas, allow for an objective and comparable assessment of enterprises' capacity for militarization. It is essential to emphasize that this assessment process cannot be of a one-off nature. Given the dynamic political, economic, and technological changes, it requires regular updating and adaptation to current national conditions and requirements. This necessity is clearly illustrated by the experience of the armed conflict in Ukraine, where the decisive factors included not only the modernity of weaponry but also, and more importantly, the resilience of supply chains, the ability to rapidly mobilize industrial resources, and the protection of critical infrastructure. The assessment model should be used as very usefull tool for government supporting both the diagnosis of industrial defense potential and the strategic planning of state development.

In conclusion, the proposed in this article assessment system poses valuable fundament for building sustainable and flexible industrial base which is capable to support the defense needs of the state. It represents only the initial stage of developing a universal and uniform evaluation framework. Thereafter, it is necessary to develop the model containing the digitalization of monitoring process, data integration in real-time and the usage of modern analitic methods including machine learning and artificial intelligence. Moreover, it will be necessary to introduce incentive mechanisms for enterprises and to propose the integration of the assessment system with national as well as allied defense capability development programs, implemented, among others, within NATO and the European Union. Such a model has the potential to provide tangible support for decision-makers and to contribute to enhancing the state's resilience to contemporary threats.

## References

1. Constitution of the Republic of Poland of 2 April 1997 (Journal of Laws of 1997, No. 78, item 483) (in polish).
2. Lewandowski, W., Fonrobert, P.: Polish defense industry-ambitions and prospects (in polish). PTU. **158**(3–4), 111–136 (2021). https://doi.org/10.5604/01.3001.0015.6792
3. Economic Mobilization Program (in polish), https://www.gov.pl/web/obrona-narodowa/pmg2, last accessed 27.10.2025.
4. Grzelak, M., Borucka, A., Guzanek, P.: Application of linear regression for evaluation of production processes effectiveness. In: Machado, J., Soares, F., Trojanowska, J., Ivanov, V. (eds.) Innovations in Industrial Engineering. ICIENG 2021 Lecture Notes in Mechanical Engineering. Springer, Cham (2022). https://doi.org/10.1007/978-3-030-78170-5_4
5. Borucka, A., Grzelak, M.: Application of logistic regression for production machinery efficiency evaluation. Appl. Sci. **9**(22), 4770 (2019). https://doi.org/10.3390/app9224770
6. Kozłowski, E., Borucka, A., Oleszczuk, P., Leszczyński, N.: Evaluation of readiness of the technical system using the semi-Markov model with selected sojourn time distributions. Maint. Reliab. **26**(4) (2024). https://doi.org/10.17531/ein/191545
7. Ruff-Stahl, H.J.K.: The strategic assessment model (STRATAM) studying and preventing strategic failure. J. Mil. Stud. **5**(2) (2014)
8. Galán, J.J., Carrasco, R.A., LaTorre, A.: Military applications of machine learning: A bibliometric perspective. Mathematics. **10**(9), 1397 (2022). https://doi.org/10.3390/math10091397
9. Ziółkowski, J., Żurek, J., Małachowski, J., Oszczypała, M., Szkutnik-Rogoż, J.: Method for calculating the required number of transport vehicles supplying aviation fuel to aircraft during combat tasks. MDPI Sustain. **14**(3), 1619 (2022). https://doi.org/10.3390/su14031619
10. Ziółkowski, J., Małachowski, J., Oszczypała, M., Szkutnik-Rogoż, J., Konwerski, J.: Simulation model for analysis and evaluation of selected measures of the helicopter's readiness.

Proc. Inst. Mech. Eng. Part G: J. Aerosp. Eng. **236**(13), 1–12 (2022). https://doi.org/10.1177/09544100211069180
11. Kozinski, O., Kotyrba, M., Volna, E.: Improving the production efficiency based on algorithmization of the planning process. Appl. Syst. Innov. **6**, 77 (2023). https://doi.org/10.3390/asi6050077
12. Pawlak, S.: The impact of selected lean manufacturing tools on the level of delays in the production process. A case study. Manag. Syst. Prod. Eng. **23**(1), 103–107 (2024). https://doi.org/10.2478/mspe-2024-0011
13. Borucka, A., Kozłowski, E., Antosz, K., Parczewski, R.: A new approach to production process capability assessment for non-normal data. Appl. Sci. **13**(11), 6721 (2023). https://doi.org/10.3390/app13116721
14. Rakyta, M., Bubenik, P., Binasova, V., Gabajova, G., Staffenova, K.: The impact of a change in maintenance strategy on the efficiency and quality of the production system. Not peer-reviewed version (2023). doi:https://doi.org/10.20944/preprints202304.0621.
15. Peter, J.: Strategic planning in production and operations management. J. Res. Int. Bus. Manag. **11**(3), 01–02 (2024). https://doi.org/10.14303/jribm.2024.026
16. Siwiec, D., Gawlik, R., Pacana, A.: Fuzzy multi-criteria decision model to support product quality improvement. Manag. Prod. Eng. Rev. **14**(2), 134–149 (2023). https://doi.org/10.24425/mper.2023.146030
17. Ziółkowski, J., Hošková-Mayerová, Š., Hutsaylyuk, V., Łukasiak, D.: Optimization of road transport within the supply network – A case study from Poland. Transp. Probl. **20**(1), 193–206 (2025). https://doi.org/10.20858/tp.2025.20.1.16
18. Andry, J.F., Nurprihatin, F., Liliana, L.: Developing a decision support system for supply chain component. Manag. Prod. Eng. Rev. **14**(2), 124–133 (2023). https://doi.org/10.24425/mper.2023.146029
19. Nurprihatin, F., Ayu, Y.N., Rembulan, G.D., Andry, J.F., Lestari, T.E.: Minimizing product defects based on labor performance using linear regression and six sigma approach. Manag. Prod. Eng. Rev. **14**(2), 88–98 (2023). https://doi.org/10.24425/mper.2023.146026
20. Tomaszewska, K.: Comparative simulation of the production flow with the implementation of Kanban and DBR. Manag. Prod. Eng. Rev. **14**(2), 79–87 (2023). https://doi.org/10.24425/mper.2023.146025
21. Chinosi, M., Trombetta, A.: BPMN: An introduction to the standard. Comput. Stand. Interfaces. **34**, 124–134 (2012)
22. Ihsane, A., Rajaa, S.: Proposal of BPMN extensions for modelling manufacturing processes. In: In: 5th International Conference on Optimization and Applications (ICOA), pp. 1–6, Kenitra, Morocco (2019). https://doi.org/10.1109/ICOA.2019.8727651
23. Arromba, A.R., Teixeira, L., Xambre, A.R.: Information flows improvement in production planning using lean concepts and BPMN. An exploratory study in industrial context. In: In: 14th Iberian Conference on Information Systems and Technologies (CISTI), Coimbra, Portugal (2019)
24. Vicens, E., Alemany, M.E., Andre, H.C., Guarch, J.J.: A design and application methodology for hierarchical production planning decision support systems in an enterprise integration context. Int. J. Production Economics. **74**, 5–20 (2001)
25. Bartosiewicz, S.: The evaluation of stock processes of the production selected company with using the indicator method. Mater. Manag. Logist. **5** (2018)
26. Koliński, A.: Review of methods and techniques for assessing the efficiency of the production process (in polish). Logistics. **5** (2011)
27. Poland is accelerating the production of artillery shells – but the road to ammunition sovereignty is still long (in polish). https://sektorobronny.pl/polska-przyspiesza-produkcje-pociskow-artyleryjskich-jednak-droga-do-amunicyjnej-suwerennosci-jeszcze-daleka/, last accessed 27.10.2025.

28. Mészáros, A.A.: Innovation in the Defence industry from the end of the cold war to the war in Ukraine. J. Reg. Secur. **19**(1), 97–118 (2024)
29. Kollias, C., Tzeremes, P.: Europe's Defence industrial strategy and the EDTIB: a connectedness-based analysis of major European Defence industries. Eur. Secur. **1–27** (2025). https://doi.org/10.1080/09662839.2025.2500296
30. Zimmer, J., Madeja, M.: Operational knowledge management in a manufacturing enterprise in the high-tech sector. Case Stud. J. Organ. Knowl. Manag. **2019** (2019). https://doi.org/10.5171/2019.458202
31. Silva, M., Rocha, C.: Assessment of industry maturity level 4.0: The case of a strategic Defense company. Fut. Stud. Res. J.: Trends Strateg. **12**(1), 31–59 (2020)
32. Robertovich Kralich, Y.: Technological innovations in increasing the efficiency of defense production. J. Econ. Reforms. **4**(56) (2024)
33. Banasik, M., Bagińska, J.: Critical considerations regarding critical infrastructure security (in polish). Int. Secur. Yearb. **13**(2) (2019)
34. Żuber, M.: Critical infrastructure of the state as an area of potential terrorist influence. Int. Secur. Yearb. **8**(2) (2014)
35. Piekarski, M.: Critical infrastructure as a target of hybrid and conventional attacks. Lessons from the Ukrainian experience. Terrorism–Studies, Analyses, Prevention Special Issue, 113–132 (2025). doi:https://doi.org/10.4467/27204383TER.25.016.21519.

# A Preliminary Study on a Proposed Energy Management System Framework for a Progressive Die Stamping SME

Dawid Bogdanowicz(✉) and Anna Woźna

Wroclaw University of Science and Technology, Wybrzeże Wyspiańskiego 28, 50-370 Wrocław, Poland
dawid.bogdanowicz@pwr.edu.pl

**Abstract.** Manufacturing industries, particularly energy-intensive sectors such as progressive die stamping (PDS), face increasing pressure to enhance energy efficiency and reduce environmental impact. This paper presents a case study of a small-to-medium enterprise (SME) in Poland, analyzing its energy consumption profile and identifying opportunities for improvement through the implementation of an Energy Management System (EMS). The study employed a mixed-methods approach combining 12 months of utility and production data with direct on-site observations. Results revealed an annual energy demand of 305.19 MWh and a facility-level Specific Energy Consumption (SEC) of 2.44 kWh per 1,000 parts. Key inefficiencies included significant idle-state energy use and limited monitoring granularity. To address these, a layered EMS framework is proposed, integrating sub-metering, real-time data processing, and visualization aligned with ISO 50001 principles. The proposed system is projected to reduce total energy consumption by 10–15%, optimize photovoltaic self-consumption, and establish a data-driven foundation for continuous improvement.

**Keywords:** Energy Management System (EMS) · Progressive Die Stamping · Specific Energy Consumption (SEC) · Sustainability · Industry 4.0 · SME

## 1 Introduction

The global imperative for sustainable development has engendered a dual challenge for the manufacturing sector: to reduce its environmental impact while concomitantly enhancing its economic competitiveness [1]. Energy consumption is a pivotal component of this challenge, representing not only a significant operational cost but also a primary source of greenhouse gas emissions [2]. The industrial sector is a major global energy consumer, and manufacturing processes are often the largest contributors to this demand [3]. Within this context, improving energy efficiency has become a critical, high-priority strategy for industrial firms seeking operational resilience, cost control, and environmental stewardship [1, 2].

Within the diverse manufacturing landscape, metal forming processes—such as progressive die stamping (PDS), press-hardening (hot stamping), and general stamping—are


A. Hamrol et al. (Eds.): MANUFACTURING 2026, LNME, pp. 149–163, 2026.
https://doi.org/10.1007/978-3-032-21701-1_11

recognized as particularly energy-intensive [3–5]. These operations often involve large hydraulic or mechanical presses to apply high forces, as well as significant thermal energy in the case of processes like hot stamping [5, 6]. The energy profile of a stamping workshop is often complex, comprising not only the core processing energy but also significant consumption from auxiliary equipment, handling robots, and long periods of non-productive machine time [7, 8]. Studies analyzing the cradle-to-gate environmental impact of stamped parts have highlighted that while the production of the raw sheet metal itself is a dominant source of embodied energy and carbon [9], the electricity consumed during the manufacturing stage is a critical lever for operational improvement [7, 9, 10]. Inefficiencies are common, particularly energy consumed while machines are in an idle state during non-productive activities like tool changeovers, maintenance, and material transport, which are characteristic of batch-oriented stamping operations [6, 8].

To manage these complex energy flows systematically, Energy Management Systems (EMS) have emerged as a robust international framework [2]. The ISO 50001 standard, in particular, provides a structured, data-driven methodology for establishing, implementing, and maintaining an EMS aimed at continuous improvement [11]. Analysis of ISO 50001 adoption reveals that its primary drivers include regulatory compliance, corporate sustainability goals, and market pressures, while its most significant benefits are consistently reported as direct cost savings, enhanced productivity, and improved operational control [11, 12].

Despite these proven benefits, the widespread adoption of comprehensive EMS faces significant hurdles, especially within SMEs [13, 14]. SMEs form the backbone of the manufacturing economy and its supply chains, yet they operate under unique constraints [13–17]. A systematic review of barriers for manufacturing SMEs identifies critical challenges, including a lack of financial resources for new technology, a shortage of dedicated personnel with technical expertise, insufficient time for planning, and a lack of awareness of the full benefits of energy-saving measures [13].

Recent research confirms that, despite significant technological advances in EMS, their adoption in small and medium-sized enterprises remains limited. A comprehensive review by Triller and Sandkuhl [18] highlights that existing EMS solutions are predominantly designed for large organizations, often characterized by high implementation complexity, substantial investment costs, and significant organizational requirements that exceed SME capabilities. The authors emphasize that SMEs face persistent barriers such as limited financial and human resources, lack of methodological guidance, and insufficient integration of EMS tools with existing operational practices, which collectively hinder effective implementation.

Empirical evidence further indicates that SMEs nevertheless perceive substantial potential benefits from EMS adoption, particularly in terms of cost reduction, supply security, and competitiveness. Based on a large-scale SME survey and expert interviews, Triller and Sandkuhl demonstrate that energy management has shifted from a peripheral activity to a strategic concern for many SMEs, driven primarily by rising energy prices and regulatory pressure [18]. However, the study concludes that the absence of tailored, low-threshold EMS solutions remains a critical gap between perceived relevance and practical implementation in SMEs.

Furthermore, the transition to Industry 4.0 presents both a solution and an additional challenge. The digitalization of manufacturing, through the integration of cyber-physical systems, the Internet of Things (IoT), and data analytics, is a key enabler for enhancing firm performance and resource efficiency [14–16]. However, many SMEs are cautious, lacking a clear digital strategy, and are concerned with the high cost, data security risks, and a lack of interoperability standards associated with new digital technologies [14, 17].

The new paradigm of Industry 4.0, however, offers unprecedented tools to overcome traditional monitoring gaps [14]. Low-cost, non-invasive IoT-based energy monitoring systems can now provide granular, real-time data for multiple machines simultaneously, often using simple smart plugs and open-source visualization platforms [19–21]. This data stream, linked to a digital twin or centralized platform, creates the transparency needed to identify energy "hot spots" (such as compressors or chillers) and quantify waste from non-productive states [19, 21]. This enables advanced monitoring strategies, such as the in-process sensor-based methods (e.g., force, temperature, acoustic) that are crucial for adaptive control and quality assurance in modern forming [22].

Recent industrial applications demonstrate that Industry 4.0 concepts are increasingly being translated into practical solutions within the stamping sector. State-of-the-art 'Smart Stamping' solutions now integrate piezoelectric sensors, load sensors, eddy currents sensors and optical sensors directly into progressive dies, enabling real-time monitoring of press force and feed alignment to detect malfunctions instantly [23–26]. Furthermore, automated quality control has advanced to include autonomous vision systems capable of high-throughput in-line inspection at production-rate speeds for dimensional and surface defects. These systems increasingly utilize OPC UA protocols to transfer data seamlessly to central Quality Assurance databases, facilitating full traceability where individual parts are laser-marked 'on-the-fly' without interrupting production [27]. However, for many SMEs, these high-end solutions remain financially out of reach, necessitating lower-cost alternatives.

This high-resolution data also fuels sophisticated analytical approaches. Non-Intrusive Load Monitoring, for instance, can use high-frequency electrical data and machine learning algorithms, like neural networks, to automatically identify a machine's operational state (e.g., Running, Idle, Starting) and even the specific part being produced, purely from its energy signature [28]. Furthermore, Discrete Event Simulation (DES) has proven to be a powerful tool for modeling the complex, dynamic nature of a stamping workshop. DES can evaluate the system-level impact of production planning, machine failures, transportation batch sizes, and mold-changing schedules on total energy consumption and makespan [6, 8]. A critical new application involves integrating DES with dynamic data on the electrical grid's CO2 intensity, allowing production to be scheduled for periods of low environmental impact, thereby actively reducing Scope 2 emissions [4].

This granular data, captured via IoT and analyzed through simulation, allows for the development of meaningful energy-related Key Performance Indicators (e-KPIs). A systematic e-KPI methodology can move a company beyond simple monthly utility bills to precisely link energy consumption (the effect) to specific manufacturing states and activities (the causes), such as "idle" or "setup," enabling targeted management interventions [2].

The scope of energy management is also necessarily expanding beyond the factory walls. As SMEs are critical links in larger supply chains, their energy performance directly impacts the Scope 3 (upstream) emissions of their customers [29]. This creates market and regulatory pressure for energy efficiency integration throughout the supply chain [12, 29]. Concurrently, on-site renewable energy generation, such as photovoltaic (PV) systems, introduces another variable to manage, requiring a cost-benefit analysis that includes optimizing self-consumption and understanding all direct and indirect benefits [30].

This confluence of factors—the high energy intensity of PDS, the specific operational and financial barriers faced by SMEs, and the transformative potential of data-driven EMS within an Industry 4.0 framework—creates a compelling case for focused research. A gap persists in the literature regarding the practical application and documented impact of a comprehensive EMS within the specific context of PDS-focused SMEs, particularly in terms of low-threshold, SME-adapted implementation approaches. These firms often operate with a mix of contemporary and legacy equipment and lack the dedicated resources for sophisticated system implementation [6, 13, 14].

This paper aims to address this gap by presenting a detailed case study of a representative PDS facility in Poland. The objectives are threefold: (1) to establish a baseline energy performance profile using facility-level utility and production data, culminating in a Specific Energy Consumption (SEC) metric; (2) to identify key sources of energy inefficiency, such as idle-state consumption and sub-optimal use of on-site PV generation, through direct qualitative observation; and (3) to propose a tailored EMS framework that leverages modern monitoring principles, quantifying its potential to enhance both the environmental and economic sustainability of the operation.

## 2 Materials and Methods

A single-case study methodology was employed at a SME specializing in progressive die stamping, located in Poland (hereafter referred to as 'the Company'). The analysis was conducted in June 2025 for period between May 2024 and April 2025 using a mixed-methods approach to ensure a comprehensive understanding of energy flows.

### 2.1 Quantitative Data Collection

A baseline for total facility energy consumption was established by analyzing monthly utility bills over a 12-month period. This was supplemented with granular generation and self-consumption data from the facility's existing 100 kWp photovoltaic (PV) system monitor. Production logs were used to determine the total annual output of parts. Electricity consumption was monitored using utility billing data and photovoltaic inverter measurements. Phase imbalance analysis was not performed at this stage; energy consumption values represent aggregated facility-level active energy (kWh) as reported by the utility provider, consistent with facility-level baseline assessment practices.

## 2.2 Qualitative Data Collection

Gemba walks (direct on-site observations) were conducted by one of the authors of this study, in cooperation with production supervisors, on the production floor to understand real-world operational patterns. This observational approach is widely recognized for its effectiveness in identifying discrepancies between expected and actual operational performance.

The direct observations focused on the operational status of the five main stamping presses, explicitly noting the duration of productive runs versus non-productive but powered-on states during activities such as tool changeovers, breaks, and material loading. This nuanced data collection is crucial, as it enables deeper insights into the operational complexities and patterns that often remain obscured in aggregate reports or utility bills alone.

Methodologies such as Single-Minute Exchange of Die (SMED) are widely recognized as effective means of reducing non-productive intervals in stamping operations, thereby improving both operational efficiency and associated energy performance [31]. In parallel, the integration of Industrial Internet of Things (IIoT) technologies has been shown to complement traditional observational approaches by enabling real-time monitoring and data-driven identification of productivity losses [32]. In this study, Gemba walks provided essential qualitative insight into operational states and energy use patterns that remain obscured in aggregated utility data, forming a necessary foundation for subsequent EMS design. By observing direct operational patterns and integrating methodologies such as SMED and IIoT, organizations can effectively improve operational performance and energy management.

## 2.3 Analytical Method

Following the collection of the relevant data, estimations were performed in order to calculate the facility's SEC. The SEC was derived by correlating the total energy consumption (in kWh) with the total pro-duction volume (in thousands of parts). This provided a normalized metric for energy efficiency. This practice is consistent with established methodologies in energy management that utilize SEC as a performance indicator, enabling facilities to assess and compare energy utilization across different operational metrics. Lawrence et al. [33] indicate that SEC serves as a critical metric in evaluating energy efficiency within industrial settings, and they emphasize its role in fostering energy conservation and operational optimization.

It is important to note that this SEC calculation represents a facility-level aver-age. It is derived from the total energy consumption of the entire site, which includes not only the stamping presses but also all auxiliary systems such as the air compressor, facility lighting, HVAC, and the in-house toolshop. Understanding the comprehensive nature of energy capture across all systems aids in identifying areas for improvement and potential savings. Zhou et al. [34] discuss how such an integrated approach is vital for achieving high energy efficiency, as focusing solely on production machinery may overlook significant energy expenditures related to auxiliary components that can either detract from or enhance overall efficiency.

A primary limitation of the current data, and a key motivator for this study, is the inability to isolate the energy consumed solely by the core production processes. This challenge is well documented in the energy management literature, where aggregated facility-level measurements are shown to obscure process-specific inefficiencies and limit the effectiveness of targeted improvement actions. Chiaroni et al. emphasize that comprehensive and disaggregated energy assessments are essential for identifying meaningful efficiency opportunities and for supporting structured energy management decisions in industrial environments [35].

Therefore, while SEC provides a valuable benchmark for energy efficiency, the complexity of energy use across various systems necessitates an integrated approach to truly enhance energy performance at the facility level.

### 2.4 Framework Validation Scope and Limitations

The proposed EMS framework was subjected to a qualitative practitioner-based review to assess its practical feasibility in an SME environment. This review did not constitute a formal Delphi study or a structured multi-round expert consensus method.

The consultation involved three practitioners with complementary expertise relevant to EMS implementation in progressive die stamping SMEs: (i) an industrial energy management specialist with experience in ISO 50001 implementation, (ii) a senior progressive die stamping engineer responsible for press shop operations, and (iii) an industrial digitalization specialist involved in energy monitoring and data acquisition systems.

The practitioners reviewed the proposed EMS architecture and the set of energy-related KPIs with respect to technical feasibility, organizational and resource constraints, and practical relevance for a resource-constrained SME. Feedback was provided in the form of open qualitative comments and suggestions, focusing on potential implementation barriers, data availability issues, and operational risks.

No quantitative scoring, statistical evaluation, or multi-round consensus procedure was applied. The purpose of this review was formative rather than confirmatory, aimed at identifying obvious feasibility gaps and implementation risks rather than establishing expert consensus or generalizable validation results.

Due to the exploratory nature of this study, the findings of the practitioner review are qualitative and non-generalizable. A formal expert validation study, potentially employing a Delphi-based approach with a larger expert panel, is planned as part of future work following full-scale EMS implementation and the availability of machine-level energy data.

## 3 Results

An in-depth analysis of Company's operations from May 2024 to April 2025 revealed a total facility energy demand of 305.190 MWh. This total demand was met by 227.1 MWh imported from the grid and 78.091 MWh of self-consumed PV energy. The on-site 100 kWp PV system generated 113.961 MWh, of which 78.091 MWh (68.5%) was consumed directly on-site, meeting approximately 25.6% of the facility's total energy needs (see Fig. 1.). With an annual production of 125,000,000 parts, the facility's SEC was calculated to be 2.44 kWh per 1,000 parts.

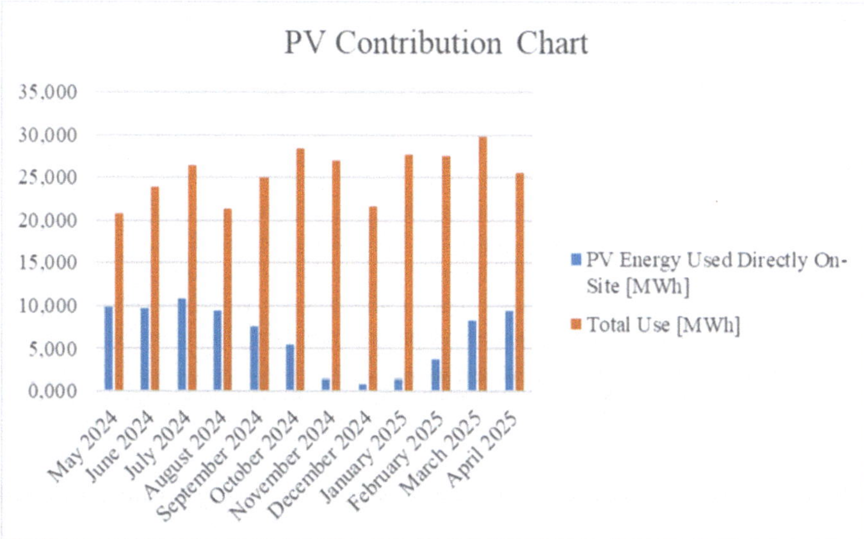


**Fig. 1.** Monthly energy demand (MWh) at the case study facility from May 2024 to April 2025, detailing the breakdown between energy imported from the grid and on-site PV self-consumption.

Despite the positive contribution of the PV system, several inefficiencies were identified.

The issue of energy consumption when the equipment is not in operation has been identified as a significant area for investigation. Direct observations have indicated that stamping presses spend a considerable proportion of their operational time in non-productive states (e.g. during tool changeovers) while remaining powered on, resulting in significant energy expenditure.

The issue of a lack of granular monitoring has been identified. The facility relies on overall utility bills for the purpose of energy tracking, as it lacks a system for real-time, machine-specific monitoring. This hinders the identification of specific inefficiencies and the accurate measurement of energy-saving initiatives.

Sub-optimal PV utilisation was observed. In instances where real-time load monitoring is not operational, the facility is unable to make dynamic adjustments to its operations with a view to optimising self-consumption of solar energy. Consequently, 31.5% of the generated PV energy is exported to the grid.

## 4 Proposed Energy Management System Architecture

To address the limitations identified in the current energy monitoring practices, a comprehensive EMS is proposed. The system is designed to provide real-time visibility of energy consumption at the process level, enabling data-driven decision-making aligned with ISO 50001 principles.

The architecture follows a layered design, integrating measurement devices, data processing infrastructure, and visualization tools into a unified management framework.

This architecture, based on non-invasive sensors and open-source platforms, is specifically designed to overcome the high-cost and lack-of-expertise barriers that typically prevent SMEs from adopting Industry 4.0 technologies. The layered structure reflects a logical flow from physical energy measurement to decision-oriented management actions, ensuring that raw electrical signals are progressively transformed into actionable information for continuous improvement.

Machine-level energy consumption data will be collected using industrial-grade, split-core current transformers (CTs) and smart sub-meters installed on the main power lines feeding the three highest-load stamping presses and the central air compressor. These devices, which do not require breaking the power connection (non-invasive), measure real-time current and power on all phases, providing the granular resolution necessary to distinguish between operating states such as Running, Idle, and Standby in an industrial environment. The data collected at this level form the foundation for subsequent processing and contextualization, enabling meaningful interpretation of machine operating states and energy use patterns.

The sub-meter signals are transmitted to an edge gateway using standard industrial communication protocols. The gateway (e.g., industrial PC or Raspberry Pi) hosts a lightweight data broker such as MQTT and a processing tool such as Node-RED to clean, format, and route the data. Once structured and time-synchronized, the processed energy data can be effectively utilized by higher-level analytical and visualization tools to support operational monitoring and analysis.

The processed data is transmitted to a centralized dashboard environment (e.g., Grafana or Power BI) where real-time and historical trends are displayed. These dashboards enable operators, maintenance personnel, and management to monitor energy performance by individual machine, shift, and production batch. These analytical outputs provide the necessary transparency and performance feedback required to support structured energy management actions and continuous improvement at the organizational level.

The platform enables the calculation of e-KPIs, supporting structured improvement initiatives (Plan-Do-Check-Act cycle) under ISO 50001. This provides a repeatable and scalable framework for reducing energy waste while improving operational control.

## 4.1 Energy-Related Key Performance Indicators

To move beyond facility-level SEC, the following process-focused e-KPIs are introduced (see Table 1). The variables and symbols used in the proposed EMS framework are summarized in Table 1.

**Table 1.** e-KPIs used for the evaluation of energy efficiency and sustainability performance in progressive die stamping operations.

| KPI Name | Formula | Purpose | Insight Enabled |
|---|---|---|---|
| Process SEC ($SEC_{press}$) | $SEC_{press} = \frac{E_{press}}{N/1000}$ | Compare efficiency across machines | Identifies the least energy-efficient press |

*(continued)*

**Table 1.** (*continued*)

| KPI Name | Formula | Purpose | Insight Enabled |
|---|---|---|---|
| Idle Energy Ratio | $R_{idle} = \frac{E_{idle}}{E_{total}}$ | Quantifies non-productive energy use | Supports idle time reduction actions |
| PV Self-Consumption Ratio | $R_{PV,self} = \frac{E_{PV,self}}{E_{PV,gen.}}$ | Evaluates renewable energy utilization | Helps optimize production during PV peak hours |
| Energy Cost per Part | $C_{part} = \frac{E_{press} \times c_{el}}{N}$ | Links energy directly to product cost | Enables cost-based production planning |
| $CO_2$ Emissions per 1,000 Parts | $EM_{press} = SEC_{press} \times EF_{grid}$ | Environmental performance indicator | Supports sustainability reporting & customer audits |

**Table 2.** List of symbols and abbreviations used in the paper, including variable definitions, units, and contextual explanations relevant to the proposed energy management framework.

| Symbol | Meaning | Unit | Explanation |
|---|---|---|---|
| $E_{press}$ | Energy consumed by stamping press | kWh | Total electrical energy drawn by the presses during production (productive and non-productive time). |
| $E_{idle}$ | Energy consumed during idle or standby states | kWh | Represents wasted or non-productive energy when machines are powered but not producing parts. |
| $E_{total}$ | Total facility energy consumption | kWh | Sum of all electrical energy used by the facility, including presses, compressors, lighting, HVAC, etc. |
| $E_{PV,gen}$ | Total photovoltaic (PV) energy generated | kWh | The amount of electricity produced by the on-site PV system over the measured period. |
| $E_{PV,self}$ | Self-consumed PV energy | kWh | Portion of PV-generated electricity consumed directly on-site (not exported to grid). |
| N | Number of produced parts | pcs | Total count of stamped components produced within the analyzed timeframe. |

(*continued*)

**Table 2.** *(continued)*

| Symbol | Meaning | Unit | Explanation |
|---|---|---|---|
| $SEC_{press}$ | Specific Energy Consumption (press-level) | kWh per 1,000 parts | Normalized measure of press energy efficiency, independent of production volume. |
| $R_{idle}$ | Idle Energy Ratio | – (dimensionless) | Share of total energy spent in idle states; indicator of operational inefficiency. |
| $R_{PV,self}$ | PV Self-Consumption Ratio | – (dimensionless) | Fraction of PV energy consumed on-site relative to total PV generation. |
| $c_{el}$ | Unit electricity cost | $PLN \cdot kWh^{-1}$ (or $EUR \cdot kWh^{-1}$) | Average cost of purchased electricity; used to convert energy consumption into cost. |
| $C_{part}$ | Energy cost per produced part | $PLN \cdot part^{-1}$ (or $EUR \cdot part^{-1}$) | Direct energy-related cost assigned to each produced part. |
| $EF_{grid}$ | Grid emission factor | kg $CO_2e \cdot kWh^{-1}$ | Represents the carbon intensity of grid electricity (varies by region). |
| $EM_{press}$ | $CO_2$ emissions per 1,000 parts | kg $CO_2e \cdot (1{,}000\ parts)^{-1}$ | Product of $SEC_{press}$ and $EF_{grid}$; quantifies environmental impact. |

## 5 Discussion

The proposed EMS framework supports Sustainable Development Goal 9 by enabling resource-efficient and digitally supported operations in energy-intensive manufacturing SMEs. In progressive die stamping, where high electrical loads and non-productive machine states are inherent, structured energy monitoring and process-level e-KPIs constitute a practical form of industrial innovation rather than a purely technological upgrade. In the current European energy context, driven by the European Green Deal strategies and characterized by price volatility and increasing regulatory pressure, such EMS-based approaches enhance operational resilience, improve transparency in sustainability reporting, and reduce exposure to energy-related risks. By relying on non-invasive sensors and open-source platforms, the proposed framework lowers the entry barrier to data-driven energy management and aligns industrial competitiveness with sustainable infrastructure development (Table 2).

The results confirm that while the case facility has made commendable investments in renewable energy, significant opportunities for efficiency gains are being missed due to a lack of data-driven energy management. This assertion is supported by studies that highlight the critical role of EMS in optimizing energy efficiency across facilities. For instance, Laskurain-Iturbe et al. emphasize that an effective EMS is key to fostering energy culture and optimizing resource usage [36]. The identified inefficiencies—idle consumption, sub-optimal PV use, and an inability to track machine-level performance—are symptoms that an EMS is designed to remedy. The implementation of an EMS is reinforced by the International Organization for Standardization's ISO 50001 standard, which provides structured frameworks to aid organizations in achieving greater energy performance [37].

An EMS represents a crucial step in the transition to a more sustainable and digitally mature manufacturing operation. Incorporating elements such as sub-metering on high-consumption assets and a centralized data visualization platform has been shown to facilitate better monitoring and optimization of energy use within facilities [38]. The proposed solution involves implementing a comprehensive EMS, structured in accordance with ISO 50001 principles, to provide clearer insights into performance and efficiency improvements.

The calculated facility-level SEC of 2.44 kWh per 1,000 parts serves as a useful baseline, but its usefulness for targeted process improvement is limited. This limitation is acknowledged in the literature, where aggregated metrics often obscure specific operational inefficiencies [39]. Thus, the composite nature of this figure may hinder effective management decision-making regarding specific product lines or machine performance, demonstrating the critical need for enhanced data granularity.

The issue under discussion is precisely the problem that the proposed EMS with sub-metering is designed to solve. By isolating the energy consumption of individual presses, the system would enable the calculation of a much more valuable metric: a process-level SEC. This approach aligns with findings from Mkhaimer et al. [40], who indicate that targeted measurement leads to enhanced energy performance monitoring capabilities. Such granular data would empower the facility to identify its least efficient machines accurately and to model return on investment for equipment upgrades.

Furthermore, the introduction of process-level energy monitoring through the proposed EMS architecture signifies a significant advancement over current practices, where energy performance is typically evaluated at a broad facility level [41]. By enabling machine-specific sub-metering and relevant e-KPIs, the facility can diagnose energy inefficiencies with greater precision. This is particularly pertinent for metrics like the Idle Energy Ratio, which will be vital in quantifying non-productive energy consumption associated with idle machinery [42]. The implementation of protocols informed by this data will facilitate improved operational practices, such as enhanced changeover coordination.

Additionally, the availability of real-time and historical machine-level consumption profiles establishes the foundational data required for systematic continuous improvement under ISO 50001. Studies have shown that organizations utilizing EMS with detailed measurement capabilities achieve significant energy reductions [43]. Such systems are essential for performing recurring energy performance reviews, benchmarking across similar machines, and verifying savings resulting from operational changes.

It is essential to note that aligning production with peak photovoltaic generation, as indicated by the proposed incorporation of the PV Self-Consumption Ratio, not only optimizes renewable energy use but also reduces dependency on grid power [44]. Similar results have been noted in studies emphasizing the benefits of integrating renewable energy management with traditional manufacturing goals [45].

The expected outcomes of implementing the EMS are multifaceted, encompassing significant quantitative energy savings projected at 10–15% annually, alongside economic and operational improvements that would enhance the facility's sustainability profile [46]. The anticipated qualitative benefits extend to fostering a culture of continuous improvement, wherein energy management becomes an integral aspect of operational performance rather than simply a line item in operational costs.

The findings of this study, while valuable, are limited to a single facility and one year of data. Ongoing research should, therefore, validate the proposed EMS framework across multiple sites and include longitudinal studies to assess both scalability and long-term sustainability of implemented efficiency gains [47].

## 6 Conclusion

This study examined the energy consumption profile of a progressive die stamping SME and identified key inefficiencies related to idle-state machine operation, limited monitoring granularity, and sub-optimal utilization of on-site photovoltaic generation. The resulting facility-level SEC of 2.44 kWh per 1,000 parts provides a baseline metric but was shown to lack diagnostic value due to its aggregation of both productive and non-productive energy use. To address these limitations, a tailored EMS architecture was proposed, integrating machine-level sub-metering, an edge-level data processing layer, and real-time visualization and analysis capabilities. The system enables the introduction of process-level energy performance indicators, such as SECpress and Idle Energy Ratio, which directly support targeted reduction of non-productive energy consumption.

The proposed EMS framework establishes the technical and organizational conditions necessary for continuous improvement of energy performance in alignment with ISO 50001. Based on efficiency improvements documented in similar industrial applications, the system is expected to facilitate a 10–15% reduction in annual energy consumption, alongside enhanced renewable energy self-consumption and improved operational decision-making. Future work should focus on extending machine-level monitoring to additional assets, integrating production scheduling with real-time PV availability, and evaluating the long-term impacts of EMS-driven interventions on cost, productivity, and emissions reduction.

The authors declare that no external funding was received for this research.

## References

1. Xia, T., et al.: Efficient energy use in manufacturing systems—modeling, assessment, and management strategy. Energies (Basel). **16** (2023). https://doi.org/10.3390/en16031095
2. May, G., Barletta, I., Stahl, B., Taisch, M.: Energy management in production: A novel method to develop key performance indicators for improving energy efficiency. Appl. Energy. **149**, 46–61 (2015). https://doi.org/10.1016/j.apenergy.2015.03.065

3. Gao, M., He, K., Li, L., Wang, Q., Liu, C.: A review on energy consumption, energy efficiency and energy saving of metal forming processes from different hierarchies. PRO. **7**(6), 357 (2019). https://doi.org/10.3390/pr7060357
4. Borbolla, A., Aboujarra, M.N., Ramanujan, D.: Investigating sustainable production planning of an industrial progressive die stamping process using discrete event simulation. In: 29th Design for Manufacturing and the Life Cycle Conference (DFMLC), vol. 5. American Society of Mechanical Engineers (2024). https://doi.org/10.1115/DETC2024-143648
5. Votava, F., Jirková, H., Kučerová, L., Jeníček, Š.: Study of transition areas in press-hardened steels in a combined tool for hot and cold forming. Materials. **16** (2023). https://doi.org/10.3390/ma16010442
6. Xiong, W., Huang, H., Li, L., Gan, L., Zhu, L., Liu, Z.: Energy consumption evaluation in stamping workshops via a discrete event simulation-based approach. Int. J. Precis. Eng. Manuf.-Green Technol. **9**, 1543–1562 (2022). https://doi.org/10.1007/s40684-021-00411-x
7. Gao, M., Huang, H., Li, X., Liu, Z.: Carbon emission analysis and reduction for stamping process chain. Int. J. Adv. Manuf. Technol. **91**, 667–678 (2017). https://doi.org/10.1007/s00170-016-9732-8
8. Xiong, W., et al.: A framework for energy consumption reduction in sheet metal forming. Procedia CIRP, 534–539 (2020). https://doi.org/10.1016/j.procir.2020.01.067
9. Cooper, D.R., Rossie, K.E., Gutowski, T.G.: An environmental and cost analysis of stamping sheet metal parts. J. Manuf. Sci. E. T. ASME. **139** (2017). https://doi.org/10.1115/1.4034670
10. Zhao, K., Liu, Z., Yu, S., Li, X., Huang, H., Li, B.: Analytical energy dissipation in large and medium-sized hydraulic press. J. Clean. Prod. **103**, 908–915 (2015). https://doi.org/10.1016/j.jclepro.2014.03.093
11. Fuchs, H., Aghajanzadeh, A., Therkelsen, P.: Identification of drivers, benefits, and challenges of ISO 50001 through case study content analysis. Energy Policy. **142** (2020). https://doi.org/10.1016/j.enpol.2020.111443
12. Zimon, D., Jurgilewicz, M., Ruszel, M.: Influence of implementation of the ISO 50001 requirements on performance of SSCM. Int. J. for Qual. Res. **15**, 713–726 (2021). https://doi.org/10.24874/IJQR15.03-02
13. Alayón, C.L., Säfsten, K., Johansson, G.: Barriers and enablers for the adoption of sustainable manufacturing by manufacturing SMEs. Sustainability (Switzerland). **14** (2022). https://doi.org/10.3390/su14042364
14. Schröder, C.: The Challenges of Industry 4.0 for Small and Medium-sized Enterprises. Friedrich-Ebert-Stiftung, Bonn, Germany (2016)
15. Kampoowale, I., Singh, H., Binti Sakka, A.A., Iwuchukwu, E., Ahmed Al-Shaikhli, E.A.: The role of digitalization on manufacturing SME firm performance in India. Int. J. Academic Res. Busi. Social Sci. **13** (2023). https://doi.org/10.6007/ijarbss/v13-i9/18060
16. Rahmani, A., Aboojafari, R., Bonyadi Naeini, A., Mashayekh, J.: Adoption of digital innovation for resource efficiency and sustainability in the metal industry. Resour. Policy. **90** (2024). https://doi.org/10.1016/j.resourpol.2024.104719
17. Khan, M.A., Saini, P., Alam, J.: Internet of things (IOT) enabled Sustainable Energy Management (SEM) for SMEs–A review. J. Electron. Electr. Eng., 133–151 (2025). https://doi.org/10.37256/jeee.4220256925
18. Triller, M., Sandkuhl, K.: Energy Management Systems in SMEs: Relevance, State of Research and Requirements. Complex Syst. Inf. Model. Quarterly, 22–42 (2025). https://doi.org/10.7250/csimq.2025-42.02
19. Azlan, A.T.B.N.N., Mativenga, P.T., Zhu, M., Mirhosseini, N.: Industry 4.0 energy monitoring system for multiple production machines. Procedia CIRP, 613–618 (2023). https://doi.org/10.1016/j.procir.2023.09.047
20. Alenazi, M.: IoT and energy. In: Internet of Things-New Insights. IntechOpen (2024). https://doi.org/10.5772/intechopen.113173

21. Uhlmann, E., Polte, J., Geisert, C.: IoT-Based energy monitoring in production. In: Measurement: Sensors (2025). https://doi.org/10.1016/j.measen.2024.101591
22. He, X., Welo, T., Ma, J.: In-process monitoring strategies and methods in metal forming: A selective review. J. Manuf. Processes. **138**, 100–128 (2025). https://doi.org/10.1016/j.jmapro.2025.02.011
23. Kolhatkar, A., Pandey, A.: Predictive maintenance methodology in sheet metal progressive tooling: a case study. Int. J. Syst. Assur. Eng. Manag. **14**, 980–989 (2023). https://doi.org/10.1007/s13198-021-01564-3
24. Skardžius, J., Nagurnas, S., Skačkauskas, P.: Modernization of the stamping process using eddy current and load sensors in the manufacturing of automotive parts. Maint. Reliab. **26** (2024). https://doi.org/10.17531/ein/189453
25. Amaral, E.S., Silva, G.C.: Preventive measures for progressive tool failure using load cell-based signal monitoring. Measurement. **231**, 114674 (2024). https://doi.org/10.1016/J.MEASUREMENT.2024.114674
26. Farioli, D., Kaya, E., Fumagalli, A., Cattaneo, P., Strano, M.: A data-based tool failure prevention approach in progressive die stamping. J. Manuf. Mater. Proc. **7**, 92 (2023). https://doi.org/10.3390/jmmp7030092
27. Kistler Group: Digital Measurement Technology For Stamping Processes. (2025)
28. Carrillo, C., et al.: Detailed energy analysis of a sheet-metal-forming press from electrical measurements. Energies (Basel). **16** (2023). https://doi.org/10.3390/en16196972
29. Xavier, B.M., Thollander, P., Hilletofth, P., Johansson, M.: Exploring energy management integration into upstream supply chains: a systematic literature review. Front. Energy Res. **12**, 1425795 (2024). https://doi.org/10.3389/fenrg.2024.1425795
30. Zhao, Y., et al.: Cost–benefit analysis of distributed energy systems considering the monetization of indirect benefits. Sustainability (Switzerland). **16** (2024). https://doi.org/10.3390/su16020820
31. Basri, A.Q., Nik Mohamed, H.N.M.Z., Nor, N.M., Yusof, Y.: SMED simulation in optimising the operating output of tandem press line in the automotive industry using witness software. Int. J. Automot. Mech. Eng. **18**, 8895–8906 (2021). https://doi.org/10.15282/ijame.18.3.2021.05.0682
32. Sundar, P.S., Chowdhury, C., Kamarthi, S.: Industrial internet of things enabled kata methodology of assembly line productivity improvement: Insights from a case study. PRO. **12**, 2611 (2024). https://doi.org/10.3390/pr12112611
33. Lawrence, A., Thollander, P., Andrei, M., Karlsson, M.: Specific Energy Consumption/Use (SEC) in energy management for improving energy efficiency in industry: Meaning, usage and differences. Energies (Basel). **12**, 247 (2019). https://doi.org/10.3390/en12020247
34. Zhou, L., Li, J., Li, F., Meng, Q., Li, J., Xu, X.: Energy consumption model and energy efficiency of machine tools: a comprehensive literature review. J. Clean. Prod. **112**, 3721–3734 (2016). https://doi.org/10.1016/j.jclepro.2015.05.093
35. Chiaroni, D., Chiesa, V., Franzò, S., Frattini, F., Manfredi Latilla, V.: Overcoming internal barriers to industrial energy efficiency through energy audit: A case study of a large manufacturing company in the home appliances industry. Clean Techn. Environ. Policy. **19**, 1031–1046 (2017). https://doi.org/10.1007/s10098-016-1298-5
36. Rotzek, J.N., Scope, C., Günther, E.: What Energy Management Practice Can Learn From Research on Energy Culture? Sustainability Accounting Manage. Policy J. **9**(4), 515–551 (2018). https://doi.org/10.1108/sampj-07-2017-0067
37. Runge, M., Reiche, L., Niemann, K.-H., Fay, A., Wurger, A.: Requirements for an Energy Data Information Model for a Communication-Independent Device Description. In: 26th IEEE International Conference on Emerging Technologies and Factory Automation (ETFA), vol. 2021, pp. 01–04, Vasteras, Sweden (2021). https://doi.org/10.1109/etfa45728.2021.9613401

38. Tauruy, A.Y., Fahmi, I., Najib, M.: Energy management system implementation strategy in the concentrating division of PT freeport Indonesia based on ISO 50001:2018. Busi. Rev. Case Stud. **3**(2) (2022). https://doi.org/10.17358/brcs.3.2.170
39. Păunescu, C., Blid, L.: Effective energy planning for improving the enterprise's energy performance. Manage. Mark. **11**, 512–531 (2016). https://doi.org/10.1515/mmcks-2016-0013
40. Mkhaimer, L.G., Arafeh, M., Sakhrieh, A.: Effective implementation of ISO 50001 energy management system. Int. J. Eng. Busi. Manage. **9** (2017). https://doi.org/10.1177/1847979017698712
41. Castrillón-Mendoza, R., Rey-Hernández, J.M., Rey Martínez, F.J.: Industrial decarbonization by a new energy-baseline methodology: Case study. Sustainability. 12(5), 1960 (2020). doi:https://doi.org/10.3390/su12051960
42. Herce, C., Biele, E., Martini, C., Salvio, M., Toro, C.: Impact of energy monitoring and management systems on the implementation and planning of energy performance improved actions: An empirical analysis based on energy audits in Italy. Energies (Basel). **14**(16), 4723 (2021). https://doi.org/10.3390/en14164723
43. Trubetskaya, A., McDermott, O., McGovern, S.M.: Implementation of an ISO 50001 energy management system using lean six sigma in an Irish Dairy: A case study. The TQM Journal. **35**(9), 1–24 (2023). https://doi.org/10.1108/tqm-08-2022-0252
44. Marimón, F., Fa, M.C.: Reasons to adopt ISO 50001 energy management system. Sustainability. **9**(10), 1740 (2017). https://doi.org/10.3390/su9101740
45. Laskurain-Iturbe, I., Saizarbitoria, I.H., Fa, M.C.: Fostering Renewable Energy Sources by Standards for Environmental and Energy Management. Renew. Sust. Energ. Rev. **50**, 1148–1156 (2015). https://doi.org/10.1016/j.rser.2015.05.050
46. Nakthong, V.V., Kubaha, K.: Development of a sustainability index for an energy management system in Thailand. Sustainability. **11**(17), 4587 (2019). https://doi.org/10.3390/su11174587
47. da Damasceno, A.J., Nakao Cavaliero, C.K., Rodrigues Souza, R.C.: The evolution of organizational energy management: Developed methodologies and the ISO 50001 standard. IOSR J. Bus. Manag. **26**(11), 10–18 (2024). https://doi.org/10.9790/487x-2611021018

# Monetisation of Environmental ESG Indicators as a Tool for Standardising Sustainability Reporting in Small and Medium-Sized Enterprises in the Metal and Automotive Industries – a New Approach

Karol Dąbrowski[1(✉)], Katarzyna Skrzypek[2], Daria Moskwa-Bęczkowska[3], Agnieszka Stachowiak[4], and Sebastian Saniuk[1]

[1] Institute of Management and Quality Sciences, University of Zielona Góra, Zielona Góra 65-417, Poland
k.dabrowski@wez.uz.zgora.pl

[2] Institute of Mechanical Engineering, University of Zielona Góra, Zielona Góra 65-417, Poland

[3] The Chair of Economics and Finance, Kielce University of Technology, Kielce 25-314, Poland

[4] Faculty of Engineering Management, Poznan University of Technology, Poznań 60-965, Poland

**Abstract.** In the face of growing economic challenges, manufacturing companies are required to adopt development models that take into account both economic information and aspects of social responsibility and environmental protection. More entities are disclosing this data in non-financial, CSR and sustainability reports, which is related to the obligation imposed by the CSRD, published in 2022. ESG reports cover environmental, social and corporate governance criteria, and their descriptive and quantitative indicators are an important tool for assessing the long-term health of an entity. Despite the assumptions of transparency and standardisation, an analysis of manufacturing companies reports shows low comparability, especially in the descriptive parts, which makes it difficult to create rankings. The article proposes simplifying selected environmental non-financial indicators by monetising them — converting them into monetary values based on the concepts of process management and activity-based costing, which link activities to the generation of costs. The aim of the research is to assess the feasibility, purposefulness and usefulness of such an approach and to identify the benefits and limitations of monetising ESG environmental indicators. This original concept may be particularly useful for SME operating simultaneously in the metal and automotive industries in EU, improving the comparability and usefulness of sustainability reports.

**Keywords:** ESG · Environmental indicators · Monetisation · Manufacturing company · Metal industry · Automotive industry · Small and medium-sized enterprises


A. Hamrol et al. (Eds.): MANUFACTURING 2026, LNME, pp. 164–174, 2026.
https://doi.org/10.1007/978-3-032-21701-1_12

## 1 Introduction

Contemporary economic, social and environmental challenges force entities and economic units in the market and public sectors to develop and implement specific models for their development. These models must ensure the ability to generate economic information, as well as information on social responsibility and environmental care.

Therefore, in order to meet the principles of corporate social responsibility (CSR), more and more entities are openly reporting on their activities in published non-financial reports, CSR reports or sustainability reports [1] in accordance with the EU's Corporate Sustainability Reporting Directive (CSRD), which replaced the 1992 Non-Financial Reporting Directive (NFRD) on the disclosure of non-financial information [2]. The analysed document imposes an obligation on a specific group of entities to prepare a sustainability report in accordance with uniform EU standards for reporting on these issues and to present the information in a structured electronic format.

So-called ESG reports must therefore include a description of the impact of entities' activities on the environment, society and corporate governance [3]. This article focuses only on the environmental aspect of ESG reports, as the authors' research shows that this is an important area of sustainable development for manufacturing companies. In this part of the report, it is necessary to indicate, among other things, whether and how the entity uses renewable and non-renewable energy sources, how it manages natural resources, or what principles it follows in waste management. All reported information in this area is expressed in terms of indicators or descriptions. These indicators are used to assess how the entity manages issues related to, for example, environmental protection, and provide important information for investors in their investment decision-making processes. However, determining specific values in this section of ESG reports requires a lot of work related to gathering additional, often unrecorded information, e.g. from outside the accounting system. This is a difficult and extremely time-consuming process.

One of the assumptions behind the creation of ESG reports is their transparency and standardisation of the information presented. An analysis of publicly available environmental sustainability reports of manufacturing companies has shown that they are not very comparable, especially in the descriptive parts or those presenting descriptive indicators.

Therefore, the premise for undertaking research in this area is the desire to simplify selected descriptive and non-financial indicators and replace them with monetary indicators based on a cost approach. This is justified by the concepts of process management and activity-based costing, according to which the processes and activities carried out in a company are the direct cause of costs [4, 5]. Therefore, every action taken by an entity subject to mandatory ESG reporting involves the necessity of incurring a specific type of cost.

In this context, the aim of this article is to analyse and assess the possibility, purposefulness and usefulness of expressing non-financial sustainability indicators in the environmental area in monetary terms.

The argument in favour of the above statement is that ESG reports are to be verified by certified auditors or other persons authorised to audit financial statements, for whom the most reliable indicator of the business performance of economic entities is monetary value. For this reason, the following sections of this article will present a new, original concept of converting a specific group of non-financial ESG environmental indicators into monetary values. This process is referred to in the literature as monetisation, selected theoretical aspects of which will be discussed later in this article [6].

According to the authors, the proposed solution may be beneficial for small and medium-sized enterprises (SMEs) operating simultaneously in the metal and automotive industries. Based on the commonly known definition of SMEs and the use of the NACE classification (Statistical classification of economic activities in the European Community) to describe enterprises, it appears that the characteristics of selected manufacturing enterprises are similar, at least in European countries. Therefore, the new, original approach to selected environmental sustainability indicators proposed in the article may be universal and utilitarian, at least in EU countries. As a result, sustainability reports will be more useful for conducting comparative analyses of the information presented in them. The selected industries are crucial for the economic development of both Poland and other European countries.

## 2 Literature Review

### 2.1 Sustainable Development

The concept of sustainable development (SD) has been the foundation of contemporary thinking on resource management and social development for years. For companies, especially those in the manufacturing sector, implementing SD principles is crucial to their long-term competitiveness and adaptability. These principles force them not only to innovate in terms of environmental efficiency and social responsibility, but also to constantly monitor and report on the impact of their activities on the environment. Interest in these aspects is also linked to growing stakeholder awareness, increasing regulatory requirements and the need for transparency, which are reflected, among other things, in the popularity of ESG reports [7]. This is because sustainable development is no longer just an optional addition to business strategy, but has become an economic and social imperative in the face of growing climate and resource crises [8]. A key element of the effective transformation of enterprises towards sustainable production and the achievement of a zero-carbon economy is transparency and accountability, which are achieved through non-financial reporting and systematic reporting on implemented measures.

Increased regulation and pressure from stakeholders (investors, consumers, society) are forcing companies to go beyond traditional financial reporting. Sustainability reporting allows for the measurement, verification and communication of progress in terms of environmental, social and corporate governance impact [9].

The literature on the subject distinguishes between manufacturing companies assessed in terms of sustainable development as [10–12]: Sustainable Development Leaders, Compliance-Driven Entities, Laggards/Uninterested Entities.

An analysis of the literature on the subject shows that while there is a growing general awareness of the importance of SD, there is still a gap between declaration and effective implementation and full transparency. The lack of standardised measurement methods, the problem of correctly interpreting the indicators used, the high initial costs of transformation and the lack of competence, e.g. in the field of circular economy (CE) and decarbonisation, are the main barriers to moving from awareness to action, especially in the small and medium-sized enterprise sector [13, 14]. Adding to the above barriers the so-called purely human factors, such as origin, experience, etc., it can be assumed that SD reports may not be objectively comparable. It is therefore advisable to develop SD assessment methods that will minimise the risk of these differences occurring.

According to the authors of the article, one possible way to overcome barriers in interpreting ESG indicators is to properly quantify non-financial or descriptive indicators, which form the basis for assessing and communicating the level of advancement of activities in areas such as environmental protection in a company. These indicators cover a wide range of variables. Their assessment and interpretation are often subjective and vary depending on the industry, institutional and cultural context, and standards applicable in a given area [15]. According to the authors of the article, a possible way to overcome these differences is to replace the values that cause the greatest discrepancies with indicators expressed in monetary terms. Money is a universal and commonly used value. The authors have not found similar analyses in the literature, which suggests that the use of such an approach in the context of sustainable development indicators is possible, justified, innovative and may lead to interesting research and results.

### 2.2 Monetisation

The essence of ESG reports boils down to preparing and publishing information on the sustainable development of a company and its impact on, among other things, the environment. This information is intended to assist potential investors and stakeholders in the company's investment decision-making processes. As mentioned above, the data presented in the reports should be comparable with other reports prepared by companies from specific sectors of the economy, industries or branches of industry. However, in practice this is not always possible, as confirmed by the results of an analysis of reports from manufacturing companies. Furthermore, for recipients of information on a company's environmental impact, monetary values provide a clearer and more reliable picture than presenting them only in descriptive form, for example. This is also consistent with the theory of behavioural economics, according to which people process information expressed in monetary units better than in abstract or descriptive units [16].

The authors of the article agree that it is appropriate and reasonable to express non-financial indicators in monetary terms, as this will significantly improve the comparability of reported data across specific types of companies operating in different countries. Monetary units are, above all, measurable, but also objective and clear to all recipients of information on sustainable development. In this case, the monetary unit is not only a measure, but also a language understandable to the global economic community.

To this end, the validity of monetising non-financial ESG indicators in order to convert physical measures (such as CO2 emissions) into monetarily measurable units, more easily expressed in monetary terms, has begun to be considered [17–19]. This also makes it possible to integrate the reported data with the company's accounting system, or more precisely, its cost accounting system.

These considerations are also confirmed by research conducted by other experts in this field. For example, N. Z. Muller points out that the current ESG indicators in the area of their impact on the environment are not sufficiently clear for investors operating on financial markets, as they do not take into account external social costs. He therefore proposes the concept of monetising these indicators based on 'a new index calculated as the ratio of each firm's contribution to total utility sector pollution damage to each firm's contribution to total utility market capitalisation. The index conveys the relative magnitude of how much each firm causes in damage to how much each firm contributes to industry value' [20].

A similar position on the issue under analysis is taken by M. Sining, A. H. Sharaai and N. K. Matthew, who point to the need to monetise non-financial indicators in their research, proposing a cost-based approach to their valuation. Their concept is based on the assumptions of product life cycle costing theory. They considered the CVM method, i.e. conditional valuation used to estimate willingness to pay (WTP), to be particularly effective. 'WTP allows for context-specific results that better reflect individual or organisational preferences' [21].

An analysis of the literature on the subject confirms the validity of the thesis adopted at the beginning of the discussion concerning the advisability and legitimacy of monetising non-financial indicators in order to facilitate the assessment of a company's impact on ESG components. A cost-based approach to the analysis of this problem is also justified, as confirmed by the authors of the report 'ESG Value Accounting Report for Chinese Listed Companies' [22]. In their opinion, the values of non-financial indicators of environmental or social impact should be determined on the basis of compensation costs, restoration costs, prevention costs, and retribution costs [22]. Thanks to this, the data for analysis will be obtainable from the traditional financial and management accounting system, obviously after appropriate modification of the level of detail of the records of key economic categories, i.e. revenues and costs. Furthermore, monetary expression will enable an objective assessment of a company's sustainability in relation to companies operating in specific sectors of the economy as well as different industries and branches of industry.

## 3 Research Problem

The aim of the article is to present the results of research conducted by the authors concerning the analysis and assessment of the possibility, purposefulness and usefulness of expressing non-financial environmental indicators of sustainable development used in ESG reports in monetary terms.

The authors analysed the results of a survey on sustainable development reporting. The survey involved 90 middle and senior managers involved in the implementation of sustainable development strategies, sustainable development reporting and CSR (Corporate Social Responsibility) in manufacturing companies operating in both the metal and automotive industries. The companies they represented could be classified as Industry 4.0 and 5.0 companies due to their technology and/or processes. Business environment organisations estimate that in western Poland (i.e. in the provinces of Lubuskie, Zachodniopomorskie and Dolnośląskie) there are approximately 300 companies belonging to the indicated industries and meeting the criteria of Industry 4.0 and 5.0. The surveyed group therefore accounted for 30% of the identified number of companies. In identifying companies and contacting them, the researchers collaborated with business environment entities from the surveyed area of Poland.

Among the surveyed companies, the authors identified a need to support selected enterprises in the area of sustainable development reporting. Respondents pointed to problems in, among other things, the interpretation, description and reporting of non-numerical indicators and the availability of such data. At the same time, they suggested that financial data is the easiest to report. The authors of the article decided to focus on environmental indicators, as the respondents indicated that the area of broadly understood environmental protection was the most important for their industries. Taking the above into account, the authors defined the following research problem: is it possible, purposeful and reasonable for manufacturing companies operating at the intersection of the metal and automotive industries to replace non-numerical environmental sustainability indicators with new financial indicators?

For this purpose, the following research hypotheses were adopted (see Fig. 1.):

H1: Replacing selected non-numerical environmental sustainability indicators with financial indicators is possible, purposeful and justified in manufacturing companies operating at the intersection of the metal and automotive industries.

H2: Financial indicators can effectively reflect the environmental aspects of sustainable development for manufacturing companies in selected industries, provided that they are appropriately selected and interpreted.

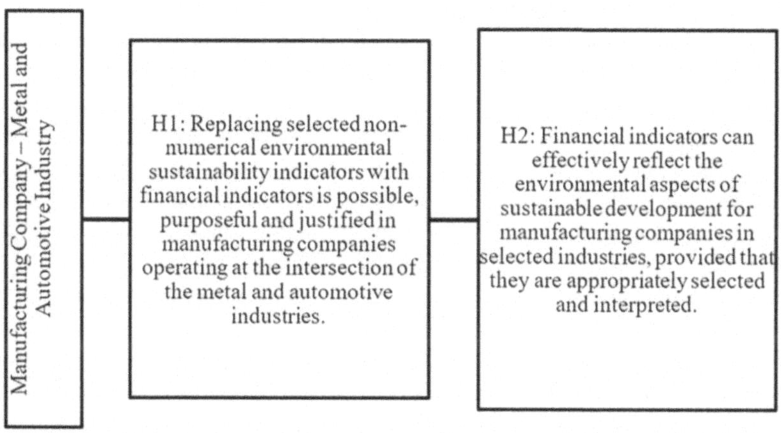


**Fig. 1** Research hypotheses. Source: Own elaborations.

For the study, the authors selected indicators from the GRI (Global Reporting Initiative) standard. This standard was chosen due to its frequent use by selected manufacturing companies. The survey also included other open-ended and closed-ended questions, which, due to the limited space available, will not be discussed in this article.

## 4 Results

The ESG report is a document containing information on a 'huge' number of financial and non-financial indicators, including those related to environmental protection in the broad sense. This data must comply with the requirements of the CSRD (Corporate Sustainability Reporting Directive) and standards such as GRI (Global Reporting Initiative) [23], presenting quantitative and qualitative data and a strategic approach to sustainable development.

The purpose of the GRI guidelines is to help entities develop transparent and consistent ESG reports so that they are readable for stakeholders and comparable with other reports [24]. These standards are divided into universal (GRI 1–3), thematic (series 200–400) and sectoral standards. The universal standards (GRI 1–3) set out general principles and requirements for their application, including how to identify material topics of an organisation's impact. The thematic standards, on the other hand, focus on specific areas of impact and are divided into series:

- series 200 economic topics,
- series 300 environmental topics,
- 400 series social topics (e.g. human rights, occupational safety).

Sector-specific standards, on the other hand, are developed for specific industries to provide more detailed and tailored guidance.

In the context of the purpose of this article, it would be reasonable to consider the possibility and advisability of expressing selected non-financial environmental sustainability assessment indicators, in accordance with GRI guidelines, in financial terms. Monetary expression is much more legible for users of the information published in these reports, while providing a clear indication for assessing the situation of the entity. Legislators also emphasise this fact, pointing out that sustainability reporting standards should promote a more integrated view of all information published by entities [25, 26].

Therefore, this article attempts to identify and analyse the possibility of replacing selected environmental non-financial indicators with monetary expressions. The study uses the GRI 300 series thematic guidelines.

Table 1 presents a list of descriptive indicators selected for analysis and proposed changes to monetisation. These indicators were considered by respondents to be the most important for their industries – they received an average rating of at least 3 in the survey – an important indicator. It is these indicators that will be the subject of an attempt at monetisation in further considerations.

It should be emphasized that the authors indicated that in some cases the transformation of an indicator into a financial one will not be easy, but it is possible.

**Table 1** GRI standards and environmental indicators expressed in financial equivalent. Source: Own eleborations

| Series 300: environment | | |
|---|---|---|
| Symbol | Indicator description according to the GRI standard | Indicator description including monetization |
| 304–1 | Organisations are located in protected areas or areas of high biodiversity value | Difficult to express in monetary values. |
| 304–2 | Significant impact of activities, products and services on biodiversity | The implementation of activities, products and services that affect biodiversity entails costs in this area, identified on thc basis of advanced calculation procedures. |
| 306–2 | Managing significant waste-related impacts - description of procedures | Difficult to express in monetary terms. However, issues related to waste management in business entities also involve the costs of securing and disposing of waste. |
| 308–2 | Negative environmental impact in the supply chain and countermeasures implemented | The cost of implementing individual measures |

## 5 Discussion

According to the research conducted, the scope of information necessary to determine the values of the 300 series indicators is in most cases associated with the need to implement measures that generate costs in business entities. However, it is worth mentioning that the introduction of the obligation to report information on sustainable development and the scope of information covered by this report will change the approach to accounting in business entities. This is therefore an excellent opportunity to provide more detailed accounting information and thus increase the reliability of the level and structure of costs disclosed in financial statements. The easiest way to collect this information would be to expand the company's chart of accounts in the area of group 4 accounts with additional analytical levels, which would also increase the scope of information on costs incurred by entities and improve the efficiency of cost management. The implementation of the activity-based costing concept could also be a helpful solution. Its main assumption is to present the entity's activities in terms of processes and, consequently, to record costs from the point of view of the activities and processes carried out within it. In summary, preparing to develop ESG reports in the context of GRI standards is a complex process that requires commitment, time, resources and consistent action. The entire entity management system should be subordinated to the reported data, with particular emphasis on the cost analysis, recording and accounting system.

There is no doubt that replacing descriptive and qualitative non-financial measures and environmental ESG indicators with monetary expressions would significantly facilitate the process of reporting on corporate sustainability. Reflecting the impact of companies on the environment in the costs of the entity's operations would significantly improve knowledge about their structure and thus improve their efficiency. However, a significant limitation is the fact that such changes would require a different approach to the standardisation of these indicators in European Union law and then its transposition into the legal systems of individual Member States. Although the literature on the subject points to the need to simplify sustainability reporting standards at the international level, due to the relatively 'young' methodology, its modification is difficult at this stage.

## 6 Conclusions

Sustainability reports are an integral strategic tool for modern enterprises, building a lasting competitive advantage for these entities. However, gathering the necessary information and compiling it is extremely difficult, both from a methodological and practical point of view. This is primarily due to the wide variety of information required to prepare these reports, which is often not collected in management systems. This information cannot be expressed in monetary terms, which makes it difficult to assess objectively and prevents the creation of industry or sector rankings in the field of sustainable development.

Therefore, the approach presented in this article, concerning the monetisation of non-financial ESG environmental indicators, should be considered fully justified. Based on an analysis of the literature on the subject and the results of their research, the authors confirm that replacing non-financial indicators with monetary expressions is possible and desirable because, as mentioned earlier, 'money' is a value that everyone understands.

The proposed solutions will also be useful from the point of view of the information needs of the interested entity and the improvement of the efficiency of its operations. The concept of monetisation of financial indicators pays particular attention to the costs of operations, their accounting details and, thus, their effective and rational management. In light of changing economic, financial, social and environmental conditions, this is extremely important for entities wishing to build a sustainable competitive advantage in the field of sustainable development.

However, it should be emphasised that although the idea of translating descriptive and quantitative non-financial (ESG) indicators into their monetary equivalents (monetisation) is attractive from the perspective of integration with traditional financial reporting, its application in highly complex sectors, such as the metal and automotive industries, may encounter a number of significant methodological and practical limitations. In the solution proposed by the authors, monetisation aims to make it easier for decision-makers, including investors, to compare environmental impacts by reducing them to a common denominator — monetary value. Nevertheless, the operational and ecological specificity of both industries means that the reliable and credible application of this method may be difficult.

A key problem may be the valuation of negative effects, such as heavy metal contamination of soil, typical of the metal industry, or nitrogen oxide emissions, typical of the life cycle of automotive products. The functioning of these sectors focuses on global and extensive supply chains. Monetising carbon footprint emissions, for example, would require collecting data and assessing the impact on thousands of third parties, from raw material extraction to product disposal. This can lead to high data uncertainty and cascading valuation errors, making the financial result unreliable. It seems that the introduction of data from ERP systems into CR reports could also be a solution to such problems.

The identified limitations prompted the authors of the article to conduct further research and analysis in this area. The proposed method is being developed and refined, and the scope of the research has been expanded to include an analysis of non-financial indicators specified in the GRI 200 and 400 series of standards. The results obtained will be discussed in a separate article.

## References


1. Czech, E., Nazarko, N.: Sustainability reporting obligations in the context of the limitation of the right to information on the environment and its protection. In: Legal Problems of Mining and Environmental Protection, pp. 1–14 (2024). DOI: 10.31261/PPGOS.2024.01.06.
2. Directive 2014/95/EU of the European Parliament and of the Council of 22 October 2014 amending Directive 2013/34/EU as regards disclosure of non-financial and diversity information by certain large undertakings and groups (OJ L 330, 15 November 2014, pp. 1–9, as amended).
3. Gillan, S.L., Koch, A., Starks, L.T.: Firms and social responsibility: A review of ESG and CSR research in corporate finance. J. Corp. Finan. **66**(C) (2021). https://doi.org/10.1016/j.jcorpfin.2021.101889
4. Piechota, R.: Designing Activity-Based Costing (in polish). Difin, Warsaw (2005)
5. Kaplan, R.S., Anderson, S.R.: Time-Driven Activity-Based Costing: A Simpler and More Powerful Path to Higher Profits. Harvard Business School Press, Boston (2007)
6. Sining, M., Sharaai, A.H., Matthew, N.K., Zainordin, N.S.: Monetization of sustainability indicators: Integrating willingness to pay into life cycle sustainability assessment. Int. J. Acad. Res. Econ. Manag. Sci. **13**(4), 155–168 (2024). https://doi.org/10.6007/IJAREMS/v13-i4/23037
7. Soyombo, O.T., Odunaiya, O.G., Okoli, C.E., Usiagu, G.S., Ekemezie, I.O.: Sustainability reporting in corporations: A comparative review of practices in the USA and Europe. GSC Adv. Res. Rev. **18**(2), 204–214 (2024). https://doi.org/10.30574/gscarr.2024.18.2.061
8. Peter, M., Peter, D., Peter, P., Yacob, P.: Pathway to net zero 2050: A conceptual evidence for smes transitioning to sustainable development. Quantum J. Soc. Sci. Humanit. **6**(2), 398–407 (2025). https://doi.org/10.55197/qjssh.v6i2.594
9. Lu, L., et al.: Green transition in manufacturing: Dynamics and simulation. PLoS One. **18**(1), e0280389 (2023). https://doi.org/10.1371/journal.pone.0280389
10. Cuhadar, S., Rudnak, I.: Literature review: Sustainable leadership. Stud. Mundi – Econ. **9**(3), 55–65 (2022). https://doi.org/10.18531/studia.mundi.2022.09.03.55-65
11. Wesseh, P.K., Lin, B., Zhang, Y., Wesseh, P.S.: Sustainable entrepreneurship: When does environmental compliance improve corporate performance? Bus. Strateg. Environ. (2023). https://doi.org/10.1002/bse.3657
12. Elsawy, M., Al-Ghurabli, Z.: Sustainable Leadership: Steering Organisations Towards a Greener Future, pp. 235–246. Springer International Publishing, Cham, Switzerland (2024). https://doi.org/10.1007/978-981-99-7798-7_20

13. Arkoh, P., Costantini, A., Scarpa, F.: Determinants of sustainability reporting: A systematicliterature review. Corp. Soc. Responsib. Environ. Manag. **31**(3), 1578–1597 (2024). https://doi.org/10.1002/csr.2645ARKOH
14. Zhao, F., Han, Z., Wang, L.: Digitization path to improve ESG performance: A study on organizational perspectives. PLoS One. **19**(12), e0315487 (2024). https://doi.org/10.1371/journal.pone.0313686
15. Fischer, D., Brettel, M., Mauer, R.: The three dimensions of sustainability: A delicate balancing act for entrepreneurs made more complex by stakeholder expectations. J. Bus. Ethics. **163**(1), 87–106 (2020). https://doi.org/10.1007/S10551-018-4012-1
16. Kahneman, D., Tversky, A.: Prospect theory: An Analysis of decision under risk. Ecnometrica. **2**(47), 263–291 (1979)
17. Krasodomska, J., Zarzycka, E.: Non-financial key performance indicators: regulations and guidelines and business practice (in polish). Theor. Account. Noteb. **180**(164), 115–132 (2020)
18. Galgani, P., Kanidou, D., Bernard, J., Mesguich, A.: Monetisation Factors for True Pricing. True Price Foundation, Amsterdam (2021)
19. Atz, U., Van Holt, T., Douglas, E., Whelan, T.: The return on sustainability investment (ROSI): Monetizing financial benefits of sustainability actions in companies. In: Bali Swain, R., Sweet, S. (eds.) Sustainable Consumption and Production, vol. II. Palgrave Macmillan, Cham (2020). https://doi.org/10.1007/978-3-030-55285-5_14
20. Muller, N. Z.: Monetization is the Key to Measuring Corporate Environmental Performance, NBER Working Paper No. w29454, https://ssrn.com/abstract=3958641, last accessed 2025/10/20.
21. Sining, M., Sharaai, A.H., Matthew, N.K., Zainordin, N.S.: Monetization of sustainability indicators integrating willingness to pay into life cycle sustainability assessment. Int. J. Acad. Res. Econ. Manag. Sci. **13**(4), 155–168 (2024)
22. China Association for Public Companies and GoldenBee (Beijing) Management Consulting Co., Ltd.: ESG Value Accounting Report for Chinese Listed Companies, https://en.goldenbeechina.com/ueditor/php/upload/file/20240520/1716193550921660.pdf. Accessed 20 Oct 2025.
23. Global Reporting Initiative: GRI 1: Foundation 2021 – GRI Standards. Global Reporting Initiative, Amsterdam (2021)
24. Polish Chamber of Statutory Auditors: G4 Reporting Guidelines. Reporting Principles and Indicators (in polish). https://www.pibr.org.pl/static/items/publishing/GRI_G4_zasady-raportowania-i-wskazniki.pdf, Accessed 21 Nov 2016.
25. European Parliament & Council of the European Union: Directive (EU) 2024/825 of the European Parliament and of the Council of 28 February 2024 amending Directives 2005/29/EC and 2011/83/EU as regards empowering consumers for the green transition through better protection against unfair practices and through better information. Off. J. Eur. Union, L 6/2024/825 (2024).
26. European Parliament & Council of the European Union: Directive (EU) 2022/2464 of the European Parliament and of the Council of 14 December 2022 on corporate sustainability reporting and amending Directives 2013/34/EU and 2006/43/EC and Regulation (EC) No 1606/2002. Official Journal of the European Union, L 3222022, pp. 15–79 (2022).

# Systematic Review of AI-based Models for Sustainable Machining Process

Hanna Łosyk[1](✉), Małgorzata Szmołda[2], and Justyna Patalas-Maliszewska[1]

[1] University of Zielona Góra, Licealna 9, 65-417 Zielona Góra, Poland
H.Losyk@iim.uz.zgora.pl

[2] The Doctoral School of Exact and Technical Sciences and Institute of Mechanical, Engineering, University of Zielona Góra, Szafrana 4, 65-516 Zielona Góra, Poland

**Abstract.** Currently, manufacturing enterprises realizing their machining processes strive to address Sustainability Development Goals (SDGs), especially SDG 9 to build sustainable industrialization. Integration artificial intelligence (AI) into production enable companies to increase productivity while reducing environmental impacts. However, monitoring and forecasting the level of sustainable development (SD) remains an area requiring further attention, as there is still a research gap in the study of the transformation of machining process into sustainable machining process (SMP). This systematic literature review analyzes a total of 8,884 articles published between 2023 and September 2025. Ultimately, 71 research articles were included in the final analysis and categorized into three thematic clusters using VOSviewer software, based on author keywords extracted from the ScienceDirect and Springer databases, and following the Preferred Reporting Items for Systematic Reviews and Meta-Analyses (PRISMA 2020) methodology and the Systematic Literature Review (SLR) approach. The research findings emphasize that AI-based technologies support SDG 9 achieving in the context reducing energy consumption and carbon emissions, improving machining performance and enabling the monitoring of changes in machining process parameters: surface roughness, depth, feed rate, and speed. Furthermore, the key challenges associated with this transformation have been identified and discussed.

**Keywords:** Machining Process · Sustainable Machining Processes · Artificial Intelligence · PRISMA Methodology · Data Analytics · Manufacturing Companies

## 1 Introduction

With the entry into force of the Corporate Sustainability Reporting Directive (CSRD), organizations are required to report on sustainable development (SD) [1]. Manufacturers are under increasing pressure to implement environmentally friendly practices, and new regulations demand tools and clear guidelines to support organizations in monitoring and evaluating SD performance. This is particularly important in the context of the post-COVID-19 recovery of the manufacturing industry and the need to reduce global carbon


A. Hamrol et al. (Eds.): MANUFACTURING 2026, LNME, pp. 175–188, 2026.
https://doi.org/10.1007/978-3-032-21701-1_13

dioxide ($CO_2$) emissions, which reached a record level of 37.6 gigatons in 2024. Manufacturers are therefore a strategic stakeholder in achieving the Sustainable Development Goals (SDGs), particularly SDG 9, which focuses on building sustainable industrialization. SDG 9 emphasizes the development of advanced technologies, the reduction of $CO_2$ emissions, and the expansion of mobile broadband availability, especially in underserved regions [2]. The implementation of this objective forms an important foundation for a sustainable future, especially since infrastructure development and technological progress directly translate into economic and social growth [3]. It is therefore crucial to increase the involvement of manufacturers, particularly in the areas of standardization and the promotion of sustainable production practices. Due to the broad interpretation of SDG 9, it is necessary to define its characteristics for a machining process in order to unify the standards for sustainability evaluation.

To monitor the level of SD within a company, it is necessary to implement and apply advanced technologies, such as Artificial Intelligence (AI), Machine Learning (ML), Machine Vision (MV), the Internet of Things (IoT), Data Analytics (DA), Additive Manufacturing (AM) [4]. In recent years, the potential of AI to raise the a satisfactory level of SD s has been developed. AI's ability to automate process monitoring and management, or the ability to predictive maintenance, has redefined process innovation and efficiency [5]. The adaptation of AI in the context of supporting the implementation of SDGs is applicable, m.in, in the field of selected sustainability materials based on a number of parameters (energy use, waste, cost, carbon footprint) [6], predictive maintenance [7], optimization of energy consumption [8], automated inspection and process monitoring [9].

Monitoring SD within a manufacturing company still require attention, as a research gap persists in transforming machining process (MP) into sustainable machining processes (SMP). SMP can be treated in the context of six key aspects [10]: (1) Environmental Friendliness, (2) Cost Reduction, (3) Operational Safety, (4) Waste Reduction, (5) Reduction of Power Consumption, (6) Preserving Operator Health. So, this study attempts to characterize a SMP, as well as the methods and techniques and challenges faced by organizations in transitioning to SMP. To this end, the following research questions (RQs) are posed in this review:

- RQ1: What advanced technologies are used to support SD in the context of the machining process?
- RQ2: What indicators/parameters characterize a sustainable machining process (SMP)?
- RQ3: What are the challenges facing the implementation of SMP?

The literature review on Sustainable Machining Process (SMP) was conducted using the Preferred Reporting Items for Systematic Reviews and Meta-Analyses (PRISMA) standard and the Systematic Literature Review (SLR) and Mapping in Literature Reviews (MLR) methodology. The aim of the review is to characterize the SMP and to review the available tools to support the monitoring process and to assess the level of sustainability for the machining process. These results of the review can provide practical support in the reporting process in accordance with the CSRD Directive.

## 2 Systematic Review – Materials and Methods

The review was conducted based on the PRISMA 2020 standard, the Systematic Literature Review (SLR) methodology, and the complementary Multi-Level Review (MLR) methodology. The MLR methodology plays a crucial role in determining the scope of the analysis, while SLR enables the identification and subsequent evaluation of all relevant literature in a given thematic area, facilitating the drawing of conclusions on the issue under consideration [11]. The integration of the PRISMA 2020 standards with the SLR and MLR methodologies helps reduce risks and systematic errors during the search and selection phase of publications for analysis.

A key element of the literature review is to clearly define the research objective (in this paper, the objectives address the research questions - RQs) and establish the inclusion (IC) and exclusion (EC) criteria for publications to be considered. This approach ensures the standardization of review protocols and enhances the credibility of the publications selected for analysis.

In first stage of the research, two databases (Springer and ScienceDirect) were selected and search criteria were adopted:

- Period time: 01.01.2023–30.09.2025.
- Type of publication: scientific articles, post-conference articles, book chapters.
- Language: English.

The novelty of this research is the choice of time period for the literature analysis. This choice is consistent with the one following the announcement of the CSRD Directive by the European Parliament (16 December 2022). Given the extended obligation to report on organizational performance in the context of sustainable development, new solutions are being sought to monitor and support organizations in the reporting process. The search was limited to materials with full access, as lack of full access would hinder the analysis of selected items at later stages of the review.

Next, search terms were defined in the selected databases, ensuring that the search results could be standardized for further comparison. The terms were chosen to capture as many synonyms of relevant concepts as possible. The analysis uses search terms such as "sustainable" and/or "sustainability" and/or "sustainable manufacturing" and "machining." Each combination was entered separately into two selected databases, which provided a comprehensive overview of the number of results for the given terms. The search results were downloaded in RIS format and stored on Google Drive to safeguard the publications accepted for analysis in case they were later removed. The results were also exported to Microsoft Excel to facilitate the removal of duplicate entries. After removing duplicates, the inclusion criteria (IC) for the publications were defined:

- IC1: The article is about mechanical processing **and** IC2: The article concerns the monitoring of SD parameters in the machining process **and/or** IC3: The article concerns the assessment of the SD level in the machining process **and/or** IC4: The article characterizes the process of mechanical machining in the context of SD **and/or** IC5: The article indicates tools/methods supporting the SD level in machining processes.

Next, the EC is defined:

- EC1: The article does not apply to mechanical processing **and/or** EC2: The article does not indicate tools/methods/technologies supporting the increase/monitoring of SD levels in machining processes.

According to the SLR&MLR methodology, publications in accordance with the adopted ICs and ECs should be excluded at this stage of the review. The selection of publications should be started by excluding them based on title, keywords and abstract. The publications accepted for further IC-compliant analysis were then subjected to a full analysis to assess them for quality based on the checklist (Table 1). The list of publications subject to full analysis is presented in Appendix 1.

The developed checklist consists of 3 questions, the aim of which is to find answers to the RQs posed in the work.

**Table 1.** Checklist for qualitative research.

| No | Question | Answer |
|---|---|---|
| 1 | Is the publication indicative of the technologies in the context of the machining process that affect the SD level? | Yes/No |
| 2 | Are the methods/techniques/technologies supporting SMP indicated in the publication? | Yes/No |
| 3 | Does the publication highlight the challenges faced by organizations in the transition to SMP? | Yes/No |

Only those publications that received at least 1 point in the qualitative assessment were included in the final results (Fig. 1). Each of the questions answers the RQs posed in this review. If the publication scored 0 points, it was not accepted for the final analysis.

Figure 1 presents the stages of searching for results in accordance with the adopted research methodology. In the first stage, a total of 8884 results were searched, of which 4480 were duplicates. On the basis of the adopted ICs, a total of 4328 publications were excluded, and 76 publications were accepted for qualitative analysis in accordance with the prepared checklist (Table 1). A total of 71 publications were accepted for final analysis.

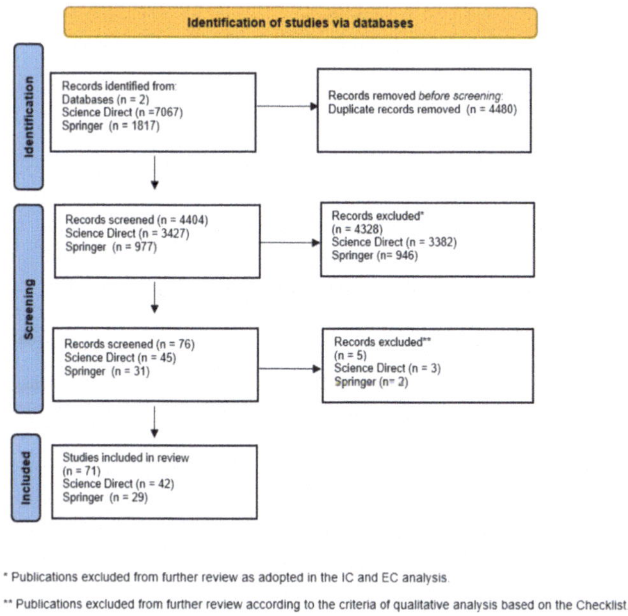


**Fig. 1.** PRISMA flow-diagram.

## 3 Research Results

In order to find answers to the RQs presented in the paper, publications were analyzed and the analysis was supported by the results obtained from the VOSviewer 13.0 software. VOSviewer allowed to generate a map of relationships between 71 publications (Fig. 2).

Map of connections created in VOSviewer 13.0. contains elements that constitute deadlines in this review. There are links between terms on the map that indicate the relationship between the two elements and the relationships of the co-occurrence of terms [2].

In the created map of connections, the strength of the link between terms is similar, which indicates a similar number of publications in which terms are present. The cluster groups in which the terms are located are called clusters, and a single term cannot be present in more than one cluster. An important part of the elements (terms) are their attributes, especially weight attributes, which are limited only to non-negative values [2]. A component with a higher weight is more important than an element with a lower weight. Terms with a higher weight in the visualization are shown more clearly than terms with a lower weight [2]. Clusters are divided by colour according to the strength of the term bindings in each group, which means that the weight attribute is the highest within a single cluster. There are also links between terms beyond the cluster in each group, but the weight attribute is smaller for them.

The map of connections (Fig. 3) consists of three clusters, where:

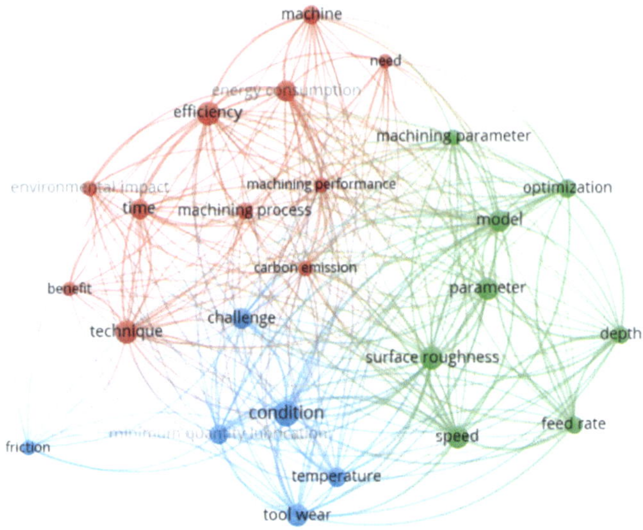


**Fig. 2.** Map of connections.

- Red cluster: machine, efficiency, energy consumption, machining performance, machining process, environmental impact, carbon emissions, benefit, need, techniques.
- Green cluster: machining parameter, optimization, model, parameter, surface roughness, depth, feed rate, speed.
- Blue cluster: challenge, condition, temperature, tool wear, friction, challenge, minimum quality lubrication.

So, answering RQ1, the results of the literature analysis (Fig. 2, Table 2) show that the parameters in the red cluster describe the expected factors that determine the increase in SD level in the context of the machining process. It is essential to implement advanced technologies that enable improvements in efficiency, reduction of energy consumption, enhancement of machining performance, and reduction of carbon emission. The key direction in the transition to SMP is also the implementation of models and solutions based on Artificial intelligence (AI). The tasks of AI-based tools for SMP are to support the automation (as well as robotization) of processes in order to increase their efficiency, efficiency, quality and safety of operators. AI is used, e.g., in the area of energy prediction [12–14], decision support and process optimization [15, 16], efficient use of resources [17] and models for predicting the non-permanent service life of tools [18, 19]. Among the methods and techniques supporting SMP, a conventional approach such as the carbon footprint calculation methodology [20] should also be indicated, LCA [21] or optical microscopy used to assess tool wear [6]. These are effective methods, but the inability to fully automate them makes them time-consuming and requires operator action. One of the limitations of this review is the lack of listing of AI-based solutions that facilitate the integration of AI into machining processes, thereby facilitating their transition towards SMP.

Next, based on the results (green cluster), the parameters of the machining process (MP) that should be examined in the context of transforming MP into a sustainable machining process (SMP) are identified (e.g. surface roughness, depth, feed rate, speed) – answering to QR2. The analysis of the literature emphasizes that the optimal selection of cutting parameters is crucial in the implementation of SMP. Appropriate selection of parameters directly affects the possibility of shortening machining time, reduces the number of used cutting tools, while maintaining the high quality of machined parts [5]. The green cluster in the link map (Fig. 2) confirms that the key trend is to look for models that optimize cutting parameters such as depth, speed, feed in terms of SMP while maintaining high quality of machined surfaces. The analysis of current solutions indicates the use of such methods/techniques in the optimization of cutting parameters as Multi-criteria Decision Making (MCDM) techniques [4, 22], multi-objective optimization (MOO) [22]. The blue cluster (Fig. 2), on the other hand, highlights the challenges of maintaining a balance between balanced process conditions and the selection of optimal machining parameters and the cooling method, temperature, friction and tool wear during machining. Among alternative cooling methods, the literature distinguishes, m.in. Nanofluid Minimum Quantity Lubrication (NMQL) [24] and Minimum quantity lubrication (MQL) [25].

Finally, the main challenges (blue cluster) facing the implementation of SMP focus on the following key terms: condition, temperature, tool wear, friction, and minimum quality lubrication. Therefore, the implementation of SMP requires support in reporting and monitoring treatment processes in the context of environmental impact, unification of norms and standards for measuring and reporting the carbon footprint [20], and coolant/lubricant consumption while maintaining a high-quality surface [26]. Moreover, the challenges involve the implementation of technologies that will enable the reduction of energy consumption in machining while maintaining process efficiency and effectiveness [16, 17, 23].

# 4 Discussion

The literature review results confirm that it is necessary to implement AI-based models/tools in order to transition from MP to SMP in the context of achieving SDG 9. Management decisions based on real data and AI-based models should provide to enhancing the level of SMP [27]. Changes in machining process parameters—such as surface roughness, depth, feed rate, and speed—should be monitored in the context of achieving results related to SD, such as: achieving in the context reducing energy consumption and carbon emissions, improving machining performance. Table 2 summarises the research results and provides the outlook into the relationships between AI-based models/tools, machining parameters, and achieving SDG 9.

The use of innovative cooling and lubrication techniques, such as MQL or cryogenic cooling, supported by decision models, significantly improves process efficiency while reducing environmental impact [23]. Especially, the combination of MQL with cryogenic $CO_2$ leads to a reduction in cutting force by nearly 20% and a reduction in tool wear and surface roughness by almost 50% compared to dry machining [28]. In the paper [29], it was indicated that large amounts of lubricants (coolants, cutting fluids) are used

**Table 2.** AI-based models/tools and machining parameter in the context of achieving SDG 9.

| AI Method | Key Machining Parameters | Effects in realising SDG 9 | Validation context |
|---|---|---|---|
| ML | Cutting speed, feed | Reduction of energy consumption | Experimental data |
| DL | Temperature, vibration | Reducing tool wear | Sensory data |
| ANFIS | Depth of Cut | Improving surface quality | Laboratory tests |
| Fuzzy MCDM (TOPSIS) | Cooling medium | Reduction of $CO_2$ emissions | Comparative analysis |
| Digital Twin + AI | Real-time process parameters | Resource optimization | Simulation + Real Data |

during conventional machining of Ti-6Al-4 V alloy. Due to the impact of lubricants on the operator's health and the environment, it has been assumed that dry machining is a sustainable alternative, but it has a negative impact on tool life and shortens surface integrity. These results confirm the potential of hybrid cooling technologies in the context of transforming MP to SMP.

Other studies point to the growing importance of the use of lubricating nanofluids, whose properties enable the reduction of friction, lowering process temperature and improving surface quality. In this case, decision support models allow not only to assess operational efficiency, but also to estimate the impact on $CO_2$ emissions and employee health [30]. A similar approach has been used in research on the use of toner waste to create eco-friendly lubricating fluids that significantly extend tool life and reduce surface roughness coefficient [31]. Decision support models for Electrical Discharge Machining (EDM) processes and Wire-electro discharge machining (WEDM) are also a significant trend. The choice of electrode (e.g. copper or brass) and dielectric parameters have been shown to have a significant impact on both process efficiency and $CO_2$ emissions. Multi-criteria models can identify the most sustainable configurations, balancing efficiency, cost, and environmental impact [32]. Additionally, fuzzy methods such as F-TOPSIS support the reduction of uncertainty in surface quality prediction, which increases the precision of process control [33].

The literature also emphasizes the growing role of computational AI-based models and digital twins. The use of fuzzy systems (FIS), adaptive neuro-fuzzy systems (ANFIS), evolutionary algorithms or machine learning techniques enables dynamic forecasting of key process parameters such as tool wear and surface roughness and their adaptive control in real time [32]. The literature indicates a number of applications of AI-based tools supporting the transition to SMP, especially through tool wear monitoring and prediction [5], surface roughness prediction [34], prediction of machining parameters [5] energy consumption prediction [16]. Integration with the Internet of Things and Industry 4.0 systems makes it possible to create intelligent, self-optimizing production lines [4]. Hybrid manufacturing processes, such as Wire Arc Additive Manufacturing (WAAM) integrated with machining, also remain an important research direction.

However, the main limitation of the research is that the selection of articles for analysis was restricted by the inclusion criteria (IC) and exclusion criteria (EC), as well as the search terms defined in the paper. Therefore, the research question that remains open is the analysis of the AI-based model for SMP integrating specific characteristics of the process and the desired effects of SDG 9 implementation. This is particularly important for further research focused on experimental studies. The next limitation of this research is a limited number of searched databases and a limited number of keywords. Finally, there is also the ethical issue of AI. Taking into account the ethical and environmental dimensions of AI is necessary due to their close link [5]. Therefore, AI systems should not only be technically efficient, but also transparent, accountable and in line with environmental protection and ethics, which would allow for the construction of a reliable and trusted system supporting the implementation of the SDGs [35].

## 5 Conclusions

This provided systematic literature review enable the identification of the expected outcomes within a manufacturing company from the implementation of advanced technologies in the context of achieving SDG 9, as well as the characteristics of MP that are crucial for the transformation from MP to SMP.

**Acknowledgments.** This work was supported by a program of the Polish Ministry of Science under the title 'Regional Excellence Initiative', project no. RID/SP/0050/2024/1.

## Appendix 1

| No | DOI | Points |
|---|---|---|
| Science Direct | | |
| 1 | https://doi.org/10.1016/j.jmapro.2024.11.057 | 0 |
| 2 | https://doi.org/10.1016/j.clet.2025.100908 | 3 |
| 3 | https://doi.org/10.1016/j.nxmate.2025.100586 | 3 |
| 4 | https://doi.org/10.1016/j.rineng.2024.103042 | 3 |
| 5 | https://doi.org/10.1016/j.measurement.2023.113937 | 2 |
| 6 | https://doi.org/10.1016/j.procir.2024.12.118 | 0 |
| 7 | https://doi.org/10.1016/j.clet.2024.100804 | 2 |
| 8 | https://doi.org/10.1016/j.procir.2024.10.080 | 2 |
| 9 | https://doi.org/10.1016/j.clet.2025.101067 | 3 |
| 10 | https://doi.org/10.1016/j.rineng.2025.106921 | 2 |

(*continued*)

(*continued*)

| No | DOI | Points |
| --- | --- | --- |
| 11 | https://doi.org/10.1016/j.rineng.2025.104457 | 3 |
| 12 | https://doi.org/10.1016/j.rineng.2025.105249 | 3 |
| 13 | https://doi.org/10.1016/j.heliyon.2024.e41027 | 2 |
| 14 | https://doi.org/10.1016/j.procir.2024.12.074 | 2 |
| 15 | https://doi.org/10.1016/j.jmapro.2024.09.120 | 2 |
| 16 | https://doi.org/10.1016/j.compositesb.2024.111247 | 1 |
| 17 | https://doi.org/10.1016/j.triboint.2024.109677 | 3 |
| 18 | https://doi.org/10.1016/j.heliyon.2024.e38299 | 2 |
| 19 | https://doi.org/10.1016/j.cirpj.2024.06.006 | 2 |
| 20 | https://doi.org/10.1016/j.wear.2025.206346 | 2 |
| 21 | https://doi.org/10.1016/j.aej.2025.07.034 | 2 |
| 22 | https://doi.org/10.1016/j.cja.2024.01.018 | 3 |
| 23 | https://doi.org/10.1016/j.jmapro.2023.08.026 | 0 |
| 24 | https://doi.org/10.1016/j.procir.2025.02.114 | 1 |
| 25 | https://doi.org/10.1016/j.jclepro.2023.138259 | 2 |
| 26 | https://doi.org/10.1016/j.matdes.2025.113690 | 2 |
| 27 | https://doi.org/10.1016/j.procir.2024.12.001 | 2 |
| 28 | https://doi.org/10.1016/j.rineng.2025.104924 | 2 |
| 29 | https://doi.org/10.1016/j.procir.2024.12.148 | 2 |
| 30 | https://doi.org/10.1016/j.heliyon.2024.e33726 | 3 |
| 31 | https://doi.org/10.1016/j.rineng.2024.102355 | 3 |
| 32 | https://doi.org/10.1016/j.procir.2023.06.086 | 3 |
| 33 | https://doi.org/10.1016/j.heliyon.2024.e34836 | 2 |
| 34 | https://doi.org/10.1016/j.procir.2023.03.059 | 2 |
| 35 | https://doi.org/10.1016/j.rineng.2025.104415 | 1 |
| 36 | https://doi.org/10.1016/j.engappai.2025.111276 | 2 |
| 37 | https://doi.org/10.1016/j.mfglet.2024.09.083 | 2 |
| 38 | https://doi.org/10.1016/j.rineng.2024.102090 | 2 |
| 39 | https://doi.org/10.1016/j.heliyon.2024.e34433 | 2 |
| 40 | https://doi.org/10.1016/j.procir.2024.12.024 | 2 |
| 41 | https://doi.org/10.1016/j.procir.2024.08.020 | 2 |
| 42 | https://doi.org/10.1016/j.cirp.2024.06.001 | 3 |
| 43 | https://doi.org/10.1016/j.susmat.2023.e00641 | 1 |
| 44 | https://doi.org/10.1016/j.procs.2025.01.211 | 1 |

(*continued*)

(*continued*)

| No | DOI | Points |
|---|---|---|
| 45 | https://doi.org/10.1016/j.susoc.2024.01.001 | 1 |
| Springer | | |
| 46 | https://doi.org/10.1007/s40684-025-00757-6 | 2 |
| 47 | https://doi.org/10.1038/s41598-024-79338-7 | 1 |
| 48 | https://doi.org/10.1038/s41598-025-18356-5 | 3 |
| 49 | https://doi.org/10.1038/s41598-025-07526-0 | 2 |
| 50 | https://doi.org/10.1007/s42452-025-07365-8 | 2 |
| 51 | https://doi.org/10.1007/s42452-024-06398-9 | 2 |
| 52 | https://doi.org/10.1007/978-3-031-77429-4_76 | 2 |
| 53 | https://doi.org/10.1007/s10010-025-00898-9 | 1 |
| 54 | https://doi.org/10.1007/s42452-025-07385-4 | 0 |
| 55 | https://doi.org/10.1038/s41598-025-01543-9 | 2 |
| 56 | https://doi.org/10.1007/s40684-024-00640-w | 2 |
| 57 | https://doi.org/10.1007/s44444-025-00030-w | 2 |
| 58 | https://doi.org/10.1007/s00170-025-16121-6 | 2 |
| 59 | https://doi.org/10.1007/978-3-031-93891-7_50 | 0 |
| 60 | https://doi.org/10.1007/s00170-025-16464-0 | 2 |
| 61 | https://doi.org/10.1007/s00170-022-10794-z | 2 |
| 62 | https://doi.org/10.1007/s00170-024-13881-5 | 2 |
| 63 | https://doi.org/10.1038/s41598-025-05181-z | 2 |
| 64 | https://doi.org/10.1007/978-3-031-28839-5_22 | 2 |
| 65 | https://doi.org/10.1007/s43939-025-00236-4 | 2 |
| 66 | https://doi.org/10.1038/s41598-025-07584-4 | 1 |
| 67 | https://doi.org/10.1007/s00170-024-14583-8 | 2 |
| 68 | https://doi.org/10.1007/s00170-024-13823-1 | 2 |
| 69 | https://doi.org/10.1007/s12008-023-01696-9 | 2 |
| 70 | https://doi.org/10.1007/s42452-025-06663-5 | 2 |
| 71 | https://doi.org/10.1186/s44147-023-00292-8 | 1 |
| 72 | https://doi.org/10.1007/s00170-023-12132-3 | 2 |
| 73 | https://doi.org/10.1007/s40684-023-00501-y | 2 |
| 74 | https://doi.org/10.1038/s41598-025-04198-8 | 2 |
| 75 | https://doi.org/10.1038/s41598-025-92713-2 | 2 |
| 76 | https://doi.org/10.1038/s41598-023-49727-5 | 1 |

## References


1. Patalas-Maliszewska, J., Łosyk, H.: Changes in sustainable development in manufacturing in cases of unexpected occurrences–A systematic review. Sustainability. **16**(2), 1–23 (2024). https://doi.org/10.3390/su16020717
2. Eck, N.S., Waltman, L.: VOSviewer Manual. CWTS. (2022)
3. The Sustainable Development Goals Report.: https://unstats.un.org/sdgs/report/2025/ 2025. Accessed 17 Nov 2025
4. Sristi, N.A., Zaman, P.B., Dhar, N.R.: Evaluation of entropy-coupled multi-criteria decision-making methods for enhancing machinability. Heliyon. **10**(19), e38299 (2024). https://doi.org/10.1016/j.heliyon.2024.e38299
5. Chung, C., Andrianto, A., Wang, P.O.: Robust integer optimization of turning parameters for cutting tool sustainability and machining economics in discrete production. Heliyon. **10**(24), e41027 (2024). https://doi.org/10.1016/j.heliyon.2024.e41027
6. Khan, M.A., Mufti, S.: Tool chip contact length analysis of dry and cryogenic turning of aerospace alloy Ti-6Al-4V. Manuf. Lett. **41**, 653–657 (2024). https://doi.org/10.1016/j.mfglet.2024.09.083
7. Soori, M., Jough, F.K.G., Dastres, R., Arezoo, B.: Sustainable CNC machining operations: A review. Sustainable Oper. Comput. **5**, 73–87 (2024). https://doi.org/10.1016/j.susoc.2024.01.001
8. Sarma, D., Ozah, R., Chandrasekaran, M., Sahoo, A.K., Kumar, R., Pattanayak, S.: Material-specific machining optimization of Ti6Al4V alloy under MQL: A sustainability-centric approach. Next Mater. **8**, 100586 (2025). https://doi.org/10.1016/j.nxmate.2025.100586
9. Gupta, M.K., et al.: Studies on geometrical features of tool wear and other important machining characteristics in sustainable turning of aluminium alloys. Int. J. Precis. Eng. Manuf.-Green Tech. **10**, 943–957 (2023). https://doi.org/10.1007/s40684-023-00501-y
10. Henkel, S., Knauf, M., Katzer, F., et al.: Development and application of material extrusion produced ultra-fine diamond grinding tools for machining hard brittle materials. Int. J. Adv. Manuf. Technol. **139**, 4631–4650 (2025). https://doi.org/10.1007/s00170-025-16121-6
11. Martinčević, I., Primorac, D., Dorić, B.: Corporate Sustainability Reporting Directive (CSRD): Obligations, challenges and requirements for companies. ENTRENOVA–ENTerprise REsearch InNOVAtion. **10**(1), 317–327 (2024)
12. Chinnathai, M.K., Alkan, B.: A digital life-cycle management framework for sustainable smart manufacturing in energy intensive industries. J. Clean. Prod. **419**, 138259 (2023). https://doi.org/10.1016/j.jclepro.2023.138259
13. Farooq, M.U., et al.: Sustainable machining of Inconel 718 using minimum quantity lubrication: Artificial intelligence-based process modelling. Heliyon. **10**(15), e34836 (2024). https://doi.org/10.1016/j.heliyon.2024.e34836
14. Gupta, M.K., et al.: Real-time monitoring and measurement of energy characteristics in sustainable machining of titanium alloys. Measurement. **224**, 113937 (2024). https://doi.org/10.1016/j.measurement.2023.113937
15. Jeong, H., Fekete, T., Bashyal, A., Wicaksono, H.: From theory to practice: Implementing causal AI in manufacturing for sustainability. Procedia Comput. Sci. **253**, 1495–1504 (2025). https://doi.org/10.1016/j.procs.2025.01.211
16. Shah, M.A., Wicaksono, H.: Leveraging machine learning for power consumption prediction of multi-step production processes in dynamic electricity price environment. Procedia CIRP. **130**, 226–231 (2024). https://doi.org/10.1016/j.procir.2024.10.080
17. Liang, Y., Wang, Y., Chiong, R., Li, A., Lu, J.: Cutting tool life prediction and extension through generative model-augmented deep learning and laser remanufacturing techniques. Eng. Appl. Artif. Intell. **158(Part A)**, 111276 (2025). https://doi.org/10.1016/j.engappai.2025.111276

18. Sen, B., et al.: AI-driven wear monitoring of PVD TiAlN coated carbide insert in sustainable machining of Hastelloy C276: An industry 4.0 perspective. Results Eng. **25**, 104457 (2025). https://doi.org/10.1016/j.rineng.2025.104457
19. Wurm, S., Tschepe, T., Petrovic, O., Herfs, W.: Developing a methodology for accurate product carbon footprint calculation in machining processes. Procedia CIRP. **135**, 1308–1313 (2025). https://doi.org/10.1016/j.procir.2024.12.148
20. Alammari, Y., Saelzer, J., Berger, S., Iovkov, I., Biermann, D.: Initial period of chip formation: Observations towards enhancing machining sustainability. In: Kohl, H., Seliger, G., Dietrich, F. (eds.) Manufacturing Driving Circular Economy. GCSM 2022 Lecture Notes in Mechanical Engineering. Springer, Cham (2023). https://doi.org/10.1007/978-3-031-28839-5_22
21. Saeed, M.A., et al.: Optimizing economic and environmental objectives in sustainable machining processes. Cleaner Eng. Technol. **28**, 101067 (2025). https://doi.org/10.1016/j.clet.2025.101067
22. Ahmad, A., et al.: Achieving sustainable machining of titanium grade 3 alloy through optimization using grey relational analysis (GRA). Results Eng. **23**, 102355 (2024)
23. Hakkı, N.R., Kaftanoğlu, B.: Enhancing sustainable machining of Inconel 718 using Multi-Walled Carbon Nanotubes and TiO2 based nanofluid minimum quantity lubrication. Procedia CIRP. **135**, 781–786 (2025). https://doi.org/10.1016/j.procir.2024.12.074
24. Ali, S.H., et al.: Recent developments in MQL machining of aeronautical materials: A comparative review. Chin. J. Aeronaut. **38**(1), 102918 (2025)
25. Elsheikh, A., et al.: A review on sustainable machining: Technological advancements, health and safety considerations, and related environmental impacts. Results Eng. **24**, 103042 (2024). https://doi.org/10.1016/j.rineng.2024.103042
26. Marconi, S., Frigerio, N., Matta, A.: Clustering-based energy modeling for sustainable CNC woodcutting machinery. Procedia CIRP. **135**, 307–312 (2025). https://doi.org/10.1016/j.procir.2024.12.024
27. Rojek, I., Dostatni, E., Mikołajewski, D., Pawłowski, L., Węgrzyn-Wolska, K.: Modern approach to sustainable production in the context of Industry 4.0. Bull. Pol. Acad. Sci. Tech. Sci. **70**(6), e143828 (2022). https://doi.org/10.24425/bpasts.2022.143828
28. Chen, X., et al.: Study on the transient thermo-mechanical coupling mechanism at tool-chip interface in ultrasonic vibration assisted chip formation process under sustainable dry machining conditions. Cleaner Eng. Technol. **25**, 100908 (2025). https://doi.org/10.1016/j.clet.2025.100908
29. Dennison, M.S., Jebabalan, S.K., Barik, D.: Applicability of nano-cutting fluids for enhanced cooling, low tool wear, and high tribological performance during machining—A review. Discov. Appl. Sci. **6**, 663 (2024). https://doi.org/10.1007/s42452-024-06398-9
30. Ross, N.S., et al.: Performance evaluation of cutting fluids formulated with recycled toner waste and rice bran oil in sustainable machining of LMD processed Ti64 alloy. Int. J. Precis. Eng. Manuf.-Green Tech. **12**, 151–167 (2025). https://doi.org/10.1007/s40684-024-00640-w
31. Hannan, A., et al.: Machining performance, economic and environmental analyses and multi-criteria optimization of electric discharge machining for SS310 alloy. Sci. Rep. **14**, 28930 (2024). https://doi.org/10.1038/s41598-024-79338-7
32. Sunder, S.C., Ramesh, A.: Digital twin-enabled surface quality prediction and optimization in dry turning of Ti6Al4V using ANFIS and genetic algorithm. J. King Saud Univ.–Eng. Sci. **37**, 17 (2025). https://doi.org/10.1007/s44444-025-00030-w
33. Sawyerr, B.A., Fasina, E., Adedeji, W.O., Adeniran, M.K., Oke, S.A., Rajan, J.: A fuzzy TOPSIS method for surface integrity criteria ranking using the wire electrical discharge machining process. J. Eng. Appl. Sci. **70**, 120 (2023). https://doi.org/10.1186/s44147-023-00292-8

34. Elangovan, M., et al.: Machine learning approach to the prediction of surface roughness using statistical features of vibration signal acquired in turning. Procedia Comput. Sci. **50**, 282–288 (2015)
35. Bansal, C.: AI ethics and sustainability: Accelerating paradigm shifts toward sustainable development. J. Strategy & Innovation. **36**(1), 200537 (2025)

# The Potential for Adopting Circular Supply Chains with Paas Models in the Apparel Industry

Dorota Filipek and Paulina Golińska-Dawson(✉)

Poznan University of Technology, Institute of Logistics, Poznan, Poland
paulina.golinska@put.poznan.pl

**Abstract.** Product-as-a-Service (PaaS) in the apparel industry is an emerging business model that aligns with the principles of the Circular Economy (CE) by promoting sustainability through the leasing, renting, or sharing of clothing items rather than traditional ownership. Currently, there are limited examples of successful implementation of circular supply chains with PaaS in consumer markets. To address this gap, the aim of this study is to assess the maturity of circular practices related to PaaS models in the apparel industry in the context of circular supply chains. Using desk research, existing practices are analysed to determine their level of maturity to support circular supply chains with PaaS. A multiple case study method was applied to compare the approaches of selected companies, allowing a detailed evaluation of the strengths and limitations in implementation. This analysis allowed for the development of a novel maturity model, which helps in identifying gaps between the current state of practice and the desired model for the circular supply chain with PaaS. The findings offer practical recommendations for both individual companies and the wider apparel sector that seek to advance the level of circularity in their supply chain.

**Keywords:** Product-as-a-Service (PaaS) · Circular supply chains · Apparel industry · Sustainability · Business model innovation · Maturity assessment

## 1 Introduction

The global apparel industry is an important economic sector valued at approximately 1.3 trillion USD and employing more than 300 million people worldwide [1]. The fashion industry alone produces 92 million tons of textile waste annually, with 85% ending in landfills or incinerators [2]. These challenges underline the need for a transition towards the adoption of the principles of Circular Economy(CE) where resources are kept in use for as long as possible, and towards circular supply chains that prioritise reuse, repair, and resource efficiency [3]. However, the circular transition in the apparel industry faces many barriers, including saturated second-hand markets, limited recycling technologies, and complex global regulations [4]. Therefore, developing of effective circular supply chains (CSCs) is essential, as they support adoption of CE practices [5]. Research in the apparel sector highlights inadequate supply chain management, as a cause of increasing


A. Hamrol et al. (Eds.): MANUFACTURING 2026, LNME, pp. 189–203, 2026.
https://doi.org/10.1007/978-3-032-21701-1_14

landfill waste [6], inefficient first-mile collection planning [7] and problems with sorting of used clothing for further recovery and reuse [8]. Logistics and SCM are widely regarded as key enablers of efficient CSCs and the creation of economic, environmental, and social value at the triple bottom line [9–11].

The aim of this paper is to investigate the maturity of circular practices within apparel supply chains operating under Product-as-a-Service (PaaS) settings, evaluating opportunities for development and improvement in the context of the circular economy. Four companies were examined using publicly available company information and industry literature. By identifying existing gaps and proposing recommendations, the research highlights the urgency of systemic change, supports wider adoption of circular supply chain, and demonstrates how leading companies can serve as examples for an industry transformation. We state the following research questions:

**RQ1:** Which circular practices can be implemented within Product-as-a-Service models in the apparel industry?

**RQ2:** How the level of maturity of the practices that support the implementation circular supply in PaaS can be measured?

The reminder of this paper is organised as follows, Sect. 2 presents the theoretical background, outlining key concepts related to circular supply chains, circular practices, and PaaS. Section 3 defines the research problem and explains the methodology used to analyse the selected companies. Section 4 provides the results of the study and discusses the findings in relation to the research questions. Finally, Sect. 0 presents the main conclusions, highlights the limitations of the research, and proposes directions for future studies to support further development of circular supply chain strategies in the apparel industry.

## 2 Theoretical Background and Research Gaps

### 2.1 Circular Supply Chain

Circular supply chains (CSCs) are designed to be restorative and regenerative, ensuring that materials circulate as long as possible to maximise value over time [12]. Within this framework, research is focused on areas such as reverse logistics, sustainable supply chain management, and closed-loop systems [11, 13]. The CSC concept advances sustainable supply chain models by integrating a regenerative approach, ideally achieving zero waste [14]. CSC involves the application of circular practices (so called R-practices) to the management of supply chains. It focuses on the restoration of technical materials and the regeneration of biological materials to achieve the zero-waste vision, supported by innovations in business models and supply chain operations from the design of products or services to product recovery [15]. This approach involves all stakeholders throughout the lifecycle, including manufacturers, service providers, consumers, and users [14]. CSC not only aims to mitigate environmental impacts, but also returning products to their original supply chain for reuse, refurbishment, remanufacturing or recycling [16]. Yang et al. [17] identify four main sources of value creation within CSCs, as follows:

- minimising material consumption and reducing costs through inner circles (production, reuse and refurbishment) and then through outer circles (recycling);
- maximising the number of circles and extending the life of the product by limiting material consumption, product longevity, reusing waste across value chains and maintaining high quality resource flows;
- enforcing of cascaded use which enables secondary applications through industrial symbiosis;
- use of an uncontaminated material streams to increase redistribution efficiency and material productivity.

Despite its potential, the transition towards CSCs presents significant operational, organisational, and strategic challenges for both companies and policymakers. From a CSC perspective, effective circularity requires enhanced product flows during the post-sale and post-consumption stages. This includes designing efficient collection networks, developing reliable systems for sorting and classifying used garments, and ensuring their recovery, refurbishment, recycling or reintegration into the market [18–20]. These activities are essential for closing material loops, but they remain underdeveloped and unevenly implemented across the apparel industry [21].

PaaS models offer new opportunities to strengthen circular supply chains by enabling companies to retain ownership or control of products and maintain ongoing relationships with customers beyond the initial point of sale [15]. Through leasing, rental, subscription, or usage-based models, PaaS can facilitate product return flows, support higher utilisation rates, extend product lifetimes, and provide valuable data for improving design, durability, and traceability. As such, PaaS has the potential to serve as a crucial mechanism for operationalising circularity in the apparel sector.

However, the existing literature on CSCs in the fashion industry tends to focus on recycling technologies, material innovation, or end-of-life management, while paying limited attention to business model innovations such as PaaS [21]. Research rarely integrates PaaS into the broader CSC framework or examines how it can enable more efficient collection, recovery, and recirculation processes. This lack of integration leaves a gap in understanding how business models and supply chain design must evolve to enable circularity at scale.

Therefore, this paper aims to address this research gap by exploring the role of PaaS in supporting and advancing circular supply chain development within the apparel industry.

## 2.2 Circular Practices in a Supply Chain

Common circular practices in circular supply chain includes [19, 21, 22]:

- developing of new business models increasing the timespan of clothing,
- environmentally safe and renewable resources, and
- recycling, reuse, refurbishing, and upcycling solutions.

These practices include the R-strategies, circular design, reverse logistics, industrial symbiosis, and PaaS models [16]. From CSC's perspective, sorting plays a pivotal role in enabling the cascade approach, which maximises retained value through successive

R-strategies such as reuse, repair and recycling [19]. Effective sorting ensures that post-consumer garments are correctly classified beginning with identifying items suitable for high-value resale, allowing reuse to be prioritised over recycling and lower waste-hierarchy options [23, 24]. Prioritising reuse, over recycling allows significantly reduces energy use and emissions [25]. Implementing PaaS in the apparel industry enables the creation and management of viable systems for reuse and resale. PaaS is servitised business model, in which ownership of the product remains with the manufacturer or retailer [26]. In this model, consumers pay for access to the product or for the results it delivers rather than purchasing the product itself, thereby shifting consumption away from ownership towards service-based use [27]. In the apparel industry, PaaS is typically implemented through two approaches: pay-per-use and subscription models. Pay-per-use, suitable for occasional or short-term consumption, allows customers to access a product for a limited period and return it after use, after which the item is assessed, repaired, or adjusted before being offered to the next user. Subscription models, by contrast, give customers ongoing access through a recurring payment, supporting continuous and extended product utilisation.

PaaS in a CSC (see Fig. 1.) requires that a single product is used sequentially by multiple customers over its lifetime. To ensure smooth circulation of products between temporary users, companies must apply strategies R-strategies, such as reuse, repair, refurbishment, remanufacturing and recycling, which serve as value retention processes (VRPs) that extend product lifetimes and preserve material and economic value while reducing environmental impact [16, 28]. Current research indicates a research gap on the limited practical application of circular practices in the apparel industry. Although CE concepts are increasingly discussed at a conceptual level, their translation into concrete industrial practices across the entire value chain, from fibre production and garment manufacturing to recycling and end-of-life recovery, remains underexplored. Many initiatives focus on individual stages, while comprehensive and system-level approaches are still limited [18–20].

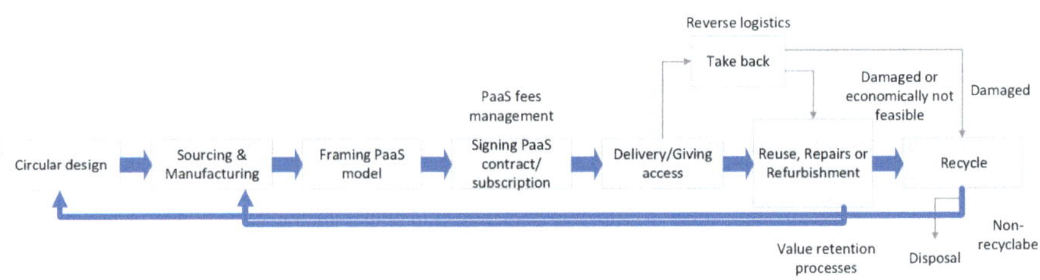


**Fig. 1** Example of circular PaaS in the apparel industry. Source: own elaboration based on [29].

## 3 Research Problem and Research Methodology

This exploratory study implements the mixed methods approach, as it combines desk research, with multiple case study methods, and a gap analysis. Methodology is summarised in Fig. 2. We applied the case study research, as it is a widely recognised qualitative approach that enables an in-depth examination of a clearly defined phenomenon within its real-life context. It involves the use of multiple sources of evidence, such as documents, observations, or reports, in order to build a comprehensive understanding of the subject under investigation [30]. A single case study allows for detailed insight into one organisation or situation, while multiple case studies facilitate comparison, enabling identification of similarities and differences across cases. Multiple case study designs are often valued for their stronger evidence base, as findings supported by several empirical examples contribute to stronger and more credible theoretical insights [30]. By contrasting or confirming the results in cases, this approach supports a wider exploration of research questions, contributing to theoretical development.

Gap analysis is a structured method used to compare the current state of an organisation with a defined target state in order to identify areas requiring improvement. The process typically involves four stages: assessing the existing situation, describing the desired future condition, determining the discrepancies between the two, and formulating actions that can minimise or ultimately eliminate these gaps [31]. In this study, multiple case analysis was applied to determine the current maturity level of implementation of circular practices in PaaS supply chains with PaaS in selected organisations.

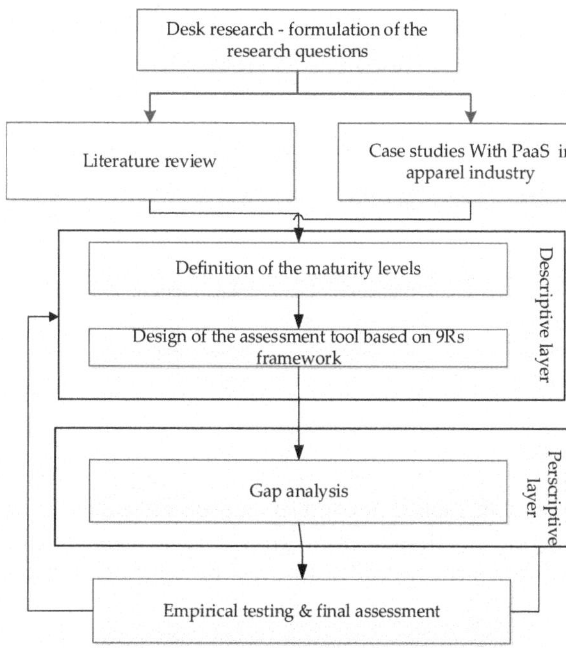


**Fig. 2** Methodology. Source: own elaboration.

To support the identification of gaps, a maturity model was employed. Such models offer a structured framework that describes the characteristics of effective processes at different stages of development and are increasingly recommended for assessing corporate sustainability performance [32]. Maturity models facilitate benchmarking and help position an organisation within a transformation pathway, illustrating the evolution from a basic level to an advanced level of practice. Their strengths lie in their clarity, ease of application, and incremental nature, which enables a step-by-step progression toward the target state [32].

## 4 Results

This paper analyses four case studies, where the circular practices are implemented across the supply chains with PaaS for consumer markets (B2C). Table 1 *shows a summary of all the cases.* Companies were selected based on the following inclusion criteria: 1) core operations in the apparel industry on international markets; 2) commitment to sustainability and environmental awareness; 3) offering of PaaS for consumer markets.

**Table 1** Summary of the analysed cases. Source: own elaboration.

| Case | A | B | C | D |
|---|---|---|---|---|
| Products | Outdoor clothing and equipment | Sports shoes and clothing | Luxury apparel and accessories | Jeans |
| Size | Big | Big | Big | Small |
| Country of origin | USA | Switzerland | USA | Netherlands |
| Area of operation | Worldwide | Worldwide | Worldwide | Europe |
| Circular strategies | Rental, repair and resale of clothes, recycling of materials, investment in sustainable raw materials | Subscription, recycling, material recovery, designing for disassembly | Subscription, resale of vintage products, investment in recycling and sustainable materials | Leasing, take-back, repair, recycling, designing with circulation in mind |

The analysis framework is based on the findings from the literature review presented, and its scope is outlined in Table 2. It considers aspects related to the implementation of 1) PaaS models, 2) circular design, and 3) product value retention processes such as repair, reuse and refurbishment, as well as recovery processes including upcycling and recycling.

### 4.1 Circular Business Model PaaS

Company A offers a pay-per-use rental service for specialised outdoor equipment, allowing customers to book items online, collect them from a single retail location, use them, and return them after use. The service supports reuse and upcycling by reducing the need for occasional-use products to be purchased, although its scope remains limited to one product category and one store.

Company B operates a subscription-based footwear service in which customers pay a monthly fee to receive fully recyclable shoes made from bio-based materials. Once worn out or when a replacement is desired, the shoes must be returned to the company, which recycles the materials and provides a new pair, creating a closed-loop cycle. The service currently covers three footwear models and is available in a wide range of countries, with plans for further expansion.

The company C offers a subscription-based clothing rental service that provides customers with a monthly selection of personalised garments, which can be worn, exchanged for new styles, or purchased at a discounted rate. Returned items are donated for second-hand use, supporting extended product life rather than material recycling. The service is limited to selected apparel categories and is currently available only in one country.

The company D allows customers to subscribe to a pair of jeans for a fixed monthly fee, covering the full range of styles offered. After the leasing period, customers may keep the jeans, exchange them for a discounted new pair or return them for reuse or recycling, with repair services included during the subscription. The system is available in many countries, and supports both product life extension and material recovery.

**Table 2** Analysis framework - scope of implementation of the circular practices.

| b | Description of practices |
|---|---|
| A. Circular business models | Increasing the timespan of clothing by implementation of Product-as-a-Service (renting, subscriptions, pay-per-use etc.) |
| B. Environmentally safe and renewable resources | Circular design: products design to last and easy to repair, with sustainable and secondary materials, and resource efficient production processes. |
| C. Product recovery & value retention processes | C1. Collection system (take-back) C2. Repair |
| | C3. Refurbish C4. Resale |
| | C5. Upcycle C6. Recycle |

Source: own elaboration.

## 4.2 Circular Design

Company A applies several circular design principles, including durability, multifunctionality, and the avoidance of obsolescence. Its products are designed to be easily repairable and made from materials that can be reused or recycled. Through these actions, the company reduces its dependence on virgin raw materials and reduces carbon emissions. It pays particular attention to the origin of materials and aims to systematically reduce its reliance on natural resources. To support this, the company has committed to using only preferred materials, such as organic and regenerative cotton, as well as recycled polyester and nylon. In addition, it plans to incorporate both synthetic and natural fibres derived from pre-consumer and post-consumer waste in the production of new garments.

Company B is in the process of transitioning low impact production methods, supported by emission reduction targets and design choices that minimise the environmental footprint of its products. The company improved its sustainable design efforts through innovation technology, which enables the creation of a complete shoe upper in a single three-minute step, reducing significantly $CO_2$ emissions compared with its conventional product. It also introduced an internal design guide and adopted the Life Cycle Assessment to support environmentally informed material selection and product development, achieving designs that are fully recyclable or incorporate recycled materials. In PaaS model, a closed material loop is implemented by reusing materials recovered from returned footwear.

Company C has set a target to establish and fully implement circular product design principles by the end of 2025, by selecting sustainable materials and processes, as well as ensuring durability, repairability and, ultimately, recyclability. In Company C, design decisions are guided by an emphasis on extending product life, enabling effective maintenance, and facilitating end-of-life recovery of materials.

In Company D, the design process also considers production methods and innovations that minimise negative environmental impacts. A key indicator of its commitment to circular design is the Cradle-to-Cradle certification it has achieved. This certification is awarded to products designed to follow a natural end-of-life cycle, meaning that they can be fully recycled, upcycled, or safely decomposed, returning to the environment without causing harm.

## 4.3 Product Value Retention Processes and Recovery

As mentioned in Sect. 2, a well-established collection system is essential to ensure a sufficient flow of materials and to enable economically viable and environmentally responsible value retention processes, such as repair or refurbishment.

Company A operates a comprehensive take-back system that accepts all products it has distributed, either through returns at physical stores or via postal return to its service centre. Returned items are assessed for reuse, in which case customers receive a voucher or otherwise recycled with no customer compensation to encourage environmentally responsible behaviour. This return scheme applies across the full product range, making it a well-established practice.

Company B operates a closed-loop take-back system in which all subscription footwear and apparel must be returned once worn out, exchanged or when the subscription ends, ensuring that no product remains with the customer beyond its intended use period. Returned items are sorted for resale, refurbishment, or full material recycling, with worn products refreshed and sold through a dedicated second-hand platform, and customers receive a voucher as an incentive. Items that cannot be resold are recycled into new products or donated, making the take-back programme comprehensive and aligned with circular recovery principles.

Company C runs a textile recycling initiative that allows customers in selected regions to return 100% cashmere garments, regardless of brand, for fibre recovery and reprocessing into new yarns and fabrics. Participants receive a shipping label, but no financial incentive, making the programme a voluntary sustainability contribution. The scheme is limited to cashmere products and relies on an external recycling partner to complete the process.

Company D offers a refund scheme for all denim products it manufactures. Customers can request a return label online and receive a discount on a new pair when donating their jeans, which are then reused, recycled, or repurposed, and each return also supporting a tree restoration initiative.

The applied practices are assessed based on their maturity (stage) of their implementation following the scoring scheme presented in Fig. 3.

| Level | Stage | Stage description | Mark |
|---|---|---|---|
| 1 | No practices | Lack of any action | □□□□ |
| 2 | Single practice | Occurrence of individual non-systematic activities, but no strategy | ■□□□ |
| 3 | Partial implementation | The practice is applied but to a limited scope, e.g. for a particular type of product or only in a particular location. The company is in the process of planning its wider implementation | ■■□□ |
| 4 | Advanced Implementation | The practice is applied for more products, on a larger scale | ■■■□ |
| 5 | Full integration | The practice is an integral part of the company applied to all products | ■■■■ |

**Fig. 3** The scoring of the maturity levels. Source: own elaboration.

The comparison of the maturity of the analysed sustainable practices (see Table 2) based on the scoring scheme (see Fig. 3) is shown in Fig. 4.

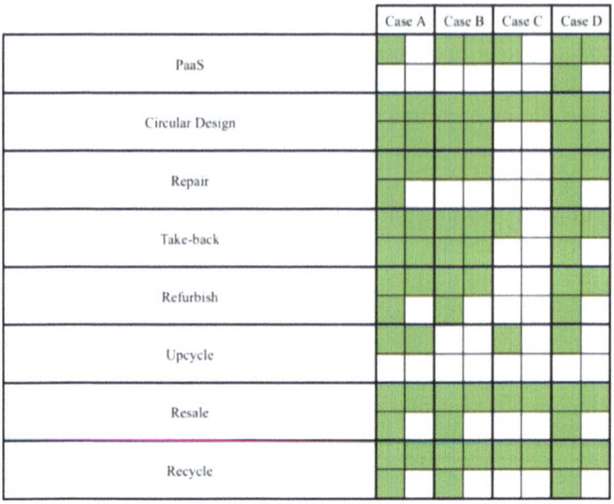


**Fig. 4** The scoring for the particular categories/subcategories of circular practices. Source: own elaboration.

The maturity of the implementation of circular practices among the analysed companies is presented in Fig. 5.

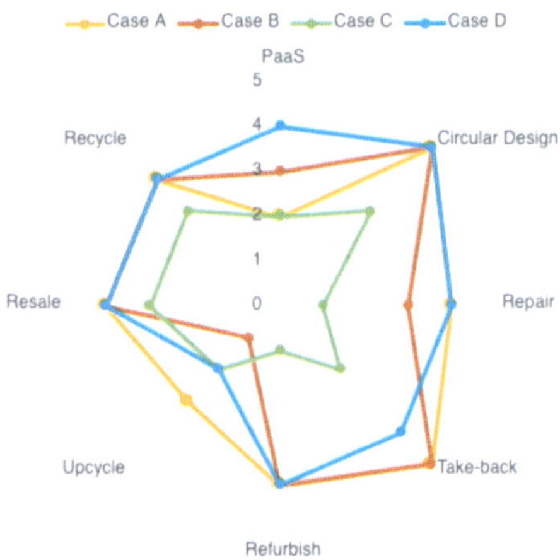


**Fig. 5** Radar chart of maturity of practice in the analysed cases. Source: own elaboration.

To provide an integrated assessment, the Circular PaaS Index is introduced. This index (see Table 3) consolidates the scores across the different levels of circular practice implementation, offering a comprehensive evaluation of the progress of each company. The index is calculated as the mathematical average of the total scores achieved across all individual circular practices, divided by the maximum possible level of implementation, and multiplied by 100 to obtain percentage values.

$S_i$ = score for practice i

n = total number of practices (here: n = 8)

$S_{min}$, $S_{max}$ = minimum/maximum possible score per practice (here: $S_{min} = 1$; $S_{max} = 5$)

$$\text{Average Score} = \frac{\sum_{i=1}^{n} S_i}{n} \tag{1}$$

$$\text{Normalised Score} = \frac{\text{Average Score}}{S_{max}} \tag{2}$$

$$\text{Circular PaaS Index} = \left( \frac{\sum_{i=1}^{n} S_i}{n \times S_{max}} \right) \times 100 \tag{3}$$

**Table 3** Scale of implementation of the circular practices.

| Case | Arithmetic Average | CircularPaaSIndex |
|---|---|---|
| A | 3.875 | 77,5% |
| B | 3.625 | 72,5% |
| C | 2.125 | 42,5% |
| D | 3.875 | 77,5% |

Source: own elaboration.

## 5 Discussion

To structure the assessment to indicate the gaps in the application of circular practices in the apparel industry, a novel maturity model was proposed. This model systemises the finding from the case studies in a more general way (see Fig. 6). The model is based on the maturity model theory [33], and presents the transition paths of a company through five constitutive levels:

- Level 1 'Initiating', when a company starts to adopt a PaaS, but without complementary other circular practices. Companies allocated at this level achieve the value of Circular PaaS Index in the range < 0;20%).
- Level 2 'Emerging', when a company begins to integrate individual circular practices into its PaaS model; Companies allocated at this level achieve the value of Circular PaaS Index in the range < 20; 40%).
- Level 3' Partially implementing', when a company implements circular practices, but only within a limited range of products or in a specific geographical region; Companies allocated at this level achieve the value of Circular PaaS Index in the range < 40; 60%).

- Level 4 'Scaling up' when a company implements multiple circular practices in a supply chain with PaaS models at a larger scale in a variety of products, regions, or operations. Companies allocated at this level achieve the value of Circular PaaS Index in the range < 60; 80%).
- Level 5 'Circular Excellence', when a company fully integrates circular practices in all product ranges within supply chains with the PaaS model. Companies allocated at this level achieve the value of Circular PaaS Index in the range <80%;100>.

The analysis (see Sect. 4), shows that all apparel companies are progressing towards the implementation of circular practices in a supply chain with PaaS models. The maturity gap (white area in circles in Fig. 6) is defined as the difference between the desired implementation at Level 5 (to be) and the current (as is) state. This gap indicates the potential to improve the circularity of the supply chain with PaaS. Company A, Company B, and Company D achieved maturity Level 4, while Company C received the lowest score (Level 3). The highest-scoring companies have established systematic progress towards fully circular supply chains with PaaS, while the lowest-scoring company remains in the early stages of development, where activities are still small-scale, isolated, or ad hoc. Despite these differences, all four companies recognise the need for circular transformation and are already taking steps to integrate circularity across their full product range or selected lines. The fact that all companies achieved relatively high maturity levels reflects their positions as leaders in PaaS within the apparel sector.

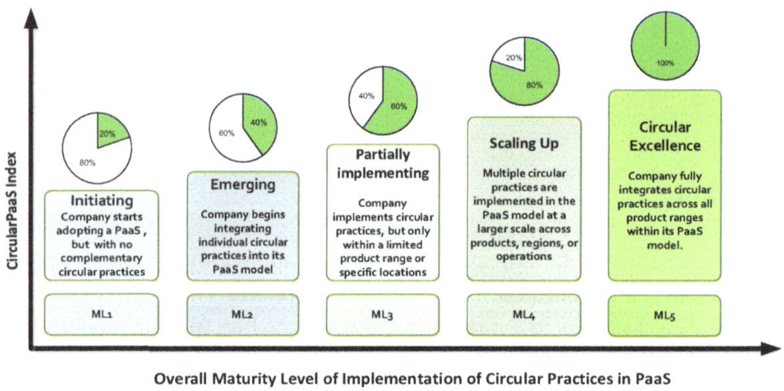


**Fig. 6** The overall level of maturity in the implementation of circular practices in PaaS. Source: own elaboration.

The proposed maturity model aims to improve the implementation of PaaS by developing practices assessed on a five-point maturity scale ranging from 1 (practice not present) to 5 (fully integrated). For each company, targeted action steps were formulated based on identified gaps, supporting a smoother transition toward effective PaaS operation and more advanced circular supply chain management. These recommended measures incorporate product, process, and organisational innovation, and several proposed solutions have the potential to improve resource efficiency, lower operational costs, and increase market competitiveness. The proposed solution lies in treating the improvement concept as an integrated framework, rather than a set of isolated practices, allowing companies to implement circularity more coherently along the entire supply chain with PaaS. Due to space limitation the details on the recommendations for each company are out of the scope of this paper.

## 6 Conclusions

The aim of this paper was to assess the maturity of circular practices related to PaaS models in the apparel industry in the context of circular supply chains. To provide an industry-relevant comparative framework for multiple case studies, first we classified the circular practices into three categories, such as: circular business models, circular design, and product value retention processes and recovery. Then we introduced a scale which reflects the scope of their implementation, ranging from 1 (practice not present) to 5 (fully integrated). The individual assessments were then translated into an aggregated Circular PaaS Index. The application of this index allowed us to propose a structured maturity model with five maturity levels. This model, together with the described assessment approach, constitutes the main theoretical contribution of this study. The proposed model supports enhanced circularity in supply chains through the development of more circular products, a broader application of value retention processes, resource recovery, waste reduction, and the adoption of the innovative PaaS business model. The proposed framework can provide companies with a clear path to achieving mature circular supply chains and fully operational PaaS models.

This study is subject to certain limitations, mainly due to the relatively small number of case studies available and the predominant reliance on secondary data. The restricted sample may limit the generalisability of the findings, while the use of secondary sources may introduce biases related to the quality, completeness, or selective presentation of publicly reported information. To mitigate these limitations, triangulation was applied using multiple types of evidence, including industry data, sustainability reports, and scientific literature, which improved the credibility and robustness of the analysis. Furthermore, internal peer review supported the validation of the interpretations and helped reduce potential researcher bias, thus strengthening the overall reliability of the study's conclusions.

Future research should examine circular supply chains in the textile industry using larger samples and longitudinal designs to capture the dynamic nature of circular transitions. Given current industry trends, future studies could explore the role of digital technologies such as digital product passports, blockchain, and automated sorting systems to enable more effective PaaS models and circular resource flows. Furthermore, more empirical research is needed to evaluate consumer acceptance, cross-sector collaboration, and policy mechanisms that can accelerate the adoption of fully circular supply chains in the apparel sector.

**Acknowledgements.** This study was supported by the internal grant from the Poznan University of Technology no. 0812/SBAD/4238.

## References


1. Ellen MacArthur Foundation: Fashion and the circular economy. https://www.ellenmacarthurfoundation.org/fashion-and-the-circular-economy-deep-dive. Accessed 13 Jan 2026.
2. United Nations Environment Programme: Unsustainable fashion and textiles in focus for International Day of Zero Waste. https://www.unep.org/news-and-stories/press-release/unsustainable-fashion-and-textiles-focus-international-day-zero, (2025). Accessed 13 Jan 2026.
3. Werner-Lewandowska, K., Golinska-Dawson, P., Mierzwiak, R.: Enablers and barriers in building the circular supply chain through remanufacturing – grey DEMATEL approach. Int. J. Prod. Econ. **284**, 109617 (2025). https://doi.org/10.1016/j.ijpe.2025.109617
4. Brydges, T.: Closing the loop on take, make, waste: Investigating circular economy practices in the Swedish fashion industry. J. Clean. Prod. **293**, 126245 (2021). https://doi.org/10.1016/j.jclepro.2021.126245
5. Dragomir, V.D., Dumitru, M.: Practical solutions for circular business models in the fashion industry. Clean. Logist. Supply Chain. **4**, 100040 (2022). https://doi.org/10.1016/j.clscn.2022.100040
6. Kamble, Z., Behera, B.K.: Upcycling textile wastes: Challenges and innovations. Text. Prog. **53**, 65–122 (2021). https://doi.org/10.1080/00405167.2021.1986965
7. Jäämaa, L., Kaipia, R.: The first mile problem in the circular economy supply chains – collecting recyclable textiles from consumers. Waste Manag. **141**, 173–182 (2022). https://doi.org/10.1016/j.wasman.2022.01.012
8. Paras, M.K., Ekwall, D., Pal, R.: Developing a framework for the performance evaluation of sorting and grading firms of used clothing. J. Glob. Oper. Strateg. Sourc. **12**, 82–102 (2019). https://doi.org/10.1108/JGOSS-11-2017-0047
9. Mishra, J.L., Hopkinson, P.G., Tidridge, G.: Value creation from circular economy-led closed loop supply chains: A case study of fast-moving consumer goods. Prod. Plan. Control. **29**, 509–521 (2018). https://doi.org/10.1080/09537287.2018.1449245
10. Hazen, B.T., Russo, I., Confente, I., Pellathy, D.: Supply chain management for circular economy: Conceptual framework and research agenda. Int. J. Logist. Manag. **32**, 510–537 (2021). https://doi.org/10.1108/IJLM-12-2019-0332
11. Sandberg, E., Pal, R.: Exploring supply chain capabilities in textile-to-textile recycling – a European interview study. Cleaner Logistics and Supply Chain. **11**, 100152 (2024). https://doi.org/10.1016/j.clscn.2024.100152
12. Batista, L., Bourlakis, M., Smart, P., Maull, R.: In search of a circular supply chain archetype – a content-analysis-based literature review. Prod. Plan. Control. **29**, 438–451 (2018). https://doi.org/10.1080/09537287.2017.1343502
13. Golinska, P., Fertsch, M., Gómez, J.M., Oleskow, J.: The concept of closed-loop supply chain integration through agents-based system. In: Gómez, J.M., Sonnenschein, M., Müller, M., Welsch, H., Rautenstrauch, C. (eds.) Information Technologies in Environmental Engineering, pp. 189–202. Springer, Berlin, Heidelberg (2007). https://doi.org/10.1007/978-3-540-71335-7_20
14. Farooque, M., Zhang, A., Thürer, M., Qu, T., Huisingh, D.: Circular supply chain management: A definition and structured literature review. J. Clean. Prod. **228**, 882–900 (2019). https://doi.org/10.1016/j.jclepro.2019.04.303
15. Hidalgo-Crespo, J., Riel, A., Golinska-Dawson, P., Peeters, J.R., Werner-Lewandowska, K., Duflou, J.R.: Facilitating circularity: Challenges and design guidelines of product-as-a-service (PaaS) business model offers for electrical and electronic equipment. Procedia CIRP. **128**, 567–572 (2024). https://doi.org/10.1016/j.procir.2024.03.037
16. Golinska-Dawson, P., Zysnarska, Z., Pender, A.: Assessment of the maturity of product-as-a-service business models for household appliances from the perspective of R strategies in

circular economy. Procedia CIRP. **122**, 1083–1088 (2024). https://doi.org/10.1016/j.procir.2024.02.038
17. Yang, M., Smart, P., Kumar, M., Jolly, M., Evans, S.: Product-service systems business models for circular supply chains. Prod. Plan. Control. **29**, 498–508 (2018). https://doi.org/10.1080/09537287.2018.1449247
18. Pal, R., Samie, Y., Chizaryfard, A.: Demystifying process-level scalability challenges in fashion remanufacturing: An interdependence perspective. J. Clean. Prod. **286**, 125498 (2021). https://doi.org/10.1016/j.jclepro.2020.125498
19. Pal, R., Sandberg, E.: Circular supply chain valorisation through sustainable value mapping in the post-consumer used clothing sector. Int. J. Logist. Manag. **35**, 1373–1416 (2024). https://doi.org/10.1108/IJLM-01-2023-0023
20. Pal, R., Sandberg, E., Paras, M.K.: Multidimensional value creation through different reverse supply chain relationships in used clothing sector. Supply Chain Manag. **24**, 729–747 (2019). https://doi.org/10.1108/SCM-12-2018-0422
21. Surjit, R., Akshatha, R., Akshya, R., Harani, C.: Rise of circular fashion for sustainable apparel retailing. In: Muthu, S.S. (ed.) Circular Business Models in the Apparel Industry, pp. 95–123. Springer Nature, Switzerland, Cham (2025). https://doi.org/10.1007/978-3-031-80193-8_6
22. Singh, N.: Circular economy in textile and apparel supply chains. In: Khan, S.A.R. (ed.) Advances in Business Strategy and Competitive Advantage, pp. 25–64. IGI Global (2024). https://doi.org/10.4018/979-8-3693-4123-0.ch002
23. Sandin, G., Peters, G.M.: Environmental impact of textile reuse and recycling – a review. J. Clean. Prod. **184**, 353–365 (2018). https://doi.org/10.1016/j.jclepro.2018.02.266
24. McCauley, E., Jestratijevic, I.: Exploring the business case for textile-to-textile recycling using post-consumer waste in the US: Challenges and opportunities. Sustainability. **15**, 1473 (2023). https://doi.org/10.3390/su15021473
25. Nørup, N., Pihl, K., Damgaard, A., Scheutz, C.: Evaluation of a European textile sorting centre: Material flow analysis and life cycle inventory. Resour. Conserv. Recycl. **143**, 310–319 (2019). https://doi.org/10.1016/j.resconrec.2019.01.010
26. Sakao, T., Wasserbaur, R., Mathieux, F.: A methodological approach for manufacturers to enhance value-in-use of service-based offerings considering three dimensions of sustainability. CIRP Ann. **68**, 33–36 (2019). https://doi.org/10.1016/j.cirp.2019.04.084
27. Kesavapanikkar, P., Amit, R.K., Ramu, P.: Product as a service (PaaS) for traditional product companies: An automotive lease practice evaluation. J. Indian Bus. Res. **15**, 40–54 (2023). https://doi.org/10.1108/JIBR-04-2022-0107
28. Hidalgo-Crespo, J., Riel, A., Duberg, J.V., Bunodiere, A., Golinska-Dawson, P.: An exploratory study for product-as-a-service (PaaS) offers development for electrical and electronic equipment. Procedia CIRP. **122**, 521–526 (2024). https://doi.org/10.1016/j.procir.2024.01.076
29. van Loon, P., Van Wassenhove, L.N.: Transition to the circular economy: The story of four case companies. Int. J. Prod. Res. **58**, 3415–3422 (2020). https://doi.org/10.1080/00207543.2020.1748907
30. Yin, R.K.: Case study research: Design and methods. SAGE (2009)
31. Kim, S., Ji, Y.: Gap analysis. In: The International Encyclopedia of Strategic Communication, pp. 1–6. Wiley (2018). https://doi.org/10.1002/9781119010722.iesc0079
32. Golinska-Dawson, P., Sakao, T., Sundin, E., Werner-Lewandowska, K.: How can OEMs scale up remanufacturing in product-as-a-service (PaaS)? – challenges and enablers from a theoretical and industrial perspective. Int. J. Prod. Res. **63**, 3854–3883 (2025). https://doi.org/10.1080/00207543.2024.2423798
33. Röglinger, M., Pöppelbuß, J., Becker, J.: Maturity models in business process management. Bus. Process. Manag. J. **18**, 328–346 (2012). https://doi.org/10.1108/14637151211225225

# Material Circularity of Selected Components in the Context of Modern Enterprise

Lucia Knapčíková(✉), Daniel Bobko, and Maksym Karakai

Department of Manufacturing Management, The Technical University of Košice, Faculty of Manufacturing Technologies with a seat in Prešov, Bayerova 1, 080 01 Prešov, Slovakia
lucia.knapcikova@tuke.sk

**Abstract.** In the steel production process, tons of scales are generated as waste materials, which contain a high proportion of iron. These scales have an undesirable effect on the quality of steel production and on the number of pieces of production equipment with which they come into contact during the production process. Since scale contains a large amount of iron oxides, heavy metals, and other various elements that can be soil and water contaminants, it is necessary to manage this waste very responsibly. Currently, it is a great challenge to turn as much of this waste as possible into a by-product and thus to find suitable processing for further use. The Slovak Republic, as a member of the European Union, is obliged to comply with European Union regulations. The European Union directive concerning waste is strictly enforced. By 2030, waste suitable for reuse or recycling should not be accepted for landfill, except where landfilling provides the best environmental result.

**Keywords:** Sustainability · Circularity · Product · Steel

## 1 The State of the Art

In the hot rolling of steel, the formation of scale is an inevitable byproduct. The primary scale is formed by heating in impact furnaces, and the secondary and tertiary scales are formed by cooling the rolled material [1]. The main objective of the dissertation will be the chemical-physical analysis of scale waste, its effects on selected components, and the investigation of the possibilities for subsequent processing as a by-product used in the preparation of new composite materials [2]. Scale formed during the forming process is removed from the surface of the rolled material using high-pressure water at 18 MPa. The surfaces of the technological hall of the hot rolling mill rolling line are sloped into a scale channel located under the rolling stands [3]. Scale is removed from the stands of the preparatory and finishing rows of the rolling line by an installed high-pressure scale-spraying system. Recirculated purified scale water is used to remove scales and cool the rolls of the preparatory and finishing rows [4]. This water is mainly contaminated with scale and oil. It is purified in sedimentation-scale pits of primary sedimentation, where coarse scale is captured, in sedimentation tanks of the UN–DORR type for secondary sedimentation, where it is also de-oiled, and in two filtration stations


A. Hamrol et al. (Eds.): MANUFACTURING 2026, LNME, pp. 204–215, 2026.
https://doi.org/10.1007/978-3-032-21701-1_15

equipped with 9 DDF type filters and 56 DynaSand DS 5000 AE-HD type filters. Part of the purified scale water is discharged for cooling. Despite multi-stage scale removal, it is found in water as microparticles of iron oxides, which cause increased wear of some components of production equipment and piping systems [5]. Scale is collected in a scale pit and then loaded onto railway wagons and transported to the foundation piles of the sintering plant for the blast furnaces, where it serves as a replacement for the iron-containing input charge in the blast furnaces within the metallurgical cycle [6]. In connection with this process, their processing and pollution, the content of petroleum products and the cleaning of cooling water, which is recirculated in the production process, appear problematic (Fig. 1). If these scales are appropriately processed, their technical capabilities and financial value would increase.

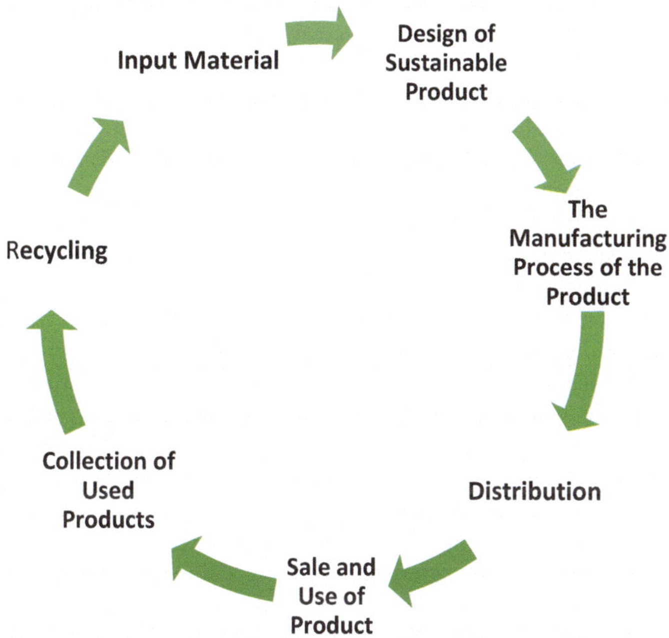


**Fig. 1.** Life cycle of the product in the context of the circular economy (Author's own processing).

Steel is one of the most recyclable materials. Its circular advantages include:

- High recyclability rate (above 90%),
- Preservation of mechanical properties after recycling,
- Low energy consumption compared to primary steel production.

Circularity plays a key role in meeting environmental, economic, and technological requirements in modern manufacturing. The main objectives include [7]:

- Reducing the consumption of primary raw materials. By reusing materials, manufacturers lower their reliance on limited natural resources.
- Minimising waste production. Waste from manufacturing is converted back into raw material through recycling or material recovery.

- Maintaining high material value. Properly designed components retain material quality throughout multiple use cycles.
- Integration with digital technologies.

### 1.1 Manufacturing Process of Steel

At the very beginning of steel production is pig iron, which is produced in a blast furnace from iron ore with the help of coke, limestone, and other additives. The pig iron is then processed into steel in a steel mill, which is then processed into semi-finished products and then into final products. Steel is traditionally produced by smelting pig iron with the addition of steel scrap or steel scrap alone in arc furnaces. During production, the carbon, phosphorus, and sulfur content of the charge is reduced. Steel preparation can be divided into two phases [7, 8]. In the first phase, the carbon content is reduced, and undesirable elements are removed by oxidation. The second phase is called refining, during which excess oxygen is removed from the steel using ferroalloys of manganese and silicon, or aluminum. Finally, the steel is desulphurized. About 70% of steel is produced today in oxygen converters, which can make all types of carbon steel. Modern steel mills use continuous casting, which is faster and less energy-intensive than traditional ingot casting [9]. Ingot casting is now used only exceptionally, for example, for large forgings or as a batch for the production of seamless pipes. Liquid steel is poured into a vertical, water-cooled crystallizer in a steel mill, where it solidifies. Below the crystallizer, in the secondary cooling zone, the center of the so-called continuous casting also solidifies. This is bent from a vertical position to a horizontal one and then cut to the required length [9]. The product of continuous casting is steel with uniform properties, while this technology also reduces waste volume. The basic shapes that can be used directly as input semi-finished products in rolling mills are obtained from continuous casting [10]. The basic products of continuous casting are slabs (wide blocks used for subsequent sheet rolling) and billets or blocks with rectangular or round cross-sections. Long products or pipes are then rolled from these. Hot rolling produces products such as bars, rails, sheets, strips, or pipes. Before entering the rolling mill, the input material (slab or billet) is heated to about 1100 °C and deformed between rotating work rolls. The material's height is reduced, its width is adjusted, and it is gradually lengthened. The speed of its passage through the next rolling mill is also increased [11]. Wires are rolled on special continuous rolling lines, while sheets are rolled on stands with smooth rolls. In the first technology, the semi-finished material is first rolled across a single rolling mill to achieve the required width, and then the sheet is rolled lengthwise to achieve a uniform thickness [12, 13]. In the second technology, the semi-finished product is gradually passed through a larger number of rolling mills and is rolled only in the longitudinal direction. Profiles with various shapes, from circular, square or hexagonal to I, U or L-shaped profiles, are also rolled in this way. Rails are also produced in the same way. In hot rolling, the material is heated above the so-called recrystallisation temperature, which is higher than 70% of the melting point of the given material [7]. Therefore, the material does not solidify during rolling. For shaping, forces up to ten times smaller than for cold rolling are sufficient. Shortly after the advent of the first hot-rolling mills, lower-capacity single-stand reversing mills, known as Steckel stands, were developed and put into operation (Fig. 2). These mills used a heated coil box on each side of the rolling stand [14]. Higher-capacity Steckel mills were also developed, incorporating an upstream reversing stand for slab reduction before the Steckel stand itself. The coilbox was developed by Stelco in the 1970s [15]. The principle of the coilbox is that the rolled strip from the reversing stand is wound into a coilbox and then unwound to the finishing stand, thereby maintaining the roll's heat and providing a more uniform inlet temperature to the finishing stand. This has led to improved strip quality and reduced motor power in the finishing stand. The thickness of the exit strip ranges from 1.0 to 25.4 mm [16].

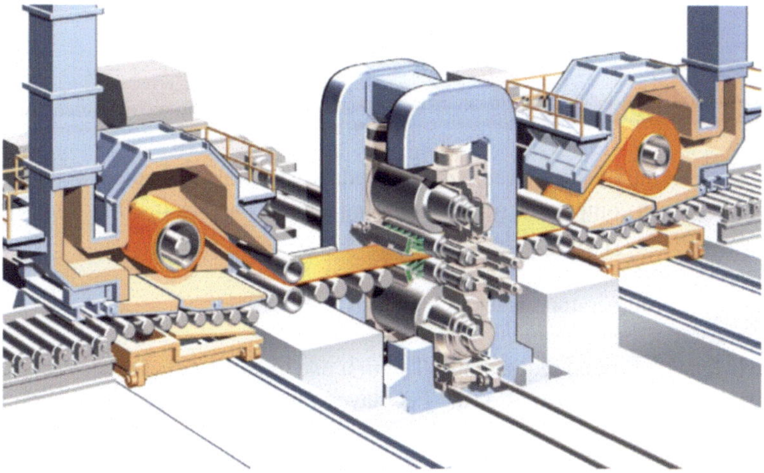

**Fig. 2.** Steckel stands [3].

Currently, there is a trend to build so-called mini mills for steel production. These are hot-rolling mills connected directly to the casting operation [17, 18]. Such operations have several competitive advantages over old types of rolling mills. One of them is the reduction of energy, since the casting process is in direct sequence with the rolling process. In this way, the energy required to heat slabs in impact furnaces is reduced [10, 19]. Another advantage, or rather a necessity due to EU legislation, is the use of arc furnaces instead of blast furnaces, which must be phased out due to high emissions (Fig. 3). Mini-mill operations are also very compact and use the most modern equipment in the hot-rolling process [20]. The output of these operations is around two million tons per year.

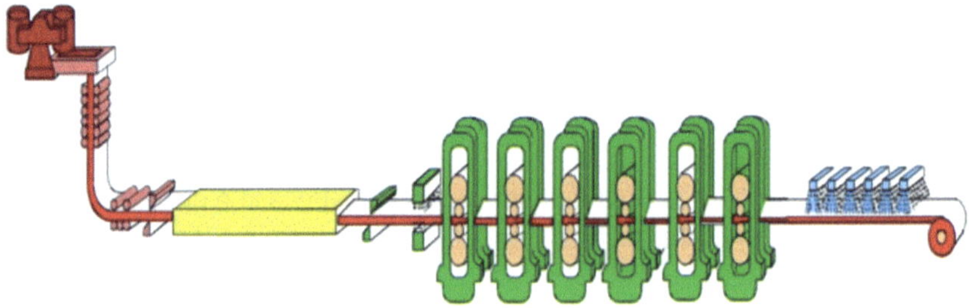

Fig. 3. Mini mill equipment [13].

## 2 Work Methodology

During hot working of steel, three types of oxidation (scale) are formed on the surface. These iron-oxygen compounds occur in varying amounts and forms (Fig. 4) [21].

There are three types of scale.

- Primary (furnace).
- Secondary (preparatory order).
- Tertiary (finished order).

The layers are formed sequentially. The first one to create is FeO. The second layer is $Fe_3O_4$, and this forms on top of the FeO layer. $Fe_2O_3$ forms the third one on top of $Fe_3O_4$ [22].

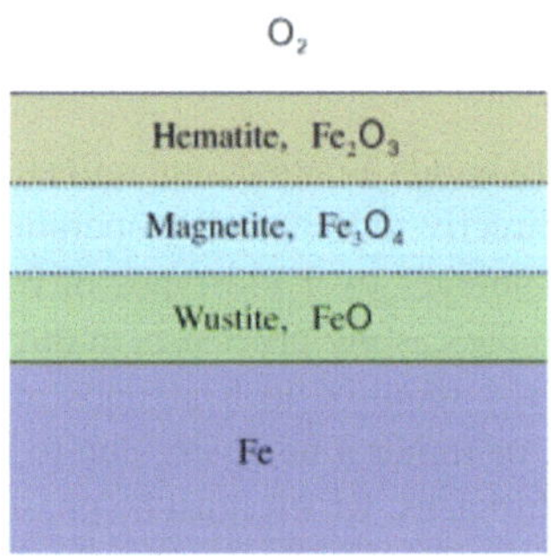


Fig. 4. Scale layers [10].

The scale's structure is time- and temperature-dependent. At hot-rolling temperatures with sufficient time, the scale forms in three layers, each with a different oxide phase [23]. The thermal stability ranges of the individual iron oxides result from the iron-oxygen phase diagram. Above 570 °C, FeO is unstable. The stable phase is $Fe_3O_4$, and the oxidation rate is controlled by the formation of this oxide [24]. At temperatures up to 500 °C, FeO forms only a thin layer immediately at the iron/oxide phase interface; above 500 °C, $\alpha$-$Fe_2O_3$ increases and dominates at 575 °C. Above 570 °C, the oxidation rate rises sharply, and the scale consists mostly of FeO, but already contains all three oxides. Above FeO lies a thin layer of Fe3O4, which is covered by an even thinner layer of

$Fe_2O_3$. FeO has a granular structure with numerous pores; Fe3O4 is even more porous, and $Fe_2O_3$ is compact. At temperatures above 1100 °C, $Fe_2O_3$ completely disappears. Scales grow linearly at the beginning because the oxidation process is regulated by oxygen absorption at the surface, which is linearly dependent on time [9].

As the scale layer increases, further fusion of $Fe_2$+ ions through the scale slows down its growth [24]. Scale growth, therefore, follows a parabolic curve as the kinetics of the scale quickly begin to outweigh the ability of Fe and O ions to diffuse through the scale [25]. Oxidation occurs on the outer surface of the scale (Fe diffusion outwards) as well as at the steel-scale interface (O diffusion inwards).

***Primary Scale***

Typical thicknesses of furnace scale are from 0.2 to 0.4 cm. Still, the thickness can be greater, representing a material loss of 1.5–2% of the slab's total mass. The amount of scale formed on the steel depends on temperature and exposure time in each environment. Scale formation is controlled by diffusion, which is exponentially dependent on temperature. Scale growth is primarily influenced by the amount of oxygen in the liver.

Low oxygen levels lead to lower scale growth [7], whereas high oxygen levels lead to higher scale growth rates. Very low oxygen levels can cause the furnace atmosphere to become reducing, resulting in sticky scale that is more difficult to remove by the descaling process. Water vapour and carbon dioxide facilitate the transport of ions through gaps and pores that would normally prevent diffusional transport. Cracks and fissures are caused by increased compressive stress in the scale as it thickens. The volume of the oxide is greater than the volume of the metal from which it is formed [7, 8]. The scale wants to expand, but the steel limits it. Porosity and voids are caused by decarburization. Internal oxidation occurs at high temperatures. Gaps between the scale and the steel are caused by the inward movement of the metal/oxide interface [9]. The scale grows more slowly than the interface recedes due to the outward diffusion of Fe. This causes the steel substrate to locally separate from the scale, forming gaps [10, 11].

***Secondary Scale***

The composition of secondary scales is uncertain because it is impossible to sample them. The exact structure/morphology of the scales is doubtful [12, 13]. Scales formed in humid air over short periods of time and at decreasing temperatures are much thinner. No visible porosity, voids or cracks are formed [14]. During the rolling in the preparatory sequence, the scales are formed from FeO, probably with some $Fe_3O_4$ present; the formation of $Fe_2O_3$ would be expected at high temperatures and long times between scale removals [13, 15]. During the rolling in the finishing sequence, the scales are composed of FeO and may again contain some $Fe_3O_4$, which forms at temperatures below the finishing and coiling temperatures. At high finishing and coiling temperatures, $Fe_3O_4$ and $Fe_2O_3$ are likely to occur. FeO is very ductile and easily rolled during rolling in both the preparatory and finishing stages [16]. There are no consequences for the surface quality during rolling. $Fe_3O_4$ is more brittle, but this can be tolerated during rolling. $Fe_2O_3$ is very brittle, fractures during rolling, and tends to roll into the steel surface, creating surface defects that increase wear on the work rolls and increase the rolling load [17].

***Tertiary Scale***

The tertiary scale is a mixture of $Fe_3O_4$ and $Fe_2O_3$ plus metallic Fe from the decomposition of FeO. A fine line of porosity along the scale lines the upper outer layer of $Fe_2O_3$. The percentage of $Fe_3O_4$ and $Fe_2O_3$ depends on the location, i.e. start/end of the strip, and edge/middle of the strip [18]. FeO is not present because it is unstable below 566 °C [11, 19]. However, low rolling temperatures can suppress the transformation, allowing some FeO to remain at ambient temperatures. The thickness can range from 8 to 15 μm and more, depending on the rolling temperature [20].

## 2.1 Problem Definition

The biggest problem with pipelines at scale is the formation of deposits, which can reach up to 5 cm thick. Such deposits then cause turbulent flow, reduced flow rates, or overheating of the pipeline and water [21]. Turbulent flow and oxidation cause uneven wear of the pipeline, leading to premature point damage and water leaks (Fig. 5). The picture shows the wall of a DN 1200 pipeline, the main supply from the pumping station to the finishing line [12]. This water cools the cylinders on the finish line. For this purpose, the so-called scale water is used, cleaned of scale and oil products, and recirculated for reuse. Spot corrosion forms under scale deposits, leading to premature pipeline damage [16, 22]. Such pipelines must be replaced or repaired, which causes production downtime and increased maintenance costs. The problem with pipelines due to scale is the formation of deposits, in some cases up to 5 cm thick. Such deposits then cause turbulent flow, reduced flow rates, or overheating of the pipeline and water [21]. Turbulent flow and oxidation cause uneven wear of the pipeline, leading to premature point damage and water leaks (Fig. 5). The picture shows the wall of a DN 1200 pipeline, the main supply from the pumping station to the finishing line [15]. This water provides cooling for the rollers of the finishing line. For this purpose, the so-called scale water is used, cleaned of scale and oil products, and recirculated for reuse. Spot corrosion forms under scale deposits, leading to premature pipeline damage [16, 22]. Such pipelines must be replaced or repaired, which causes production downtime and increased maintenance costs.

The CND 5 feed pump is the pressure source for the scale removal system. Reliability is key for this equipment [19]. Without scale removal from the roll, production of the required quality is not possible (Fig. 6). Therefore, the availability of this pump is essential for 24 h operation [23]. Scales smaller than 200 μm on the equalisation pipe reduced the pipe's cross-section and thus decreased the flow rate.

The pump rotor is axially balanced through the balancing pipe. This axial balancing was insufficient, resulting in damage to the thrust bearing. In this case, thanks to the high temperature and timely inspection, the fault was identified in time, preventing an accident that could have caused the pump to seize completely. Such faults can cause significant downtime and high repair costs.

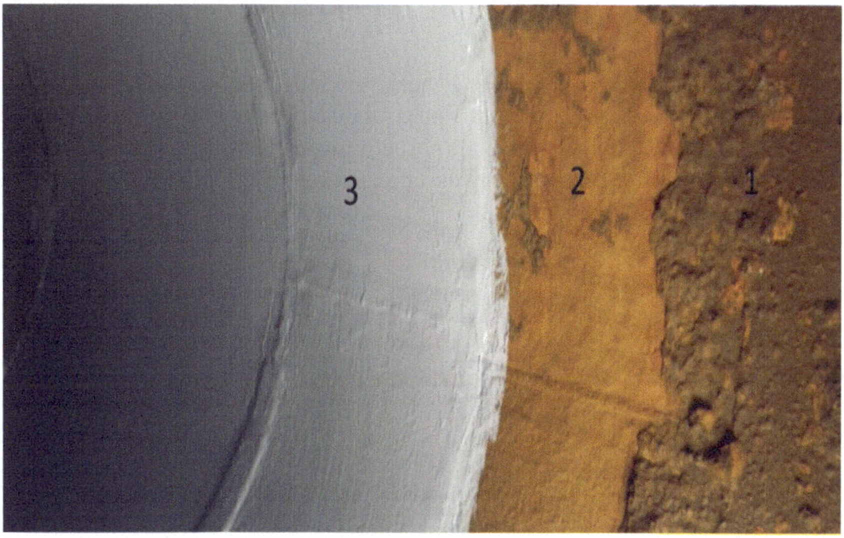


**Fig. 5.** Repair of a pipe damaged by pitting corrosion [25].

**Fig. 6.** Rotor axial balancing compensation pipe [22].

## 3 Results and Discussion

The problem with all valves, gate valves, check valves and shut-off valves with fittings is that the places where the valve seat contacts the body of the fitting cause wear and deposits [17]. In these cases, the fittings do not perform their closing function sufficiently, and leakage and overflow occur through the closed fittings (Fig. 7). These then lose their function and need to be replaced [21].

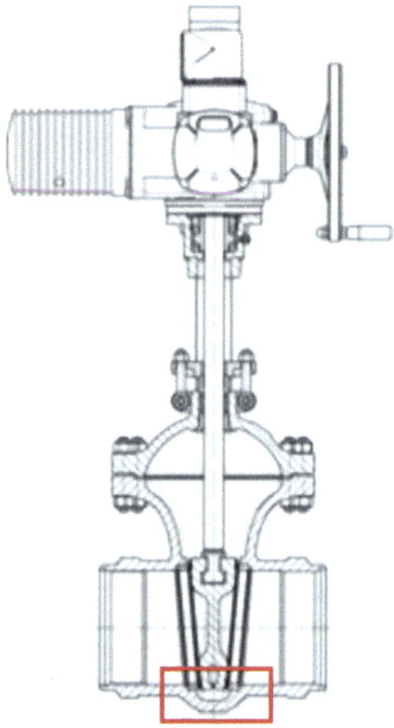

**Fig. 7.** Shut-off valve – place of scale deposition [17].

**Collectors and Nozzles**

As in pipework, scale in cooling collectors causes deposits and reduces flow rates. In addition, in some cases it acts as an abrasive, causing nozzle wear [20]. In the hot-rolling process, a large amount of water is required for the necessary cooling of the material and of all equipment that meets it. For this reason, hundreds of different collectors are part of the production equipment in a hot rolling mill [19]. A laminar cooling collector was chosen as an example. In Fig. 8, scale, specifically hematite, reduced the hole's cross-section. Laminar cooling collectors ensure strip cooling in accordance with requirements for achieving the mechanical properties of the produced material [12, 24]. The control model determines how much water is needed to reach the desired temperature and, therefore, how many collectors are switched on during cooling [23]. If these collectors are clogged, it causes problems with the theoretical calculation and the actual flow rate (Fig. 8). These collectors must be cleaned to avoid differences between the desired and actual finishing temperatures [26].

**Fig. 8** Laminar cooling collector [25].

In applications where high pressure is used, such as high-pressure scale blasting (18 MPa), scale acts as an abrasive in the water, increasing wear on nozzles and equipment in contact with it [18, 24]. This concerns parts of spray boxes and water collectors. When worn, the water flow changes and the impact decreases, which is undesirable because it reduces scale removal from the surface of the rolled material. Such nozzles must be replaced. Research into the effects of $Fe_2O_3$ on the wear of selected components primarily focuses on mitigating the negative effects of scale on production equipment and on exploring new applications for composite materials derived from iron oxide waste [25].

- It is necessary to investigate the scale that is generated in the hot rolling production process. This waste will be analyses to create a new material by combining two components, one of which is a waste raw material in the form of iron oxides.
- It is necessary to monitor its adaptability in a demanding working environment
- From a research perspective, it is very important to pay attention to the waste that is generated in the form of scale during the hot rolling process and to find new possibilities for its use as an input raw material for the production of new materials and the possibilities of their application and applicability of these materials [26].

## 4 Conclusion

The vision is to reduce the amount of waste disposed of to 10% or less of total municipal waste generated by 2035. Companies that create waste during production should consider implementing reverse logistics. It is through reverse logistics that we are approaching the sustainability of the production process and the state's economy. Sustainability is quickly becoming a global priority, with significant efforts to drive positive change. Currently, there are strict legislative targets in the European and global economy that mandate zero-carbon production across all sectors to reduce their environmental impact. An important task is to monitor selected parameters for specific mechanical engineering products, compare their current and proposed states, and subsequently evaluate the feasibility of applying the newly developed material to the chosen mechanical engineering products from a technical and economic perspective. Finally, it is important to address the challenge of finding an environmentally acceptable way to recycle the produced material within the circular economy. At the end of the life cycle, the product and its application automatically cease to fulfil their primary function for which they were manufactured. It is important that, even after use, it is recycled as best and as carefully as possible, in accordance with currently valid legislation, with priority given to material recycling.

**Acknowledgements.** This work was supported by the project VEGA 1/0210/25 granted by the Ministry of Education, Research, Development and Youth of the Slovak Republic.

## References

1. European Parliament: Circular economy: Definition, importance and benefits. https://www.europarl.europa.eu/news/en/headlines/economy/20151201STO05603/circular-economy-definition-importance-and-benefits, last accessed 2025/10/10.
2. Wang, Y.Q., Hu, J., Dai, Z.Y., Li, J.L., Huang, J.: In vitro assessment of physiological changes of watermelon (Citrullus lanatus) upon iron oxide nanoparticles exposure. Plant Physiol. Biochem. **108**, 353–360 (2016)
3. Goslinska, M., Selmer, I., Kleemann, C., Kulozik, U., Smirnova, I., Heinrich, S.: Novel technique for measurement of coating layer thickness of fine and porous particles using focused ion beam. Particuology. **42**, 190–198 (2019)
4. Eichner, E., Salikov, V., Bassen, P., Heinrich, S., Schneider, G.A.: Using dilute spouting for fabrication of highly filled metal-polymer composite material. Powder Technol. **316**, 426–433 (2017)
5. Elsner, P., Eyerer, P., Hirth, T.: Plastics: Properties and Applications, 8th edn, pp. 401–412. Springer, Berlin, Heidelberg (2012) (in German)
6. Wittenberg, E., Meyer, A., Eggers, S., Abetz, V.: Hydrogen bonding and thermoplastic elastomers–A nice couple with temperature-adjustable mechanical properties. Soft. Matter. **14**, 2701–2711 (2018)
7. Kumar, S., Singh, K., Singh, T.P., Batish, A.: Thermosetting polymers for 4D printing. Ref. Modul. Mater. Sci. Mater. Eng. **1-4**, 596–602 (2020). https://doi.org/10.1016/B978-0-12-820352-1.00026-2
8. Sun, W.: A study of the characteristics of oxide scale in hot rolling of steel. PhD thesis, University of Wollongong (2005). https://ro.uow.edu.au/cgi/viewcontent.cgi?article=1440&context=theses, last accessed 2025/09/15.

9. Ivanovic, D., Zivic, F.: Hardness and non-destructive testing (NDT) of ceramic matrix composites (CMCs). In: Encyclopedia of Materials: Composites. **2**, 183–201 (2021)
10. General Electric Company: Inspection technology: SEIFERT XRD 3003/PTS. http://www.spectro.co.th/images/1214052326/GEIT-30020US_xrd-3003, last accessed 2025/09/20.
11. ASTM D638: Standard test method for tensile properties of plastics. ASTM International, last accessed 2025/10/20.
12. Polymer Shapes: High impact polystyrene – properties and uses. https://www.polymershapes.com/high-impact-polystyrene-properties-and-uses, last accessed 2025/10/22.
13. TESCAN: SEM solutions for materials science. https://www.tescan.com/product/sem-for-materials-science-tescan-vega/, last accessed 2025/10/30.
14. STN EN ISO 6507-1: Metallic materials – Vickers hardness test (in Slovak).
15. STN EN ISO 868: Plastics and ebonite – Determination of indentation hardness by means of a durometer (Shore hardness) (in Slovak).
16. Mitutoyo: Hardness measuring–Durometers series 811. https://www.mitutoyo.com/webfoo/wp-content/uploads/1801.
17. Olympus: Innov-X DS-2000 Delta XRF analyzer. https://ndtadvancedtools.com/analyzer/olympus-innov-x-ds-2000-delta-xrf-analyzer.
18. Rokidi, S.G., Koutsoukos, P.G., Perez, L.A., Amjad, Z., Zuhl, R.W.: Iron oxide colloidal suspension stabilization by polymeric dispersants. NACE-2016-7530 (2016).
19. Maxwell Additives Pvt. Ltd.: Selecting a corrosion inhibitor/dispersant in water treatment. https://maxwelladditives.com/selecting-a-corrosion-inhibitor-dispersant-in-water-treatment/, last accessed
20. Behari, M., Das, D., Mohanty, A.M.: Influence of surfactant for stabilization and pipeline transportation of iron ore water slurry. ACS Omega. **7**(33), 28708–28722 (2022)
21. Jayaweera, C.D., et al.: Assessing the feasibility of using a data-driven corrosion rate model for optimizing dosages of corrosion inhibitors. NPJ Mater. Degrad. **8**, 127 (2024)
22. Finšgar, M., Milošev, I.: Inhibition of copper corrosion by 1, 2, 3-benzotriazole: A review. Corros. Sci. **52**, 2737–2749 (2010)
23. Mandičák, T., Mesároš, P., Tkáč, M.: Impact of management decisions based on managerial competencies and skills developed through BIM technology on performance of construction enterprises. Pollack Periodica. **13**(3), 131–140 (2018)
24. Kanáliková, A., Behúnová, A., Mandičák, T., Mesároš, P.: New technologies for sustainable development in infrastructure, logistics and construction industry. Acta Logistica. **6**(4), 171–177 (2019)
25. Martiček, M.: Study of the Influence of $Fe_2O_3$ on the Wear of selected types of components under hot rolling mill conditions. Dissertation thesis, FVT TUKE (2025).
26. Cvitić, I., Peraković, D., Periša, M., Botica, M.: Novel approach for detection of IoT generated DDoS traffic. Wirel. Netw. **27**(3), 1573–1586 (2019). https://doi.org/10.1007/s11276-019-02043-1

# A Way for Improving the Efficiency of Energy Media Use in the Propylene Glycol Production Process

Mateusz Duraj and Andrzej Loska(✉)

Silesian University of Technology, Faculty of Mechanical Engineering, 18a Konarskiego st., 44-100 Gliwice, Poland
andrzej.loska@polsl.pl

**Abstract.** This article analyzes the effective use of energy media generated during the production of propylene glycol from distilled glycerin, with particular focus on thermal energy losses arising from process steam released in the hydrogenation stage. In the current system, this steam is vented to cooling towers, causing significant energy losses and high electricity consumption by the cooling infrastructure. The aim of the study was to identify and assess solutions enabling the recovery and reuse of process steam within the plant's internal energy system. Two variants were examined: injecting steam into the main steam pipeline and using it in a heat exchanger to preheat raw materials. For each option, mass and energy balances as well as technical and economic analyses were performed. The results indicate that injecting steam into the main pipeline is the most advantageous solution. It is characterized by simple implementation, low investment cost (about PLN 48,000), and a very short payback period of approximately two to three weeks of plant operation. The solution also reduces electricity consumption and $CO_2$ emissions. The proposed modernization demonstrates a practical approach to improving energy efficiency in chemical production and supports sustainable development objectives related to rational energy use and environmental protection .

**Keywords:** Production efficiency · Continuous production installations · Propylene glycol production process · Energy efficiency of the process · Excploitation of apparatus systems

## 1 An Introduction

The modern chemical industry faces the need to continually improve the energy efficiency of technological processes, driven by both rising energy costs and increasingly stringent environmental requirements. One crucial area of optimization is the management of energy utilities, including process steam, cooling water, compressed air, and process gases. Efficient use of these utilities, including their recovery and reuse within the production plant, significantly reduces primary energy consumption and greenhouse gas emissions, while simultaneously lowering operating costs.

The production process of propylene glycol, an important intermediate in many industries–from pharmaceuticals to food and chemical–is characterized by significant


A. Hamrol et al. (Eds.): MANUFACTURING 2026, LNME, pp. 216–229, 2026.
https://doi.org/10.1007/978-3-032-21701-1_16

demand for thermal and cooling energy. The energy media generated during this process, such as waste heat from reactors, steam condensates, and process gases, constitute a potential source of energy that, if properly managed, can be reused within the process plant. In industrial practice, solutions integrating heat recovery systems, regenerative heat exchangers, and energy media management systems are becoming increasingly important, enabling dynamic balancing of energy flows depending on process conditions.

Analyzing the effective use of energy media in the propylene glycol production process is therefore a crucial element in assessing the energy efficiency of the entire plant. It not only allows for the identification of losses and potential reductions, but also for developing directions for modernizing technological systems. Implementing modern energy management methods based on process data and energy flow analysis aligns with the concept of sustainable development and the chemical industry's energy transformation strategy.

The aim of the research is to analyze the methods of effective use of energy media generated in the propylene glycol production process and to assess the potential of their recovery in the context of improving the energy balance and reducing heat losses in technological systems.

## 2 Propylene Glycol Production Process–Main Assumptions

The production process of propylene glycol (1, 2-propanediol, $C_3H_8O_2$) is a crucial element of the modern chemical industry due to its widespread use in many sectors of the economy. Propylene glycol is used in the production of polyester resins, antifreeze, cosmetics, pharmaceuticals, and as an additive in the food industry (E1520). In recent years, the importance of production methods based on renewable raw materials has been growing, aligning with the principles of a circular economy and sustainable development. One such direction is the production of propylene glycol from glycerin–a byproduct of biodiesel production.

The diagram of the propylene glycol production process is shown in Fig. 1.

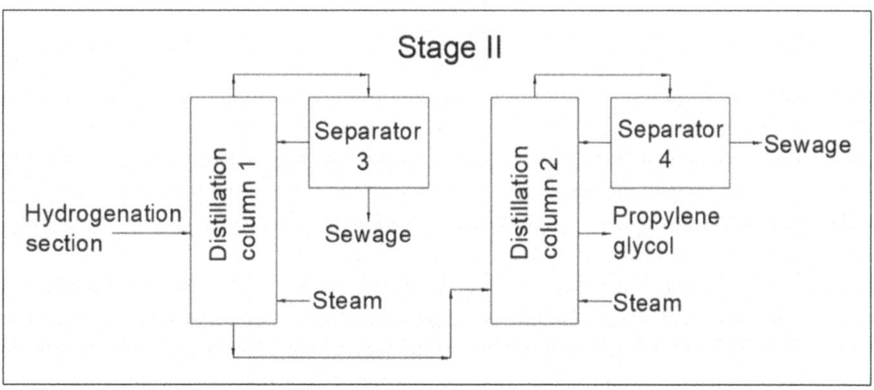


**Fig. 1.** The diagram of the propylene glycol production process.

The input to the propylene glycol production process is distilled glycerin, dosed as a precise injection into the recycled glycol stream returning from the reactor system. At this stage, the raw material is pre-mixed with the partially reacted process medium, which stabilizes feed parameters and limits fluctuations in the glycerin concentration. Process hydrogen, necessary for the hydrogenation reaction, is introduced into the prepared mixture. The resulting reaction mixture—containing glycerin, hydrogen, and recycled glycol—is then directed to Reactor 1, where the first stage of hydrogenation occurs. After leaving the reactor, the product is transferred to Separator 1, where phase separation occurs. The light stream containing hydrogen and volatile fractions is directed to the appropriate collection point, while the liquid stream containing glycol passes through a cooling exchanger. Temperature reduction and thermal separation occur in the exchanger, allowing for further separation of the process. After cooling, the stream divides into two sets: the first sets is returned to Reactor 1 as a recirculation stream, stabilizing process conditions and improving reaction efficiency, the second sets is directed to Reactor 2, where the next stage of hydrogenation takes place. From Reactor 2, the product then passes to Separator 2, where phase separation occurs again. Finally, the liquid glycol fraction is directed to the distillation section, the final stage of the process, where propylene glycol of the required purity is separated.

After the hydrogenation process in the reactors is completed, the homogenized post-reaction mixture is first directed to the pre-distillation section. The reactor stream first enters distillation column 1, which cooperates with separator 3. This is where the first stage of hydrogenation product purification occurs: low-boiling compounds, primarily water, and light by-products formed during the process are separated. From column 1, the steam is directed to a condenser and then to separator 3, where the condensate is stratified and the light phase is removed. The purified, pre-liquid glycol is then sent to the next distillation stage. The stream from column 1 is then fed to distillation column 2, which operates in conjunction with separator 4. This section of the system removes remaining organic impurities and higher, less readily evaporating fractions that could compromise the quality of the final product. Column 2 operates at a higher separation purity, thus defining the final quality of the propylene glycol. The steam from column 2 goes to the condenser and the condensate is separated in separator 4, from where the waste fractions are directed to further neutralization.

After the purification process in the second distillation column and separator 4, the resulting propylene glycol constitutes a finished product stream that meets the process's quality requirements. This product is then sent to the final cooling system and storage tanks, from where it can be further transferred for packaging or shipment. The stability of process parameters in the distillation section is crucial to maintaining uniform quality of the final product, therefore, numerous control points and continuous temperature, pressure, and flow measurements are used at each stage.

During distillation, waste streams are also produced – mainly aqueous solutions of organic pollutants and small amounts of heavy fractions. In accordance with applicable environmental requirements, these streams are sent to the plant wastewater treatment system for further treatment. This waste is not recycled back into the process cycle, which is intended to prevent recontamination of the product stream and maintain the high quality of the propylene glycol.

The resulting reaction product is then subjected to fractional distillation, which separates the reaction water, organic impurities, and byproducts from the main propylene glycol fraction. The distillation system consists of two columns operating at different temperatures and pressures, enabling the production of a high-purity product ($\geq$99.5%). The distillation step is one of the most energy-intensive unit operations in the entire process, requiring both significant amounts of thermal energy to evaporate the light fractions and cooling during the condensation and rectification processes.

## 3 Energy Media System in the Propylene Glycol Production Process

The propylene glycol production system from distilled glycerin utilizes a complex system of energy media necessary to maintain stable process conditions in individual process nodes. Proper management of these media is a key element in ensuring the safety, efficiency, and reliability of the installation.

The key types of energy media used in the processes of this type include [1, 2]:

- process steam, used to heat reactors, distillation columns, and heat exchangers; it is the primary thermal energy carrier in the system,
- steam condensates, generated during heat exchange, are a potential source of reusable waste energy,
- cooling water, used to condense vapors and stabilize product temperatures in reactors and heat exchangers,
- compressed air, used in automation systems, valve control, and safety devices,
- process gases, primarily process hydrogen, which serves a dual function: as a chemical reagent in the glycerin hydrogenation reaction and as an energy medium with high thermodynamic potential.

The complexity of the system stems from the need to simultaneously manage the flows of heating, cooling, and gaseous media [3]. Each section of the plant – from hydrogenation reactors to distillation and condensation systems – requires a precise energy balance to ensure stable process conditions with minimal media consumption.

The steam cycle utilizes several pressure and temperature levels, enabling cascading energy utilization [4, 5]. The steam with the highest parameters (from the boiler room or recovered from the reactors) is directed to heat the hydrogenation reactors, while the steam with lower parameters is used to preheat raw materials or regenerate distillation columns. This arrangement maximizes the use of steam enthalpy, reducing heat losses [6].

Condensates generated in the heat exchangers and reactors are cooled and recycled to the water-steam cycle, which not only reduces the consumption of fresh feed water but also lowers the energy demand for its reheating [6, 7]. The cooling water system is organized in a closed manner, with recirculation through fan coolers and intermediate heat exchangers [8, 9].

From a methodological perspective, the main direction of development for propylene glycol from glycerine production plants is the implementation of integrated energy systems – encompassing energy flow analysis, heat exchanger modernization, heat pumps, and automated media control systems. Implementing such solutions reduces primary media consumption, increases energy recovery, and improves the overall energy efficiency of the plant.

Integration of the energy media system in the propylene glycol plant is crucial for the overall energy efficiency of the plant. Optimized steam and condensate circulation, along with control of heat flows between individual process sections, allows for a significant reduction in energy consumption per unit of product.

Optimizing energy use in the glycerin hydrogenation and distillation process is therefore not only an economic but also an environmentally beneficial measure – it contributes to reducing $CO_2$ emissions and improving waste management. Analysis of the research object in this context provides the basis for developing modern energy management strategies in the chemical industry based on renewable raw materials [10–13].

Implementing solutions that enable energy recovery from steam from hydrogenation reactors not only reduces thermal losses but also reduces the demand for electricity used to operate cooling towers. This approach increases the plant's energy stability and reduces its carbon footprint, thus contributing to the chemical industry's sustainable development strategy [14–16].

## 4 The Energy Problem of the Propylene Glycol Production Plant

The primary industrial problem of the analyzed system is the insufficient utilization of waste energy potential and the lack of full thermal integration between individual process stages. The large temperature difference between the hot and cold streams in the system creates the potential for effective heat recovery, but technological, design, and process safety limitations often prevent its full utilization. Furthermore, the intermittent operation of some nodes, variability in reactor load, and heterogeneity in media parameters lead to difficulties in dynamic energy balancing in real time.

The propylene glycol production process using distilled glycerin generates a significant amount of process steam as a byproduct of the hydrogenation reaction. In the current process system, this steam is directed to cooling towers, where it condenses and disperses into the atmosphere, leading to significant thermal energy losses and increased electricity consumption by the fan system. Despite the stability of the technological process and maintaining the required product quality parameters, low efficiency of energy utilization in the production system is observed.

Figure 2 shows a simplified diagram of the hydrogenation process. The analyzed diagram shows the technological system for producing propylene glycol from distilled glycerin, consisting of two reactors, two separators, and a cooling exchanger. Vapors containing process heat are removed from the reactors and connecting streams; some of this heat is passed through the cooling exchanger, while the condensed condensate and cooled streams are returned to the system or further passed to the distillation section. The diagram shows the hydrogen and glycerin feeds to Reactor 1 and the connections between the separators and reactors.

The analyzed system reveals the following specific industrial problem: the inappropriate discharge and loss of process steam (an energy stream). The steam (energy) released in the hydrogenation reactors is directed to a cooling exchanger and then - without effective return to the plant steam pipeline is discharged. This results in irreversible

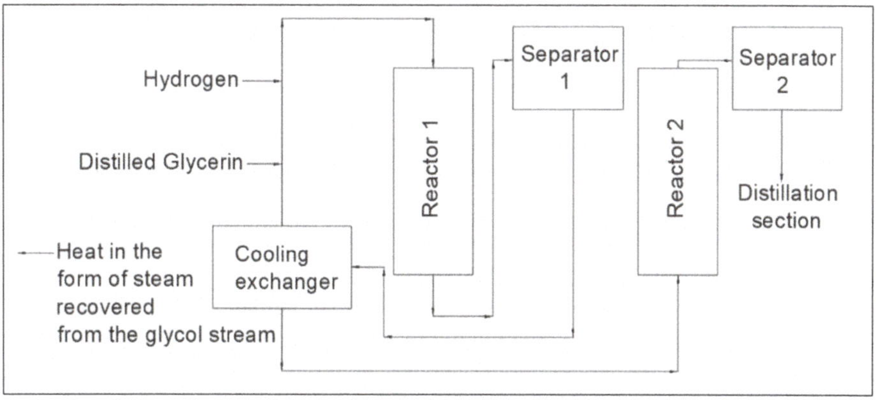


**Fig. 2.** Simplified diagram of the hydrogenation process.

energy loss, generating high operating costs and excessive steam production in the boiler room. This phenomenon results in steam losses of hundreds of kg/h (significant costs), the risk of process disruptions in the reactors, and operational problems (condensate, water hammer, corrosion).

**Fig. 3.** The part of technological system for producing propylene glycol from distilled glycerin - cooling exchanger.

The identified research problem therefore concerns the optimization of waste steam management generated in glycerine hydrogenation reactors by developing solutions that enable its recovery and reuse in the plant's energy system (Fig. 3). The aim of the research is to determine the most technically and economically effective variant of this stream management, which will reduce heat losses, reduce electricity consumption, and improve the environmental balance of the plant.

## 5 Research on the Energy and Economic Efficiency of Process Steam Recovery in a Propylene Glycol Production System

The main objective of the research was to assess the energy and economic efficiency of recovering process steam generated in a propylene glycol production plant from distilled glycerin. The work focuses on analyzing the feasibility of utilizing this stream to reduce thermal energy losses and the consumption of electricity used in the reactor cooling system. The ultimate goal is to develop a technical solution that will increase the plant's energy self-sufficiency and improve its environmental performance.

The analysis utilized balance and simulation methods, taking into account both thermal energy flows and the consumption of electrical energy by auxiliary devices. An analysis of the existing state was conducted, including measurements of temperature, pressure, and steam flow rate at key points in the system.

Based on the process data, an energy balance of the installation was developed, which was used to determine the steam recovery potential and identify possible places for its reuse.

The Table 1 gives a summary of current electricity costs and losses of steam sent to the cooling tower.

**Table 1.** Summary of current electricity costs and losses of steam sent to the cooling tower.

| Position | Value | Unit cost [PLN] | Month cost [PLN] | Year cost [PLN] |
|---|---|---|---|---|
| Cooling capacity (2 × 25 [kW]) | 50 [kW] | – | – | – |
| Electricity consumption | 36000 [kWh/month] | 0,65 | 23 400 | 246 000 |
| Steam loss from reactors (525 [kg/h]) | 378000 [kg/month] | 0,1 | 37 800 | 397 000 |
| Total cost of energy loss | – | – | 61 200 | 643 000 |

The Table 1 shows, that the current method of operating the plant generates very high costs resulting from inefficient use of electricity and losses of process steam from the hydrogenation reactors. The largest cost item is steam loss—approximately 378,000 kg per month, which, at an assumed unit price of PLN 0.10/kg, translates to PLN 37,800 in losses per month, or PLN 397,000 per year. This means that the unused steam energy potential alone accounts for over 60% of the total energy costs incurred in the current system configuration.

The second significant factor is the cooling tower's electricity consumption (two 25 [kW] motors each), which generates a load of around 36,000 [kWh] per month, generating a cost of PLN 23,400 per month, or PLN 246,000 per year. As a result, the total monthly cost of energy losses is PLN 61,200, and the annual cost reaches PLN 643,000, clearly indicating a very large savings potential.

The analysis confirms that the current energy management system in the propylene glycol installation is burdened with significant losses that economically and technically justify the implementation of modernization measures - especially those enabling steam recovery and eliminating the operation of the cooling tower.

## 6 A Method of Improving the Energy and Economic Efficiency of the Tested Installation

The method for improving the energy and economic efficiency of the propylene glycol production plant was directly related to the management of recovered steam, which was described in the previous section. In this regard, two variants were proposed and analyzed:

1. Injecting steam into the main steam pipeline (variant no. 1).
2. Using steam to heat raw materials (variant no. 2).

The first variant analyzed involves injecting steam from the hydrogenation reactors into the plant's main steam pipeline. Instead of being directed to the cooling tower, this steam is used as an additional source of thermal energy in the existing steam distribution system. The implementation of variant no. 1 is shown in Fig. 4.

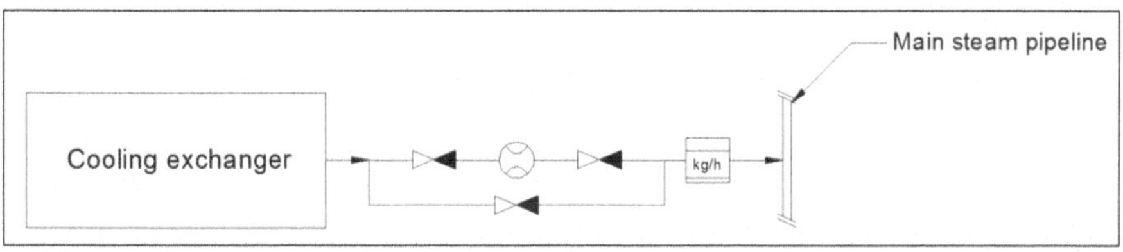


**Fig. 4.** Diagram of variant no. 1 - injecting steam into the main steam pipeline.

Steam generated during the hydrogenation process is directed through check valves, a flow meter, and a counter to the plant's main steam pipeline, where it is then reused in the steam cycle. This makes it possible to:

- complete shutdown of the cooling tower for this part of the installation,
- reduction of electricity consumption by two 25 [kW] fan motors,
- recovery of thermal energy in the form of steam at usable pressure,
- improvement of the plant's energy balance and reduction of operating costs.

This solution requires minimal investment, as existing pipelines and valves can be partially reused. Simply connect to the main pipeline, install appropriate check valves, a steam flow meter, and a steam meter for the steam entering the main pipeline.

The Table 2 gives a summary of current electricity costs and losses of steam sent to the cooling tower.

**Table 2.** Cost estimate for modernization of the installation (variant no. 1 - steam pipeline).

| Position | Quantity | Unit cost [PLN] | Value [PLN] |
|---|---|---|---|
| Steam pipeline DN80 (stainless steel) 40 [m] | 40 [m] | 380 | 15200 |
| Shut-off and check valves Electricity consumption | 1 [set] | 3 600 | 3 600 |
| Steam flow meter (with pressure sensor) | 1 [piece] | 7 500 | 7 500 |
| Steam volume meter (with integration with DCS system) | 1 [piece] | 4 200 | 4 200 |
| Assembly work and thermal insulation | 1 [service] | 12 000 | 12 000 |
| Modification of the measurement system software | 1 [service] | 5 500 | 5 500 |
| Total investment costs | | – | 48 000 |

The second variant under consideration involves using recovered steam to preheat distilled glycerin prior to the hydrogenation process. For this purpose, the steam would be directed to a heat exchanger located in the raw material preparation section.

The advantage of this solution is the ability to reduce the amount of energy needed to preheat the raw material in the reactors. However, it requires redesigning the plant and installing additional equipment: a heat exchanger, steam pipelines, control valves, and a condensate drainage system.

The Table 3 gives a summary of current electricity costs and losses of steam sent to the cooling tower.

The result of the comparison of the developed variants is presented in the Table 4.

From an economic perspective, variant no. 1 (injecting steam into the plant pipeline) is a more cost-effective and technically simpler solution. It requires significantly less installation modifications and does not require additional heat exchangers or complex automation. Variant no. 2, although theoretically improving the energy efficiency of the raw material preheating process, is associated with higher investment costs and a longer payback period. Due to the simplicity of implementation, low cost of modification and high energy efficiency, variant no. 1 was considered recommended.

Next, a time-based simulation economic analysis of variant no. 1 was carried out, including investment costs, savings resulting from the reduction of electricity consumption and the value of recovered steam.

It assumed operation of the installation for 50 weeks a year (3 shutdowns of 2 weeks each for inspections and operational replacements), which corresponds to approximately 8,400 h of operation per year.

The graph of the increasing savings values, as a result of implementing variant no. 1 is presented in Fig. 5.

**Table 3.** Cost estimate for modernization of the installation (variant no. 2 - heating of raw materials).

| Position | Quantity | Unit cost [PLN] | Value [PLN] |
|---|---|---|---|
| Steam pipeline DN100 (acid-resistant steel, with insulation) | 70 [m] | 950 | 66 500 |
| Plate heat exchanger (industrial type, stainless steel) | 1 [piece] | 85 000 | 85 000 |
| Control, safety and check valves | 1 [set] | 11 000 | 11 000 |
| Temperature and pressure sensors (with DCS integration) | 1 [set] | 9 500 | 9 500 |
| Flow meters (steam, condensate, feed water) | 1 [set] | 12 000 | 12 000 |
| Thermal energy meter with data archiving system | 1 [piece] | 8 200 | 8 200 |
| Water piping with valves (secondary installation) | 60 [m] | 450 | 27 000 |
| High-temperature circulation pump | 1 [piece] | 18 500 | 18 500 |
| Design work, technical documentation and supervision | 1 [service] | 14 000 | 14 000 |
| Equipment assembly and pressure testing | 1 [service] | 24 000 | 24 000 |
| Modification of the control software and DCS system | 1 [service] | 9 000 | 9 000 |
| Total investment costs | | – | 284 700 |

**Table 4.** The result of the comparison of the developed variants for modernization of the tested installation.

| Parameter | Variant No.1: Injecting steam into the main pipeline | Variant No.2: Heating raw materials Value [PLN] |
|---|---|---|
| Investment cost | 48 000 [PLN] | 284 700 [PLN] |
| Payback time | 1 [month | 7–12 [months] |
| Risk of failure | low | high (exchanger, pumps, automation components |
| Implementation difficulty level | low | high |
| Additional maintenance costs | insignificant | significant |
| Economic assessment | very profitable | unprofitable |

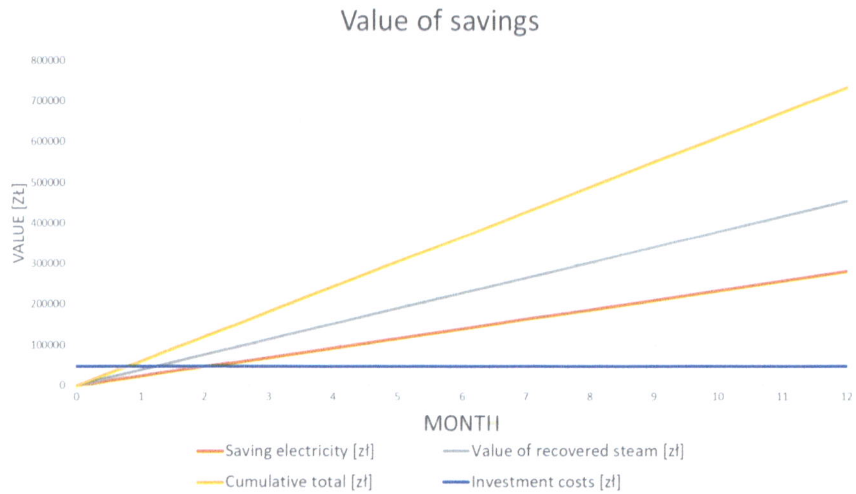


**Fig. 5.** Increasing savings values, as a result of implementing variant no. 1.

## 7 Interpretation of the Results of Analysis of the Introduced System Modifications

The economic analysis conducted for the selected variant, which involves injecting recovered steam from the hydrogenation section reactors into the main steam pipeline, clearly confirms the validity and profitability of this solution. Based on the data summarized in Table 6, the cumulative value of savings, including reduced electricity costs and process steam recovery, shows a dynamic increase in subsequent months of the installation's operation.

After just the first month of system operation, the savings achieved are significant, and on an annual basis, they exceed several hundred thousand zlotys. Assuming three maintenance shutdowns per year (six weeks in total), the investment achieves a full return on the modernization costs within a very short time of implementation. This means, that compared to variant no. 2, variant no. 1 is characterized by several times lower investment costs and a short payback period.

The savings are due to two main factor:

1. Elimination of the electricity requirement for the cooling tower motors (2 x 25 [kW]), allowing for a significant reduction in operating costs.
2. Increasing the energy efficiency of the steam system by using waste steam as a full-fledged process medium in the main circuit, without the need for additional condensation and cooling.

The implementation of variant no. 1 allows for achieving the highest economic efficiency with minimal investment outlays, which clearly confirms the validity of its selection as the target solution.

The results of the analysis conducted and described in this article led to the implementation of variant 1 on a real propylene glycol production line.

Figure 6 shows the cooling tower, previously operating continuously, now completely out of exploitation.

**Fig. 6.** Closed cooling tower shutters - cooling tower out of exploitation.

This is a direct result of the modernization described in variant 1, which involves recovering heat from the reactors and reducing the need for air cooling. Shutting down the cooling tower not only confirms that the device is no longer needed for energy stabilization of the process, but also clearly indicates that the regeneration system is operating stably and the heat recovery circuit is fulfilling its operational function. The installation maintains the required parameters without the use of an additional heat sink, which in practice is the best proof that the implemented solution is functioning properly and delivering the expected energy savings.

## 8 Summary

Implementing a modernization project that involves recovering steam from the glycol stream exiting the hydrogenation section reactors and injecting it into the main steam pipeline brings not only tangible economic benefits but also a number of significant environmental benefits. This solution aligns with the sustainable development of the chemical industry and the strategy of increasing the energy efficiency of production plants, in line with the European Union's climate and energy policy guidelines [17].

The primary positive environmental impact is the reduction in electricity consumption. Eliminating the operation of cooling towers, whose 50 [kW] motors previously operated continuously, translates into a reduction in annual electricity demand by several hundred megawatt-hours. At current emission rates, this translates to a reduction in indirect carbon dioxide emissions by approximately 300–400 tons of $CO_2$ per year.

The second important aspect is increasing the efficiency of primary energy use in the technological process. Recovering steam from the hydrogenation reactors, which was previously lost in the cooling process, allows for the use of this thermal energy in the plant's main steam cycle. This reduces the amount of steam generated in the steam boiler, leading to lower fuel consumption (e.g., natural gas or fuel oil) and related $CO_2$ emissions. It is estimated that reducing steam generation demand by approximately 500–600 kg/h reduces fuel consumption by several thousand cubic meters of gas annually, further reducing the plant's carbon footprint.

It's also worth noting that the modernization doesn't require expanding the existing refrigeration infrastructure or auxiliary systems, which reduces the consumption of consumables and process water. In practice, this also means a reduction in the amount of industrial wastewater generated in refrigeration systems and a reduced risk of failures resulting from corrosion or leaks in the water-air systems. From an environmental perspective, this is an additional advantage of the implemented solution, as it reduces not only atmospheric emissions but also the impact on water resources.

Another significant positive factor is the reduction of industrial noise. Deactivating cooling towers reduces noise levels in the plant's surroundings, improving working conditions for staff and reducing acoustic nuisance to the external environment. For chemical plants with stringent noise emission standards, this may be an additional argument for modernization.

From the point of view of sustainable development and a low-emission economy, this solution is an example of effective integration of technological and energy processes, leading to minimization of energy and resource losses while maintaining high production efficiency [17].

The analyses carried out confirm that the recovery and reuse of water vapor in the propylene glycol production process is one of the most effective and economically justified modernization activities in the chemical industry.

Ultimately, it can be stated that the implementation of the solution consisting in injecting steam from the reactors into the main steam pipeline is not only technically justified, but also brings measurable economic and environmental benefits, and its implementation is an example of good practice in the field of energy efficiency in the chemical industry.

## References

1. Matsumoto, T.: Industrial heat recovery optimization. Energy Rep. **6**, 845–854 (2020). https://doi.org/10.1016/j.egyr.2020.03.013
2. Wu, H., et al.: Energy efficiency improvement through waste heat recovery. Appl. Energy. **238**, 1032–1045 (2019). https://doi.org/10.1016/j.apenergy.2019.01.189
3. Patel, V., Gupta, R.: Integration of renewable energy and waste heat recovery. Energy Procedia. **159**, 271–277 (2019). https://doi.org/10.1016/j.egypro.2019.01.037
4. Gao, X., Liu, Y.: Sustainable energy management in chemical processes. Renew. Sust. Energ. Rev. **132**, 110059 (2020). https://doi.org/10.1016/j.rser.2020.110059
5. Santos, A., Oliveira, A.: Steam network optimization in process industries. Energy. **257**, 124673 (2022). https://doi.org/10.1016/j.energy.2022.124673
6. Lee, C., et al.: Catalyst design for efficient glycerol hydrogenolysis. Catal. Today. **320**, 160–172 (2019). https://doi.org/10.1016/j.cattod.2018.06.034

7. Chen, Y., et al.: Propylene glycol production from glycerol hydrogenolysis: A review. Chem. Eng. J. **425**, 131537 (2021). https://doi.org/10.1016/j.cej.2021.131537
8. Li, D., et al.: Catalytic hydrogenolysis of glycerol to propylene glycol. Appl. Catal. B Environ. **278**, 119314 (2020). https://doi.org/10.1016/j.apcatb.2020.119314
9. Tan, J., Zhang, Z.: Energy recovery strategies in industrial distillation systems. Energy Convers. Manag. **246**, 114676 (2021). https://doi.org/10.1016/j.enconman.2021.114676
10. Piechowski, M., Wyczółkowski, R., Paszkowski, W.: The concept of a system supporting the implementation of an intelligent lubrication strategy within the company using advanced information technologies. In: International Scientific-Technical Conference Manufacturing, pp. 136–150. Springer Nature, Switzerland, Cham (2024)
11. Singh, S., et al.: Integration of heat recovery in propylene glycol production. J. Clean. Prod. **200**, 1023–1034 (2018). https://doi.org/10.1016/j.jclepro.2018.07.234
12. Timofiejczuk, A., et al.: The concept of a system for monitoring the condition of gas transmission pipelines and their surroundings. Nafta-Gaz. **79**(1), 52–60 (2023). https://doi.org/10.18668/ng.2023.01.06. (in Polish)
13. Zhou, J., Gao, L.: Recycling of process steam in chemical plants. Ind. Eng. Chem. Res. **60**(38), 13862–13871 (2021). https://doi.org/10.1021/acs.iecr.1c02016
14. Loska, A.: Variant assessment of exploitation policy of selected companies managing technical network systems. Manag. Syst. Prod. Eng. **3**(19), 179–188 (2015). https://doi.org/10.12914/MSPE-13-03-2015
15. Loska, A., Paszkowski, W., Waszkowski, R.: The scenario approach to the concept of maintenance of technical systems of urban engineering. In: Trojanowska, J., Kujawińska, A., Pavlenko, I., Husar, J. (eds.) Advances in Manufacturing IV. Manufacturing 2024 Lecture Notes in Mechanical Engineering, pp. 84–97. Springer, Cham (2024). https://doi.org/10.1007/978-3-031-56444-4_7
16. Sun, B., et al.: Failure-based sealing reliability analysis considering dynamic interval and hybrid uncertainties. Eksploatacja i Niezawodność–Maintenance and Reliability. **23**(2), 278–284 (2021). https://doi.org/10.17531/ein.2021.2.7
17. Janik, A., Ryszko, A.: Open innovation and environmental sustainability–A systematic literature review from the firm-level perspective. Technol. Forecast. Soc. Chang. **218**(124196), 1–31 (2025). https://doi.org/10.1016/j.techfore.2025.124196

# Ergonomics in the Light of ESRS Standards. Analysis of ESG Reports from Selected Wood Based Panel Manufacturers

Fryderyk Wachowiak(✉)

Poznan University of Technology, Rychlewskiego 2, 60-965 Poznan, Poland
Fryderyk.wachowiak@put.poznan.pl

**Abstract.** The article presents the role of ergonomics in ESG reporting, resulting from the CSRD [1], in line with the ESRS [2], with particular emphasis on the S1–14 requirements regarding health and safety at work. This paper checks (1) how ergonomics appears in six ESG reports and (2) proposes simple indicators aligned with ESRS S1–14. The author analyzed the content of the regulation in question for its references to ergonomics and human factors. The article then identifies disclosures in the field of ergonomics - good practices and reporting gaps - based on an analysis of selected, available ESG reports from six selected wood-based panel manufacturers. A proposed set of ergonomic indicators, compatible with ESRS S1–14, was presented. The results showed that while most of the companies analyzed included references to occupational health and safety and preventive measures in their reports, only a few addressed ergonomics. The information provided was very general, failing to provide detailed data and indicators on their ergonomic activities. In the last part, after a discussion on the advisability of disclosing ergonomic data in ESG reports, the author points to the role of an ergonomic audit, which could successfully strengthen ergonomics reporting in the field of sustainable development.

**Keywords:** CSRD · ESRS S1–14 · Ergonomics · Wood-Based Panels

## 1 Introduction and Regulatory Context

In the face of growing stakeholder pressure and ESG regulations – in the area of environmental, social and governance (ESG), the CSRD introduces new requirements for companies to disclose non-financial data in a credible, comprehensive and verifiable manner [3]. The CSRD (Corporate Sustainability Reporting Directive) replaced the previous NFRD (Non-Financial Reporting Directive), expanding reporting obligations in the scope of disclosing non-financial information relating to the activities of companies in the environmental, social and governance context. The reporting obligation according to the amended regulations will apply to a much wider group of companies. The new regulations aim to increase the transparency and comparability of ESG reports. One of the key assumptions of the CSRD is the concept of double significance – Companies must assess both the impact of ESG issues on their financial operations and the impact of


A. Hamrol et al. (Eds.): MANUFACTURING 2026, LNME, pp. 230–239, 2026.
https://doi.org/10.1007/978-3-032-21701-1_17

their own activities on people and the environment. Importantly, from the perspective of both companies and their stakeholders, the reports are intended to enable comparisons between companies in a given industry, as well as cross-sectoral comparisons. It's therefore not surprising that every company is motivated to outperform its competitors, at least in selected, business-critical metrics. Crucially, the significance of these metrics is determined by taking into account the voices of interested parties – the aforementioned stakeholders – as required by law.

On the other hand, apart from the imposed legal regulations, companies are increasingly setting their own additional business goals, which are not limited to profit-oriented goals, but also include a number of benefits for employees related to building a broadly understood safety culture [4], oriented towards sustainable social development [5] and well-being [6].

It should be noted, therefore, that both the direction of self-development of enterprises and EU social guidelines are heading in the same direction. Key factors in the social area, according to the ESRS requirements, include aspects related to working conditions, human capital, employee well-being, and psychosocial issues. This is largely the area covered by ergonomics, whose importance in relation to sustainable development has been and continues to be the subject of research by scientists [7, 8].

The requirements for disclosing information relating to an organization's own employees are set out in ESRS S1–14 under "*occupational health and safety measures*".

## 2 Characteristics of ESRS S1–14

ESRS S1–14 "Occupational Health and Safety Measures" requires an organization to disclose information on the extent to which its workforce is covered by an occupational health and safety management system. This includes indicators related to the number of injury incidents, work-related ill health, and fatalities, as well as the number of days lost due to the aforementioned causes. Additionally, an organization may include information on the percentage of its workforce covered by an occupational health and safety management system based on legal regulations or recognized standards or guidelines that have been internally or externally audited or certified.

Significantly, the ESRS requirements do not include a direct reference to ergonomics in the strict sense. Only in the list of definitions (which, incidentally, are not mentioned anywhere in the document) are work-related hazards identified as having an ergonomic nature, providing examples such as "incorrectly positioned workstations and chairs, unnatural movements, vibrations." It is also worth noting that some of the other hazard descriptions presented in Table 1 fall within the scope of ergonomics.

Ill health, defined in ESRS is described as recurrent and chronic health problems caused or exacerbated by working conditions or work practices. These include musculoskeletal disorders, skin and respiratory diseases, malignant neoplasms, diseases caused by physical factors, and mental illness.

Ergonomics affects employee health and safety, absenteeism, turnover, productivity, benefit costs, and work quality [9] – all of these factors are important and are cited as indicators in the "S" - social area of ESG. Good ergonomic practices reduce musculoskeletal complaints [10, 11] and occupational health and safety incidents, which is reflected in

**Table 1.** Description of defined types of threats. Source: own study based on ESRS S1–14.

| Type of threat | Examples indicated |
|---|---|
| Physical | Radiation, extreme temperatures, constant exposure to noisy environments, spilled substances on floors or trip hazards, unattended machinery, faulty electrical equipment. |
| Chemical | Exposure to carcinogenic, mutagenic, reprotoxic substances, solvents, carbon monoxide or pesticides. |
| Biological | Exposure to blood and body fluids, fungi, bacteria, viruses, or insect bites |
| Psychosocial | Verbal abuse, harassment, stalking. |
| Related to the organization of work | Excessive workload demands, shift work, overtime, night work, workplace violence. |

the disclosed information described in reporting standards. Organizations implementing ergonomic policies gain a reputational and financial advantage thanks to lower injury costs and improved operational efficiency, while demonstrating that they take actions that go beyond the standard approach to occupational health and safety requirements. In the area of sustainable development, there is a growing interest in integrating human factors and ergonomics with the concept of sustainable production, where employee health is perceived as an element of long-term organizational value [12, 13].

What does the issue of reporting in relation to ergonomics look like among representatives of one of the most important industries in Poland – producers of wood-based panels?

## 3 Ergonomics in Reports from Wood-Based Panel Manufacturers

The wood-based products industry is part of the wood sector. Poland ranks second in Europe in terms of the quantity of wood-based panels produced, behind Germany (according to Polish Central Statistical Office – 2024). Because the vast majority of producers of these products are large enterprises, they are subject to the non-financial ESG reporting requirement. Although the entry into force of legal requirements for ESG reporting has been postponed until 2027 by a decision of the European Parliament, many representatives of this industry have already published their reports. Some are based on ESG requirements, in accordance with the CSRD directive, others refer to GRI indicators[1] [14], and still others provide information in relation to SDG reporting.[2] Regardless of the model by which companies present information in their reports, the author analyzed selected reports for disclosure of information regarding ergonomics.

[1] The Global Reporting Initiative (GRI) is an independent not-for-profit organization that leads a globalmulti-stakeholder process to develop and refine practical yet practical sustainability reporting.

[2] SDGs (Sustainable Development Goals) are 17 Sustainable Development Goals that were adopted by theUN in 2015 as part of the 2030 Agenda.

## 4 Materials and Methods

A qualitative-quantitative approach was used to evaluate the reports. The content of non-financial reports (ESG/CSRD/GRI) of wood-based panel manufacturers was analyzed – a full-text reading – to assess the scope and quality of ergonomic disclosures and their compliance/functional compliance with the ESRS S1–14 requirements. The unit of analysis is a single annual report (or sustainability report) from a given issuer for year 2023 or 2024.

Reports from six companies were included in the study[3]: Egger (2023/2024), Ikea (2024), Kronospan Polska (2023/2024), Pfleiderer (2023/2024), Sonae Arauco (2023), Swiss Krono (2023). Inclusion criteria: (a) publicly available ESG/CSRD/GRI/SDG report published on the official corporate website; (b) report language: Polish/English/German; (c) reference year 2023 or 2024; (d) the report covers operations in Poland or a consolidated group report with disclosures covering plants in Poland. Marketing presentations, HR/OHS brochures that are not reports, and documents without a reporting methodology section were excluded from the analyzed materials.

Reports were downloaded directly from corporate websites. For each issuer, the following information was archived: document title, publication year, scope (company/group), reporting framework (ESRS/GRI/SDG/other), language, version number, and any errata. The files were indexed in the research repository along with metadata (file hash, download date) to ensure replication.

Ergonomics-related content was defined as content containing references or descriptions that met one of the following criteria: substantive scope or reference to indicators. Content considered ergonomics included:

a) workstation and/or tool design,
b) musculoskeletal system (MSD) loading,
c) biomechanics,
d) cognitive and/or psychosocial ergonomics,
e) employee participation in workstation improvement,
f) ergonomic audits and or ergonomic risk assessments,
g) ergonomic training,
h) intervention programs (e.g., exoskeletons, line redesign), and references to ergonomic principles.

The second criterion (2) was the relationship to outcomes:

a) reported operational indicators: MSD incidence rate, days of absence due to MSDs, LTI/TRIR indicators with ergonomic causes highlighted,
b) procedural systemic evidence: regularity of ergonomic risk assessments, program coverage, internal/external audits.

References to occupational health and safety in the strict sense were not counted as ergonomic issues.

[3] EGGER Group (2024). Sustainability Report 2024, Pfleiderer Group (2024). ESG Report 2024; Sonae Arauco (2024). Sustainability Report 2024; IKEA Industry (2024). Sustainability & ESG Overview; Swiss Krono (2024). ESG Progress Report; Kronospan (2023). Sustainability Overview.

For each company, the following were recorded: (1) meta (reporting framework, audits), (2) ergonomic policies and objectives, (3) processes (risk assessment, programs, training, participation), (4) resources/technologies (e.g., exoskeletons), (5) indicators (MSD, LTI/TRIR with reasons, absence days, program coverage), (6) results and trends ($\geq$3 years, if available), (7) compliance with ESRS S1–14 and mapping to other ESRS/GRI points.

The limitations identified were (a) heterogeneity of reporting frameworks, (b) lack of full integration of ergonomics in the ESRS, (c) differences in language and level of detail that may reduce comparability.

## 5 Results

The analysis revealed that references to ergonomics appear with varying frequency in the reviewed ESG reports, unfortunately without any specific, measurable indicators (KPIs). Egger and Kronospan Polska refer to ergonomics most frequently, each three times. Sonae Arauco has two references. Pfleiderer - with a single reference to ergonomics. The IKEA and Swiss Krono reports did not include any references to ergonomics, nor to any indicators or related terms, as defined in the content considered ergonomics (given in material and method part).

The above may lead to the conclusion that comparisons between companies in the same industry regarding ergonomic activities are currently impossible due to the heterogeneity of the disclosed data and information. This may translate into the inability to perform comparative assessments by stakeholders, which is a key assumption of ESG reporting.

The lack of clear requirements for ergonomics disclosure in legal requirements – the CSRD Directive and the related ESRS and GRI indicators – may prove crucial.

The author presents a table below (Table 2) summarizing the references to ergonomics in the analyzed reports, including the report content and ergonomics-related issues. The table also includes the number of references and the number of Key Performance Indicators (if any).

Why should companies raise ergonomics issues in their reports? How can they reliably, credibly, and verifiably assess their commitment to ergonomics and the resulting outcomes through disclosed metrics?

## 6 Discussion

The analysis of selected ESG reports revealed that ergonomics is generally not presented in the form of measurable indicators – zero examples were found in all checked reports. Some of the organizations refer to ergonomics in analyzed ESG reports in a general content (see Table 2).

Disclosure of ergonomic data – i.e. information on ergonomic risks, the occurrence of musculoskeletal disorders (MSDs), preventive measures, and the effects of ergonomic interventions – should become an integral part of transparent and responsible corporate social communication. Incorporating this data into ESG reports will enable stakeholders

**Table 2.** Ergonomics in ESG reports. Source: own study based on the analyzed reports.

| **Organization**/Number of references/Key Performance Indicators (KPI) | Examples indicated |
|---|---|
| **Egger**<br>3 references<br>0 KPI | Measurements from the past: an increase in the number of **ergonomic** workstations as part of the construction or renovation of new offices.<br>...<br>At all our sites, we offer: (...); **technical and ergonomic measures** for physically demanding work; (...) and **ergonomically** adapted devices and work surfaces in plants and offices. |
| **Ikea**<br>0 references<br>0 MI | *Not occur* |
| **Kronospan Polska**<br>3 references<br>0 KPI | Our commitment to achieving these goals includes: compliance with legal regulations and industry standards applicable within the company; ensuring safe **and ergonomic** working conditions [...]<br>In accordance with the Occupational Health and Safety Policy, we focus our efforts on achieving these goals by providing a safe **and ergonomic** working environment (...). Our primary goal is to ensure safe **and ergonomic** working conditions (...). |
| **Pfleiderer**<br>1 reference<br>0 KPI | In addition, the occupational health service assesses various workstations – including in terms of **ergonomic factors** – in order to detect health risks early and take preventive measures. |
| **Sonae Arauco**<br>2 references<br>0 KPI | We made progress in implementing measures to strengthen the detection and prevention of occupational diseases through inspections, training, and programs focused on specific areas (hearing protection, respiratory protection, hydration, and **ergonomics**).<br>We implemented comprehensive medical and environmental monitoring programs, including **ergonomic assessments** of workstations, control of harmful factors (noise, formaldehyde), and vaccination campaigns. |
| **Swiss Krono Polska**<br>0 references<br>0 KPI | *Not occur* |

(investors, employees, regulators) to assess whether the organization is actively managing one of the key determinants of employee health and productivity [7, 15]. Among the objectives of ergonomic data disclosure, the author indicates the following: (a) health and financial risk significance. Ergonomic problems (e.g., musculoskeletal disorders (MSDs) generate significant direct costs (sick pay, absenteeism, compensation) and indirect costs (loss of productivity, loss of knowledge, turnover). Through the lens of ESG significance, ergonomic data can be of significant importance for assessing long-term risk and the operational resilience of the enterprise [15, 16, 18]; (b) compliance with reporting standards expectations. The European Sustainability Reporting Standards (ESRS) and the ESG "S" guidelines require disclosure of information on working conditions and employee health – which should include indicators related to ergonomics and MSDs prevention programs [3, 14]; (c) synergy with well-being and sustainability. Ergonomics supports long-term sustainability goals by maintaining employee fitness, improving working conditions, and reducing health inequalities in the workplace – which is increasingly being assessed by stakeholders. Research indicates that companies reporting KPIs related to health, ergonomics, and well-being are perceived more positively in terms of social responsibility [7, 17].

From the author's perspective – a short recommendation – the next step for organizations publishing reports, as well as for requirements authors, could focus on the following aspects.

For ergonomic data, that could be given in the reports; to be informative, disclosures should be measurable, comparable, and contextualized (e.g., industry, number of employees, geography). Example categories and KPIs may include: Outcomes: number of MSDs cases (absolute and per 1,000 employees), rate of days of absence related to MSDs, number of ergonomic events reported in a given period [15]; Process: number of ergonomic risk assessments conducted (e.g., RULA/REBA), percentage of workstations covered by an ergonomics program, number of ergonomic interventions implemented in a given period [18]; Input: budget allocated to the ergonomics program, number of people trained in ergonomics, presence of a formal ergonomics policy [17].

For disclosures to be useful, organizations should use standardized ergonomic risk assessment methods (e.g. REBA, RULA, OWAS) [19, 20], report both absolute data and standardized indicators [17], and combine quantitative data with qualitative descriptions of procedures and action plans [15].

Challenges and limitations related to comparability between industries: different work environments make direct comparisons difficult [7]. Data reliability: ergonomic events are sometimes underreported by employees [15]. Fears of "social washing" – superficial disclosure without real action – may discourage the publication of additional indicators [18].

The author also points out the need for a clear articulation of ergonomics in ESG reporting requirements, which would force reporting of ergonomic indicators. The proposed example of recommendations include:

a) ergonomic integration with the "S" policy and the occupational health and safety management system [3];
b) use of unified metrics [14];
c) publishing case studies of effective ergonomic interventions [7];

d) engaging stakeholders in defining relevant ergonomic indicators [15].

A sample set of ergonomic indicators for an ESG report, as example, could include:

a) % of workstations assessed for ergonomics in a given year;
b) % of workstations meeting ergonomics requirements;
c) % of employees receiving ergonomics training;
d) mean/median ergonomics risk assessment using recognized ergonomic tools;
e) number of cases of identified musculoskeletal complaints;
f) number of sick days/employees related to musculoskeletal complaints.

The presented indicators should be measurable and verifiable, for example, through audits. An audit documents and confirms ongoing processes, illustrating how the company identifies, assesses, and manages ergonomic risks—which aligns with sustainability reporting requirements. It generates data that can be standardized and therefore compared, which also aligns with ESG reporting expectations. The results of an audit conducted by independent, competent auditors constitute a source of credible, reliable, and verified information, which is valuable in itself in the era of social-washing.

An audit can therefore be a practical and essential tool for industrial companies seeking to demonstrate and disclose information that exceeds the minimum occupational health and safety requirements. The results of an ergonomics audit will provide data for assessing the materiality of the organization's impact on employee well-being. This impact may be positive or negative, depending on the audit findings.

As a next step, incorporating ergonomic audits into formal reporting processes, standardizing indicators, and providing external confirmation of the tasks performed will increase the value of information and, crucially, will also contribute to improving employee well-being – which would seem to be one of the key elements of the S – social aspect in ESG reporting.

## 7 Conclusions

As demonstrated in this publication, ergonomics is not explicitly included in companies' ESG reporting obligations in accordance with the applicable disclosure criteria according to ESRS standards. It is also not widely discussed in the analyzed reports of the six representatives of wood-based panel producers. However, it is worth noting that both Kronospan Polska and Egger most frequently (each three times) refer to ergonomics in their reports, which may indicate interest in the topic and ongoing activities in this area. It seems obvious that disclosing ergonomic data in ESG reports should become not only a voluntary best practice but also an ESRS requirement and a strategic tool for risk management, improving operational efficiency, and building stakeholder trust. Reliable, comparable, and contextualized ergonomic disclosures, along with the publication of structured ergonomic indicators, will enable understanding how companies invest in ergonomics – and consequently, in employee health and well-being, which translates into its long-term value and resilience [7, 15]. The last stage – no less important – is the confirmation of the disclosed information in the form of an independent audit conducted by ergonomics experts with competences such as Euro Ergonomist status.

## References


1. Directive (EU) 2022/2464 of the European Parliament and of the Council of 14 December 2022 amending Regulation (EU) No 537/2014, Directive 2004/109/EC, Directive 2006/43/EC and Directive 2013/34/EU with regard to corporate sustainability reporting.
2. Commission Delegated Regulation (EU) 2023/2772 of 31 July 2023 supplementing Directive 2013/34/EU of the European Parliament and of the Council with regard to sustainability reporting standards.
3. EFRAG: European Sustainability Reporting Standards (ESRS): Draft standards S1–S2. European Financial Reporting Advisory Group, Brussels (2023)
4. Jasiulewicz-Kaczmarek, M., Antosz, K., Wyczółkowski, R., Sławińska, M.: Integrated approach for safety culture factor evaluation from a sustainability perspective. Int. J. Environ. Res. Public Health. **19**(19), 1–30 (2022)
5. Butlewski, M., Czernecka, W.: Social sustainability in practice: Bridging the gap from declarations to real world scenarios on sustainability driven by ergonomics. Sustainability. **16**(14), 6–19 (2024)
6. Butlewski, M.: Well-being through design for dynamic diversity: The voice of minorities in design for all. In: Advances in Social and Occupational Ergonomics: Proceedings of the AHFE 2019 International Conference on Social and Occupational Ergonomics, Washington DC, USA, July 24–28, pp. 49–56. Springer International Publishing, Cham, Switzerland (2020)
7. Firescu, V., Suciu, M., Năstase, M.: Human factors and ergonomics in sustainable corporate practices: Integrating HF/E into ESG frameworks. Sustainability. **17**(2), 554–567 (2025)
8. Wróbel, K., Hoffmann, T.: Ergonomic Interventions in Shaping the Sustainable Development of Organizations, vol. No. 200. Scientific Papers Of Silesian University of Technology., Organization and Management Series (2024)
9. Kujawińska, A., Vogt, K., Wachowiak, F.: Ergonomics as significant factor of sustainable production. In: Golińska, P., Kawa, A. (eds.) Technology Management for Sustainable Production and Logistics. Eco Production. Springer, Berlin, Heidelberg (2015)
10. Attia, R.M.: Relationship between ergonomic awareness and work musculoskeletal disorders. J. Ergon. **10**(2), 45–62 (2023)
11. Jin, X., Dong, Y., Yang, L., et al.: Ergonomic interventions to improve musculoskeletal disorders among vehicle assembly workers: A one-year longitudinal study. BMC Public Health. **25**, 824 (2025)
12. Dul, J., Neumann, W.P.: Ergonomics contributions to company strategies. Appl Ergon. **40**(4), 745–752 (2009)
13. Hasanain, B.: The role of ergonomic and human factors in sustainable manufacturing: A review. Machines. **12**(3), 159 (2024)
14. GRI 403: Occupational Health and Safety 2018 (updated 2021). Global Reporting Initiative (2023)
15. Bazaluk, O.: Ergonomic risk management process for safety and health. Front. Public Health. **11**, 1120 (2023)
16. Sohrabi, M.S.: Ergonomics role in sustainable development: A review article for updates the recent knowledge. In: Congress of the International Ergonomics Association, pp. 588–602. Springer International Publishing, Cham, Switzerland (2021)
17. Health and Well-Being KPIs Report: Measuring Employee Health and Well-Being in ESG Disclosures. World Benchmarking Alliance (2023)
18. Zhang, L., Li, P., Xu, J.: Workplace ergonomics, health risk and ESG performance: Evidence from manufacturing companies. J. Clean. Prod. **355**, 131823 (2022)

19. McAtamney, L., Corlett, E.N.: RULA: A survey method for the investigation of work-related upper limb disorders. Appl. Ergon. **24**(2), 91–99 (1993)
20. Hignett, S., McAtamney, L.: Rapid Entire Body Assessment (REBA). Appl. Ergon. **31**(2), 201–205 (2000)

# Industry 4.0 & 5.0

# Influence of Focusing Tube and Process Variables on Wear Rates in Abrasive Waterjet Machining

Akash Nag(✉), Prasath Govindaraj, Jan Zelinka, Jana Petrů, and Sergej Hloch

Faculty of Mechanical Engineering, VŠB–Technical University of Ostrava, Ostrava, Czech Republic
akash.nag@vsb.cz

**Abstract.** Abrasive waterjet (AWJ) machining is a versatile technique for cold disintegration of various materials, with its performance significantly influenced by various process parameters. This study examines the relationships between input variables and critical output wear rates, including material removal rate (MRR), groove depth, and groove width, during the machining of AISI 316 L stainless steel. Utilizing a design of experiments employing focusing tube diameters of 0.89 mm and 0.5 mm, traverse speeds of 125, 250, and 500 mm/min, abrasive flow rates ranging from 5 to 50 g/min, and abrasive mesh sizes of 80 and 200, a dataset is generated. While existing literature explores the individual and combined effects of these parameters, a holistic statistical approach to simultaneously predict material removal performance across this parameter domain remains underexplored. The results showed a direct relationship between abrasive flow rate and depth, as well as MRR, and an inverse relationship with traverse speed. However, the groove width was unaffected by these parameters and only depended on the focusing tube diameter. The results obtained contribute to a deeper understanding of interactions within the AWJ process and provide valuable guidance for industrial applications seeking to maximize the material removal rate.

**Keywords:** Abrasive Water Jet · Material Removal Rate · MESH · Focusing Tube · ANOVA

## 1 Introduction

Abrasive waterjet (AWJ) machining is a highly versatile and increasingly vital non-traditional manufacturing process capable of disintegrate practically all materials regardless of their mechanical properties [1]. Currently, AWJ is used for the disintegration of a wide spectrum of materials, including metals, composites, ceramics, and even brittle substances, with minimal thermal impact [2]. This capability, natures from its fundamental principle of material erosion through a high-velocity jet of water containing abrasive particles, distinguishes it from traditional thermal cutting methods [3], making it particularly advantageous for materials sensitive to heat-affected zones (HAZ) [4] and for


A. Hamrol et al. (Eds.): MANUFACTURING 2026, LNME, pp. 243–258, 2026.
https://doi.org/10.1007/978-3-032-21701-1_18

applications requiring intricate geometries and near-net-shape manufacturing [5]. The "cold cutting" nature of AWJ technology eliminates thermal distortion, metallurgical alterations, and the generation of hazardous fumes often associated with laser or plasma disintegration, rendering it an environmentally conscious and technologically significant alternative. In addition, this method can also be used to disintegrate thermolabile or heat-sensitive materials [6].

The AWJ process technology is controlled by a complex interplay of numerous parameters [7], which are categorized into jet generation parameters (supply pressure p (MPa), d (mm) orifice diameter), abrasive characteristics (abrasive type, grain size distribution denoted by MESH number, abrasive mass flow rate $m_a$ (g/min)), and operational parameters (stand-off distance z (mm), v (mm/min) traverse speed, focusing tube diameter $d_f$ (mm)). Among these, the focusing tube diameter, a critical component responsible for collimating the abrasive-laden water jet into a coherent and high-energy stream, experiences significant wear due to the abrasive nature of the particles propelled at high velocities [8]. The wear of the focusing tube not only affects jet coherence and cutting efficiency over time but also contributes to operational costs and necessitates periodic replacement. This wear progression inherently influences the jet's geometry; a less worn, more collimated jet, often achieved with smaller initial focusing tube diameters, generally translates to increased geometrical precision in the cut kerf. Consequently, understanding and predicting focusing tube wear is crucial not only for cost management and operational planning but also for maintaining the dimensional accuracy of the machined parts [9].

Extensive research efforts have been devoted to understanding and modeling the mechanisms governing material removal in AWJ machining, both analytically [10] and statistically [11]. Analytical models, often based on energy transfer and erosion mechanics, have been proposed to predict material removal rates (MRR) and cutting depth as a function of process parameters [12]. These models, while providing valuable theoretical insights, often rely on simplifying assumptions and may struggle to accurately capture the nonlinear and synergistic effects of multiple interacting parameters across a wide range of operational and technological conditions [13]. Empirical modeling approaches, utilizing statistical techniques and the design of experiments, have also been widely adopted to establish relationships between input parameters and output responses. These methods offer practical predictive capabilities within the tested parameter space but may lack the generalizability and ability to extrapolate beyond the experimental boundaries. This is because statistical datasets and their correlations do not represent the natural response.

Despite the significant effort in the field of AWJ modeling, a significant research gap exists in the comprehensive and simultaneous prediction of wear rates (MRR, depth of cut, width of cut), specifically considering the combined influence of focusing tube diameter, abrasive flow rate, and abrasive MESH size for AISI 316 L stainless steel. Furthermore, the specific interaction between a smaller focusing tube diameter (0.5 mm), often employed for enhanced precision, and finer abrasive particles (200 MESH), utilized for improved surface finish, on both tool wear and cutting efficiency warrants dedicated investigation. The interaction between abrasive grain size [14] and mass flow rate [15] presents a critical area of investigation, especially considering the overall consumption of abrasive materials. Utilizing very fine abrasives at low flow rates offers a potential solution for resource conservation [16]. While such combinations are well-established in micro-machining applications, their efficacy in disintegrating materials like stainless steel, particularly in achieving specific thicknesses, could lead to significant savings in abrasive materials. From this perspective, this study specifically focuses on the cuttability of stainless steel under significantly reduced abrasive mass flow rates and with high MESH numbers, aiming to achieve both effective material removal and relatively good surface quality.

Therefore, this study is motivated by the need to develop more accurate and comprehensive models for AWJ machining of AISI 316 L, a material widely used in demanding applications requiring both precision and material integrity. This research explores the relationships between focusing tube diameter ($d_f = 0.5$ and 0.89 mm), abrasive flow rate ($m_a$), abrasive mesh size ($MESH$), and the traverse speed ($v$) on material removal rates, cutting depth, and width. This research aims to provide valuable insights for optimizing processes and managing tool life. The findings will enable manufacturers to select appropriate process parameters to maximize productivity, minimize operational costs associated with focusing tube replacement, and achieve desired cutting quality, thereby enhancing the overall efficiency and sustainability of AWJ machining for this important engineering material.

## 2 Material and Methods

### 2.1 Material

In the present study, stainless steel AISI 316 L was used which is commonly used in various industrial and medical applications. The material was supplied as a plate measuring 100 x 25 x 20 mm with a thickness of 20 mm. In addition to its application in various fields, it demonstrates resistance to corrosion in humid tap water environments, thereby improving the repeatability and reliability of experiments conducted using AWJ technology. The surface of all samples tested was mechanically ground to eliminate surface roughness and ensure consistent surface properties for all experimental runs. Mechanical grinding was performed using a Saphir 230 (ATM) metallographic grinding and polishing machine, employing progressively finer abrasives. The chemical composition of AISI 316 L in wt. % is as follows: Cr 16.63, Ni 10.00, Mo 2.04, Mn 1.26, Si 0.38, C 0.02, N 0.04, P 0.032, S 0.001, Fe rest [17].

### 2.2 Experimental Setup

The comparative tests were conducted on WJ1010-1Z-PJ5AX-45° - PreciseJet using the pump PTV JETS 2.2/60 for testing configuration $df = 0.5$ mm and $MESH$ 200, as well as on the high-pressure abrasive water jet device COBRA WJ4025-1Z-COBRA-PJ5AX-60° from PTV company with testing configuration $df = 0.89$ mm and MESH 80. The technological conditions are listed in Fig. 1 and Table 1. The study (Fig. 1) systematically varied focusing tube diameters (0.5 mm and 0.89 mm), traverse speeds (125, 250, and 500 mm/min), abrasive flow rates (5 to 50 g/min), and abrasive mesh sizes (80 and 200), while maintaining a constant supply pressure of 400 MPa and a cutting length of 25 mm. A structured experimental design (DoE) (Table 1), involving factorial combinations of these parameters as independent variables. Figure 2 shows the workpiece samples after the experiment for visual comparison.

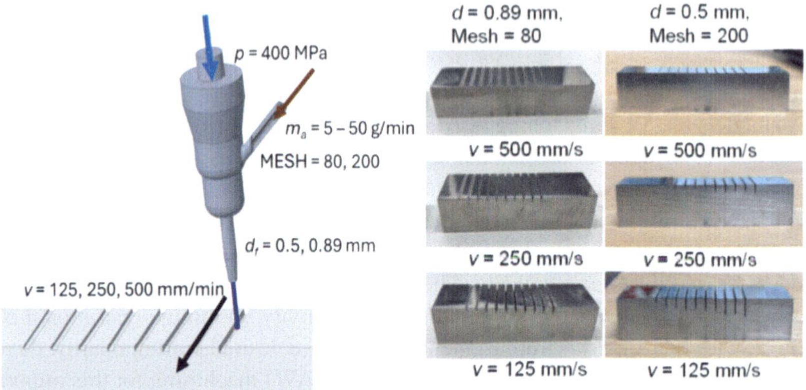


**Fig. 1.** Technological experimental setup and Pictorial representation of the generated groove on AISI 316 L sample with different experimental conditions as mentioned in Table 1.

## 2.3 Measurements

An Alicona Infinite Focus G5 optical non-contact 3D microscope was used to measure dependent variables such as groove depth (mm), width (mm), and material removal rate ($mm^3$/min). A total of 60 grooves generated under different experimental conditions, as shown in Fig. 1, were scanned using an ALICONA optical microscope. The entire groove length of 25 mm was scanned to obtain the volume of material removed during each experimental run. All the scanned data were imported into MountainsMap software, where data extraction from the scanned images was carried out. The groove depth and width were calculated from five equidistant cross-sectional profiles of the generated groove. This was done to ensure the statistical correctness of the measured values, as well as the deviation in the generated contours along the groove length. For the calculation of material removal rate, the volume removed, as measured by MountainsMap, was divided by the time taken for the jet to remove this volume. The time taken by the jet to generate the erosion groove of 25 mm length was 3 s, 6 s, and 12 s with traverse speeds of 500, 250, and 125 mm/min, respectively. The measured mean data, along with their deviations, were then plotted to understand the effect of the input parameters on the output responses.

**Table 1.** Experimental conditions.

| Constant Factors | | |
|---|---|---|
| Factors | Symbol | Level |
| Pressure | $p$ | 400 MPa |
| Standoff distance | $z$ | 2 mm |
| Angle of attack | $\phi$ | 90° |
| Abrasive type | | Barton Garnet |
| Variable factors | | |
| Configuration 1 | | |
| MESH | – | 80 |
| Focusing tube diameter | $d_f$ | 0.89 mm |
| Abrasive mass flow rate | $m_a$ | 5, 10, 15, 20, 25, 30, 35, 40, 45, 50 g/min |
| Traverse speed | $v$ | 125, 250, 500 mm/min |
| Configuration 2 | | |
| MESH | – | 200 |
| Focusing tube diameter | $d_f$ | 0.5 mm |
| Abrasive mass flow rate | $m_a$ | 5, 10, 15, 20, 25, 30, 35, 40, 45, 50 g/min |
| Traverse speed | $v$ | 125, 250, 500 mm/min |

# 3 Results and Discussion

## 3.1 Effect of Input Parameters on Groove depth

Figure 2 presents the influence of $m_a$ (5–50 g/min) and $v$ (125, 250, and 500 mm/min) on the groove depth achieved during AWJM of AISI 316 L stainless steel, using two different configurations of focusing tube diameter and abrasive MESH size: (a) $d_f$ = 0.89 mm with $MESH$ = 80, and (b) $d_f$ = 0.5 mm with $MESH$ = 200. In both cases, increasing the $m_a$ resulted in a corresponding rise in groove depth across all traverse speeds. This trend can be ascribed to the increased number of abrasive particles entering the stream, which in turn increases the kinetic energy of the jet, thereby generating deeper erosion grooves, as also reported by Hashish [18] and Momber and Kovacevic [19]. The material is removed through the impact of the abrasive-laden jet on the workpiece surface, which, upon impact, induces plastic deformation and localized work hardening, finally leading to the material failure. Each impact of high-velocity abrasive particles causes localized plastic indentation and produces microcraters. This continuous impact leads to strain accumulation and the initiation of microcracks at the edges of these craters, ultimately resulting in the detachment of small fragments from the surface. The observed mechanism is consistent with Finnie's [20] normal-impact erosion model and later AWJ studies by Hashish [18], which describe ductile erosion at 90° incidence as a process dominated by deformation and fatigue-driven material removal rather than shear cutting. Approximate 8.15- and 7.34 times increases in groove depth were achieved with $d_f = 0.5$ mm using MESH 200 and $d_f = 0.89$ mm with *MESH* 80, respectively, corresponding to a tenfold increase in the $m_a$ with a $v = 125$ mm/min. However, for a higher traverse speed ($v = 500$ mm/min), such a significant increase in the groove depth was not observed for either of the testing configurations. However, at higher abrasive flow rates, the increase in penetration depth tends to plateau, indicating that particle–particle collisions within the jet result in energy dissipation before impacting the work surface, a phenomenon known as the 'critical mass flow rate'.

Further, the influence of $v$ is also evident; for a constant $m_a$, the groove depth decreases as traverse speed increases. At lower traverse speeds ($v = 125$ mm/min), the jet interacts with a given region for a longer duration, resulting in a higher number of abrasive impacts per unit length and greater local erosion. Conversely, increasing the $v$ significantly reduces the groove depth. This is because a higher $v$ results in a shorter interaction time between the abrasive jet and any given point on the material surface. The material is exposed to fewer abrasive impacts per unit length, thereby decreasing the cumulative energy transfer and limiting the progression of material removal, a key factor in ductile material erosion. This relationship aligns with observations by Arola and Ramulu [21] and Akkurt et al. [22], who demonstrated that traverse speed significantly influences cutting depth through its impact on local energy input. Furthermore, a 6.7 times and 4.65 times increase in groove depth is observed with $d_f = 0.5$ mm using MESH 200 and $d_f = 0.89$ mm with *MESH* 80, respectively. This corresponds to a 4 times increase in $v$ while maintaining a constant $m_a = 50$ g/min.

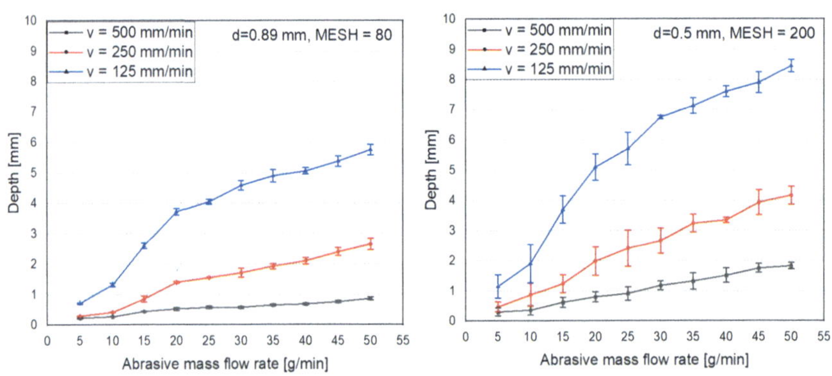


**Fig. 2.** Influence of abrasive mass flow rate ($m_a = 5$–$50$ g/min) and traverse speed ($v = 125, 250$ and $500$ mm/min) on groove depth with a) $d_f = 0.89$ mm, *MESH* = 80 and b) $d_f = 0.5$ mm, *MESH* = 200.

A comparison between the two configurations used in the current study reveals that the grooves generated with a smaller $d_f = 0.5$ mm and finer abrasives (*MESH* = 200) consistently produced deeper grooves under identical conditions. This improvement results from the higher kinetic energy density and improved focusing efficiency of the smaller jet, which concentrates the erosive action over a smaller area. The larger diameter provides greater stability to the jet; however, the coarser *MESH* 80 particles, though possessing greater individual kinetic energy attributable to their mass, are also more prone to fracture within the mixing tube and may lead to a slightly less focused jet due to the larger tube diameter [23]. Furthermore, the smaller $d_f$ produces a more coherent and higher-velocity jet, concentrating the kinetic energy over a smaller area [24]. However, the finer *MESH* 200 particles have lower individual kinetic energy but provide a higher number of cutting edges per unit of mass flow rate, resulting in more uniform momentum transfer and increasing the overall erosion rate. The deeper groove depths observed, compared to the previous configuration, demonstrate this trade-off. The fine abrasive and smaller tube may create a more precise groove, but the reduced momentum of the finer particles could lead to slightly shallower depths, especially at high traverse speeds where impact duration is limited. Similar effects of nozzle diameter and abrasive size on groove depth have been reported by Wang [25] and Hashish [26].

### 3.2 Effect of Input Parameters on Groove Width

Following the analysis of groove depth, Fig. 3 shows the variation of groove width with respect to $m_a$ (5–50 g/min) and $v$ (125, 250, and 500 mm/min) for the two experimental configurations. Unlike the depth behavior, the groove width remained nearly constant across the entire range of $m_a$ and $v$. For the configuration with $d_f = 0.89$ mm, the groove width was approximately 1.0 mm, while for $d_f = 0.5$ mm, it was about 0.6 mm, indicating that the width is primarily governed by the jet diameter rather than the process parameters. This observation agrees with prior findings by Hashish [18] and Wang [25], who noted that the lateral extent of material removal in AWJM is predominantly determined by the jet's hydrodynamic envelope and focusing tube geometry. The minimal influence of $m_a$ on groove width can be attributed to the fact that variations in abrasive loading mainly affect the jet's erosive intensity rather than its spread. Increasing abrasive concentration enhances particle density but does not significantly alter the overall jet structure, which remains confined by the focusing tube. Similarly, v has little effect on the width, as it primarily dictates the interaction time along the cutting direction, rather than the lateral jet distribution. The mean widths across the entire $m_a$ range (5–50 g/min) were 1.023 mm, 1.017 mm, and 1.013 mm for $v = 500$, 250, and 125 mm/min, respectively, with $d_f = 0.89$ mm and *MESH* 80 configuration. However, for $d_f = 0.5$ mm and *MESH* 200 configuration, mean widths were 0.627 mm, 0.619 mm, and 0.606 mm at $v = 500$, 250, and 125 mm/min, respectively. Interestingly, in both testing configurations, an inverse relationship between the $v$ and the width is observed, which can be attributed to the imperfect interaction of the abrasive jet at higher $v$ compared to lower $v$, resulting in a wider groove. However, as Momber and Kovacevic [19] suggested, at normal incidence, the jet maintains an axisymmetric erosion footprint, and the resulting width corresponds closely to the effective jet diameter at the standoff distance used.

A comparison between the two configurations further confirms that narrower nozzles produce proportionally smaller groove widths. The reduction from approximately 1.02 mm for $d_f = 0.89$ mm to 0.62 mm for $d_f = 0.5$ mm is consistent with the geometric scaling of the focusing tube outlet. The use of finer abrasives (*MESH* = 200) also contributes marginally to maintaining a more coherent and concentrated jet, which minimizes lateral divergence. The mean width ratio between the two configurations is ≈1.66, demonstrating that the nozzle geometry is the dominant factor controlling lateral spread. However, the measured ratio is somewhat smaller than the simple geometric ratio of the focusing tube (0.89/0.50 = 1.78), which is consistent with known jet expansion and standoff effects that reduce strict geometric scaling. This behavior also aligns with observations by Hashish [26], who reported that finer abrasive particles improve jet focus and stability, leading to enhanced dimensional accuracy. Finally, this difference in the mean width highlights that groove width is primarily determined by the $d_f$, making it a predictable and consistent geometric feature that is largely independent of the kinetic parameters ($m_a$ and $v$) once the jet is stable. This finding provides a straightforward way to control precision, where using a smaller focusing tube configuration is clearly beneficial for creating narrower features.

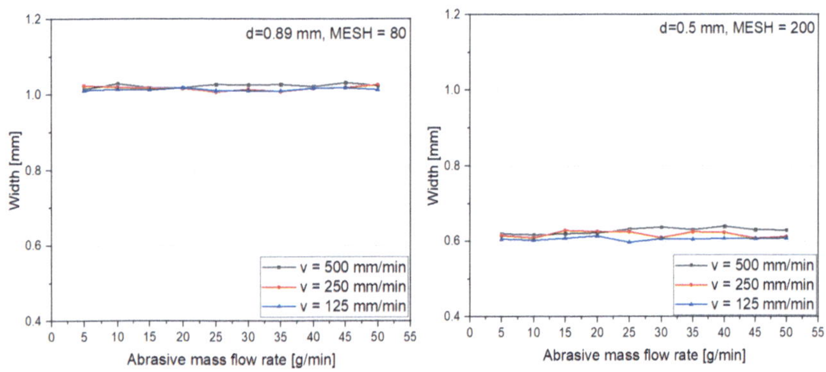


**Fig. 3.** Influence of abrasive mass flow rate ($m_a$ = 5–50 g/min) and traverse speed ($v$ = 125, 250 and 500 mm/min) on groove width with a) $d_f$ = 0.89 mm, *MESH* = 80 and b) $d_f$ = 0.5 mm, *MESH* = 200.

### 3.3 Effect of Input Parameters on MRR

The influence of $m_a$ and $v$ on the MRR for AISI 316 L is presented in Fig. 4 for the two nozzle configurations: (a) $d_f$ = 0.89 mm with *MESH* = 80, and (b) $d_f$ = 0.5 mm with *MESH* = 200. The results indicate that MRR increases almost linearly with the $m_a$ for all traverse speeds. At lower $m_a$ (below ~10 g/min), MRR values tend to be relatively low due to inadequate abrasive loading within the jet, which consequently leads to limited erosion energy per unit time. As the $m_a$ exceeds this threshold, there is a sharp increase in MRR, indicating that more abrasive particles are involved in the cutting process and that there is a more efficient transfer of kinetic energy to the surface, thereby accelerating the erosion mechanism of plastic deformation and micro-chipping. However, at the higher end of the tested range (≥40 g/min), the growth rate of MRR begins to taper slightly, suggesting that excessive abrasive concentration leads to particle–particle collisions within the focusing tube and partial energy dissipation prior to impact, consistent with the findings of Hashish [18] and Momber and Kovacevic [19].

Traverse speed inversely affects MRR at any given $m_a$, with lower $v$ resulting in higher MRR values. This can be attributed to the increased exposure time of the jet over a given area, which allows more abrasive impacts and cumulative plastic deformation, leading to higher localized erosion. At a slower $v$, even though the jet contacts a unit area for a shorter duration, the resulting groove depth is significantly greater (as shown in Fig. 3), which leads to a higher MRR [21]. Conversely, at higher $v$ with faster linear movement, the substantially reduced groove depth negates the advantage of increased speed, resulting in a lower MRR. Quantitatively, at $d_f = 0.89$ mm and $MESH = 80$, the maximum MRR achieved at $m_a = 50$ g/min was approximately 478 mm$^3$/min for $v = 125$ mm/min, compared to 445 mm$^3$/min for $v = 250$ mm/min and 359 mm$^3$/min for $v = 500$ mm/min. A similar but slightly enhanced trend was observed for the finer abrasive configuration ($d_f = 0.5$ mm, $MESH = 200$), where MRR reached nearly 589 mm$^3$/min (23% increase) at the same flow rate ($m_a = 50$ g/min) and lowest traverse speed ($v = 125$ mm/min). These results indicate that the finer abrasive particles ($MESH = 200$) and narrower focusing tube promote better momentum transfer and more efficient material dislodgment, as also reported by Wang [25]. The observed MRR behavior can be explained mechanistically by considering the energy distribution within the abrasive water jet under perpendicular impact. When $m_a$ is increased, the total mass of particles per unit time grows, enhancing the overall kinetic energy flux striking the surface. For ductile materials such as AISI 316 L, which undergo predominantly deformation-driven erosion at normal incidence, this additional energy contributes to more intense plastic deformation and subsequent fatigue failure of surface asperities, as described by Finnie [20]. At the same time, excessively high abrasive concentrations reduce particle acceleration efficiency within the mixing chamber, resulting in slightly lower particle velocities and diminishing returns in MRR at the upper flow limits.

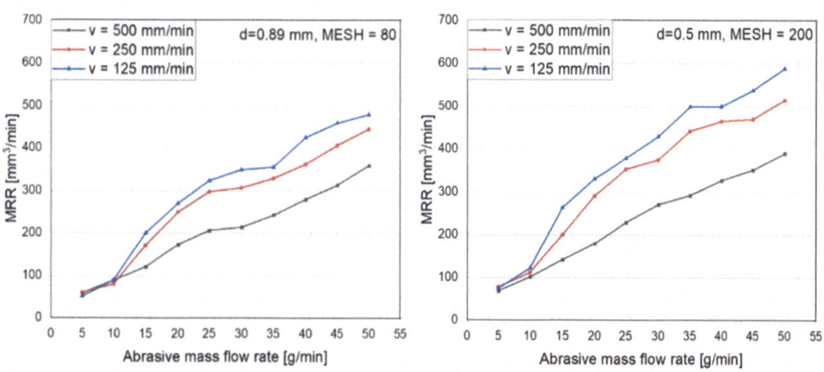


**Fig. 4.** Influence of abrasive mass flow rate ($m_a = 5$–$50$ g/min) and traverse speed ($v = 125$, 250 and 500 mm/min) on groove width with a) $d_f = 0.89$ mm, $MESH = 80$ and b) $d_f = 0.5$ mm, $MESH = 200$.

Between the two tested configurations, the $d_f = 0.5$ mm, $MESH = 200$ combination consistently achieved higher MRR values across all flow rates and speeds. The MRR is the product of depth, width, and speed. While with $d_f = 0.89$ mm and $MESH$ 80, a wider and slightly deeper groove is achieved, the significantly higher depth provided by $d_f = 0.5$ mm with *MESH* 200 ultimately results in a configuration that yields a greater total volume removed per minute. This makes the setup with the finer abrasive and smaller nozzle the more efficient choice for maximizing volumetric removal on AISI 316 L. This improvement can be attributed to the superior focusing of the smaller-diameter jet, resulting in higher power density and more effective utilization of impact energy. Moreover, the finer abrasives produce smoother energy distribution and reduce rebound losses upon impact. Together, these factors enhance erosion efficiency despite the jet's smaller cross-sectional area. This finding reinforces earlier conclusions that nozzle design and abrasive size optimization are key to maximizing process efficiency in AWJM of stainless steels.

Overall, the MRR trends confirm that process productivity in AWJM is primarily influenced by $m_a$, $v$, and nozzle configuration. While groove width remains nearly constant and depth increases with higher energy input, the combined effect of these parameters is most evident in the MRR. As a result, achieving high MRR without compromising dimensional accuracy requires a balanced approach that involves moderate v, an optimized $m_a$ (typically around 30–40 g/min for these conditions), and a well-focused jet produced by smaller $d_f$ and finer abrasives. These findings are consistent with the optimization criteria proposed by Hashish [18] and later validated experimentally by Wang [25] for ductile alloys under normal jet impingement.

## 4 Statistical Analysis

### 4.1 Results Obtained with Configuration 1 ($d_f$ = 0.89 mm and *MESH* 80)

Table 2 presents the ANOVA analysis for all measured responses, including groove depth, width, and MRR, to validate the experimental observations and quantify the influence of the input parameters. This formal statistical proof is essential for confirming the reliability and predictability of the AWJ model for AISI 316 L material within the experimental framework domain. The ANOVA specifically determined the statistical significance and the percentage contribution of independent variables such as v and $m_a$ on the main output responses. The analysis used a 95% confidence level, with statistical significance set at a P-value less than 0.05. ANOVA analysis confirms the strong statistical dependency of the input parameters on the selected responses and also closely aligns with the trends discussed in the previous sections.

For groove depth, the model explains 87.70% of the total variance ($R^2 = 87.70\%$), with $v$ showing the highest contribution of 62.71% ($p < 0.001$) and $m_a$ contributing 24.99% ($p = 0.005$). The dominance of $v$ reflects its strong control over jet-material interaction time and, consequently, the erosion depth. This statistical trend is consistent with the experimental results (Fig. 2a), where a significant increase in depth was observed with decreasing $v$. The influence of $m_a$, although secondary, still plays a statistically significant role by increasing the number of effective abrasive impacts, thereby enhancing localized micro-cutting and plowing actions. The small model error (12.30%) and an adjusted $R^2$ of 80.18% demonstrate a good model fit, while the predicted $R^2$ of 65.84% indicates moderate predictive reliability, indicating that the experimental variation was well captured.

In contrast, the groove width shows weaker sensitivity to the process parameters, with the model explaining 40.64% of the variation ($R^2 = 40.64\%$). This indicates that the groove width generated remains nearly constant regardless of $m_a$ and $v$. Only $v$ shows a statistically significant influence ($p = 0.001$), although its contribution remains modest. This behavior is consistent with the experimental data (Fig. 3a), where the groove width remained ~1.0 mm across all parameter combinations. The limited variability suggests that the lateral spread of the jet is primarily governed by the $d_f$ and standoff distance rather than the process parameters. The lower $R^2$ value indicates that a large portion of the variance arises from inherent experimental variability and the natural stability of the jet diameter. These results reinforce the earlier conclusion that, under perpendicular impact conditions, the groove width remains almost constant, being a geometrically controlled characteristic of the jet rather than a process-driven response, for perpendicular AWJ impacts.

The MRR exhibits a strong dependence on the selected parameters, with the model accounting for 96.88% of the variance ($R^2 = 96.88\%$) and an adjusted $R^2$ of 94.97%. The $m_a$ is the most dominant factor, contributing 87.12% of the total variance, while v contributes 9.76% (both $p < 0.001$). This high degree of dependence aligns well with the experimental MRR results shown in Fig. 4(a), where the MRR increases almost linearly with $m_a$ across all v, indicating that the abrasive particle flux and kinetic energy density predominantly control volumetric material removal. These findings align with the mechanistic framework proposed by Arola and Ramulu [21], where greater abrasive flux leads to increased impact frequency and cumulative energy transfer, resulting in higher erosion rates. The small residual error (3.12%) and high predicted $R^2$ (91.32%) demonstrate excellent model reliability and predictive accuracy.

### 4.2 Results Obtained with Configuration 2 ($d_f = 0.5$ mm and *MESH* 200)

ANOVA was also performed for experimental observations for the configuration with $d_f = 0.5$ mm and $MESH = 200$ for all the studied responses. The statistical results, summarized in Table 3, indicate that the model accurately reflects the impact of $m_a$ and $v$ on all machining responses, exhibiting consistency with the experimental trends outlined in the preceding sections.

For groove depth, the ANOVA results indicate that the model explains 89.41% of the total variance ($R^2 = 89.41\%$), demonstrating a slightly higher predictability compared to the previous experimental configuration, with $v$ and $m_a$ both exhibiting highly significant effects ($p < 0.001$). Among these, $v$ contributes 58.20% of the total variation with a high F-Value of 49.49, confirming its strong statistical significance and dominant role in controlling the jet exposure time and, consequently, the penetration depth. The $m_a$ contributes 31.21%, reflecting its influence on the number of active cutting edges and the local energy density of the jet. These statistical results are consistent with the previously observed trend in Fig. 2(b), where the groove depth increased as the $v$ decreased and the $m_a$ increased. Moreover, it also strongly validates that while both parameters are important for increasing groove depth, the duration of jet-material interaction, controlled by the $v$, primarily determines the final depth of the cut. The small residual error (10.59%) and adjusted $R^2$ value of 82.95% further affirm the model's reliability and its strong

**Table 2.** ANOVA analysis for groove depth, width, and MRR for experimental configuration 1.

Groove Depth

| Source | DF | Seq SS | Contribution | Adj SS | Adj MS | F-Value | P-Value |
|---|---|---|---|---|---|---|---|
| Model | 11 | 77.94 | 87.70% | 77.94 | 7.0854 | 11.67 | 0.000 |
| Linear | 11 | 77.94 | 87.70% | 77.94 | 7.0854 | 11.67 | 0.000 |
| $v$ | 2 | 55.73 | 62.71% | 55.73 | 27.8646 | 45.89 | 0.000 |
| $m_a$ | 9 | 22.21 | 24.99% | 22.21 | 2.4678 | 4.06 | 0.005 |
| Error | 18 | 10.93 | 12.30% | 10.93 | 0.6072 | | |
| Total | 29 | 88.87 | 100.00% | | | | |

Model Summary

| S | R-sq | R-sq(adj) | PRESS | R-sq(pred) | AICc | BIC |
|---|---|---|---|---|---|---|
| 0.779251 | 87.70% | 80.18% | 30.3616 | 65.84% | 103.60 | 99.06 |

Groove Width

| Source | DF | Seq SS | Contribution | Adj SS | Adj MS | F-Value | P-Value |
|---|---|---|---|---|---|---|---|
| Model | 2 | 0.00047 | 40.64% | 0.000476 | 0.000238 | 9.24 | 0.001 |
| Linear | 2 | 0.00047 | 40.64% | 0.000476 | 0.000238 | 9.24 | 0.001 |
| $v$ | 2 | 0.00047 | 40.64% | 0.000476 | 0.000238 | 9.24 | 0.001 |
| Error | 27 | 0.00069 | 59.36% | 0.000695 | 0.000026 | | |
| Total | 29 | 0.00117 | 100.00% | | | | |

Model Summary

| S | R-sq | R-sq(adj) | PRESS | R-sq(pred) | AICc | BIC |
|---|---|---|---|---|---|---|
| 0.0050740 | 40.64% | 36.24% | 0.0008582 | 26.71% | −225.44 | −221.44 |

Material removal rate

| Source | DF | Seq SS | Contribution | Adj SS | Adj MS | F-Value | P-Value |
|---|---|---|---|---|---|---|---|
| Model | 11 | 465863 | 96.88% | 465863 | 42351.2 | 50.75 | 0.000 |
| Linear | 11 | 465863 | 96.88% | 465863 | 42351.2 | 50.75 | 0.000 |
| $v$ | 2 | 46927 | 9.76% | 46927 | 23463.7 | 28.11 | 0.000 |
| $m_a$ | 9 | 418936 | 87.12% | 418936 | 46548.4 | 55.78 | 0.000 |
| Error | 18 | 15022 | 3.12% | 15022 | 834.6 | | |
| Total | 29 | 480885 | 100.00% | | | | |

Model Summary

| S | R-sq | R-sq(adj) | PRESS | R-sq(pred) | AICc | BIC |
|---|---|---|---|---|---|---|
| 28.8888 | 96.88% | 94.97% | 41728.2 | 91.32% | 320.37 | 315.83 |

agreement with the physical erosion mechanisms of perpendicular AWJ impingement on ductile materials such as AISI 316 L.

In contrast, the groove width again shows relatively low statistical sensitivity to process parameters. The regression model for groove width explains 63.18% of the variation ($R^2 = 63.18\%$), with $v$ as the only statistically significant factor ($p < 0.001$). The smaller focusing tube yielded an average groove width of ~0.6 mm with minimal variation across all parameter combinations (Fig. 3b). However, considering the previously discussed physical constraint that the $d_f$ inherently determines groove width, the large residual error percentage (Error contribution of 36.82%) indicates that a significant portion of the variation in observed groove width results from random and unmodeled factors rather than a strong, predictable relationship with $v$ and $m_a$. This statistical result supports the conclusion that the groove width is primarily a function of the fixed hardware ($d_f$) and less sensitive to the kinetic process parameters, as noted by Hashish [18] and Momber and Kovacevic [19]. The limited influence of abrasive flow rate and traverse speed on width aligns with the confined jet structure produced by the $d_f = 0.5$ mm and finer abrasives ($MESH = 200$), which maintain a coherent and narrowly focused jet stream. However, the slightly higher $R^2$ compared to the earlier configuration suggests improved jet coherence and reduced turbulence, which produce more repeatable and stable widths.

The final studied response, which is the MRR, exhibits the strongest correlation with the process parameters, as reflected by the very high coefficient of determination ($R^2 = 96.48\%$) and an adjusted $R^2$ of 94.33%. Both $m_a$ and $v$ are statistically significant ($p < 0.001$), with $m_a$ contributing 82.27% of the total variation and $v$ accounting for 14.21%. This quantifies the earlier observed behavior in Fig. 4(b), where MRR increased almost linearly with $m_a$ and decreased with increasing $v$. The large contribution of $m_a$ highlights its dominant influence on process productivity, as higher abrasive loading increases the number of impacting particles and the overall kinetic energy flux. The low residual error (3.52%) and high predicted $R^2$ (90.22%) confirm the model's robustness and predictive capability. These findings agree with the conclusions drawn by Wang [25] and Hashish [18], who observed that MRR in AWJM is primarily controlled by the abrasive feed rate and energy density, particularly in finely focused jets.

The comparative statistical analysis of both configurations clearly indicates that the fine focusing tube ($d_f = 0.5$ mm, $MESH = 200$) configuration exhibited higher statistical sensitivity and stronger linear correlations for depth and MRR ($R^2 \approx 89$–$96\%$), indicating more stable and predictable cutting behavior due to enhanced jet coherence and energy concentration. Conversely, the larger nozzle ($d_f = 0.89$ mm, $MESH = 80$) exhibited slightly lower predictability in depth ($R^2 = 87.7\%$) and weaker width dependence ($R^2 = 40.6\%$), which can be attributed to the broader jet and higher turbulence near the impact zone. Across both setups, MRR consistently exhibits the strongest linear dependence on $m_a$, confirming that the abrasive particle flux and velocity distribution govern the overall erosion volume. Groove depth is primarily dictated by the combined effects of $v$ and $m_a$, while groove width remains largely invariant, being determined by nozzle geometry and hydrodynamic jet spread. These outcomes align well with established erosion models and provide quantitative validation for the mechanistic interpretation proposed earlier in this study.

Overall, the ANOVA results substantiate that the process parameters exert statistically significant effects on material removal and penetration depth during perpendicular AWJ machining of AISI 316 L, while lateral dimensions remain geometrically constrained. This separation of control variables offers valuable guidance for process optimization, allowing precise control of depth and MRR through $m_a$ and $v$, without compromising dimensional accuracy in groove width. These differences highlight that finer abrasives and smaller nozzles result in more controlled and repeatable erosion, which is particularly beneficial for precision machining applications.

**Table 3.** ANOVA analysis for groove depth, width, and MRR for experimental configuration 2.

Groove Depth

| Source | DF | Seq SS | Contribution | Adj SS | Adj MS | F-Value | P-Value |
|---|---|---|---|---|---|---|---|
| Model | 11 | 162.48 | 89.41% | 162.48 | 14.771 | 13.82 | 0.000 |
| Linear | 11 | 162.48 | 89.41% | 162.48 | 14.771 | 13.82 | 0.000 |
| $v$ | 2 | 105.77 | 58.20% | 105.77 | 52.883 | 49.49 | 0.000 |
| $m_a$ | 9 | 56.72 | 31.21% | 56.72 | 6.302 | 5.90 | 0.001 |
| Error | 18 | 19.24 | 10.59% | 19.24 | 1.069 | | |
| Total | 29 | 181.72 | 100.00% | | | | |

Model Summary

| S | R-sq | R-sq(adj) | PRESS | R-sq(pred) | AICc | BIC |
|---|---|---|---|---|---|---|
| 1.03376 | 89.41% | 82.95% | 53.4326 | 70.60% | 120.55 | 116.02 |

Groove Width

| Source | DF | Seq SS | Contribution | Adj SS | Adj MS | F-Value | P-Value |
|---|---|---|---|---|---|---|---|
| Model | 2 | 0.00223 | 63.18% | 0.002231 | 0.001116 | 23.17 | 0.000 |
| Linear | 2 | 0.00223 | 63.18% | 0.002231 | 0.001116 | 23.17 | 0.000 |
| $v$ | 2 | 0.00223 | 63.18% | 0.002231 | 0.001116 | 23.17 | 0.000 |
| Error | 27 | 0.00130 | 36.82% | 0.001300 | 0.000048 | | |
| Total | 29 | 0.00353 | 100.00% | | | | |

Model Summary

| S | R-sq | R-sq(adj) | PRESS | R-sq(pred) | AICc | BIC |
|---|---|---|---|---|---|---|
| 0.0069395 | 63.18% | 60.46% | 0.0016052 | 54.55% | −206.66 | −202.65 |

Material removal rate

| Source | DF | Seq SS | Contribution | Adj SS | Adj MS | F-Value | P-Value |
|---|---|---|---|---|---|---|---|
| Model | 11 | 673591 | 96.48% | 673591 | 61236 | 44.86 | 0.000 |
| Linear | 11 | 673591 | 96.48% | 673591 | 61236 | 44.86 | 0.000 |
| $v$ | 2 | 99195 | 14.21% | 99195 | 49597 | 36.33 | 0.000 |
| $m_a$ | 9 | 574397 | 82.27% | 574397 | 63822 | 46.76 | 0.000 |
| Error | 18 | 24570 | 3.52% | 24570 | 1365 | | |
| Total | 29 | 698162 | 100.00% | | | | |

Model Summary

| S | R-sq | R-sq(adj) | PRESS | R-sq(pred) | AICc | BIC |
|---|---|---|---|---|---|---|
| 36.9461 | 96.48% | 94.33% | 68250.6 | 90.22% | 335.13 | 330.59 |

## 5 Conclusion

The present study examines the influence of input parameters, including traverse speed, abrasive flow rate, focusing tube diameter, abrasive MESH size, and their interplay, on the cutting groove depth, width, and Material Removal Rate in stainless steel AISI 316 L samples. Two distinct experimental configurations, (a) with $d_f = 0.89$ and $MESH = 80$, and (b) with $d_f = 0.5$ and $MESH = 200$, were evaluated. The key conclusions of this investigation are summarized as follows.

Groove depth and MRR increased with higher ma and lower v, while groove width remained nearly constant. ANOVA confirmed that v was the most influential factor for depth (≈58–63% contribution), and ma was dominant for MRR (≈82–87%), both statistically significant ($p < 0.001$).

1. Groove width showed negligible sensitivity to selected process parameters, remaining primarily governed by nozzle geometry. Mean widths closely matched the focusing tube diameters (≈1.0 mm for df = 0.89 mm and ≈0.6 mm for df = 0.5 mm), indicating high dimensional consistency.
2. The finer focusing tube (df = 0.5 mm, MESH 200) produced deeper grooves, higher MRR, and better statistical predictability ($R^2$ up to 96%), owing to improved jet coherence and energy concentration. In contrast, the larger nozzle (df = 0.89 mm, MESH 80) exhibited greater turbulence and lower precision.
3. The combined experimental and statistical findings demonstrate that groove depth and MRR can be effectively controlled via ma and v, whereas groove width is mainly geometrically fixed and governed by df.

Future work can focus on developing predictive models using machine learning to optimize AWJ parameters for various materials. Detailed surface integrity and microstructural studies can further clarify the influence of perpendicular jet erosion on subsurface deformation and residual stresses. Extending the study to angled jet impacts and multi-pass machining could also enhance process versatility and industrial applicability.

**Acknowledgement.** The authors extend their acknowledgment to the financial support of the European Union under the REFRESH-Research Excellence For Region Sustainability and High-tech Industries project number CZ.10.03.01/00/22_003/0000048 via the Operational Programme Just Transition

## References

1. Fard, M.G., Nag, A., Petru, J., Hloch, S.: Towards sustainable precision: A review of water jet meso and micromachining. Results Eng. **27**, 106447 (2025)
2. Perzel, V., et al.: Measurement of thermal emission during cutting of materials using abrasive water jet. Therm. Sci. **21**, 2197–2203 (2017)
3. Harničárová, M., et al.: Techno-economical comparison of cutting material by laser, plasma and oxygen. Tehnički vjesnik. **19**, 813–817 (2012)
4. Krajcarz, D.: Comparison metal water jet cutting with laser and plasma cutting. Procedia Eng. **69**, 838–843 (2014)
5. Brandstätter, M., Wöhrle, A., Kaufeld, M., Pude, F., Nag, A., Hloch, S.: The quantitative evaluation of the cutting surface quality levels in abrasive water jet cutting by measurement of the representative striation mark displacement. Int. J. Adv. Manuf. Technol. **120**, 1625–1642 (2022)
6. Ramulu, M., Arola, D.: Water jet and abrasive water jet cutting of unidirectional graphite/epoxy composite. Composites. **24**, 299–308 (1993)
7. Perec, A.: Research into the disintegration of abrasive materials in the abrasive water jet machining process. Materials. **14**, 3940 (2021)

8. Hreha, P., et al.: Monitoring of focusing tube wear during abrasive waterjet (AWJ) cutting of AISI 309. Meta. **53**, 533–536 (2014)
9. Nguyen, T., Wang, J.: A review on the erosion mechanisms in abrasive waterjet micromachining of brittle materials. Int. J. Extreme Manuf. **1**, 012006 (2019)
10. Hlaváč, L.M.: Investigation of the abrasive water jet trajectory curvature inside the kerf. J. Mater. Process. Technol. **209**, 4154–4161 (2009)
11. Perec, A., Radomska-Zalas, A., Fajdek-Bieda, A., Pude, F.: Process optimization by applying the response surface methodology (rsm) to the abrasive suspension water jet cutting of phenolic composites. Facta Universitatis: Series Mechanical Engineering (2022)
12. Hashish, M.: On the modeling of surface waviness produced by abrasive-waterjets. In: Jet Cutting Technology, pp. 17–34. Springer (1992)
13. Ganovska, B., et al.: Design of the model for the on-line control of the AWJ technology based on neural networks. Indian J. Eng. Mater. Sci. **23**(4), 279–287 (2016)
14. Khan, A.A., Haque, M.M.: Performance of different abrasive materials during abrasive water jet machining of glass. J. Mater. Process. Technol. **191**, 404–407 (2007)
15. Hreha, P., et al.: Impact of abrasive mass flow rate when penetrating into a material on its vibration. Tehnički vjesnik. **17**, 475–480 (2010)
16. Sabarinathan, P., Annamalai, V.E., Rajkumar, K.: Sustainable application of grinding wheel waste as abrasive for abrasive water jet machining process. J. Clean. Prod. **261**, 121225 (2020)
17. Hloch, S., et al.: Erosion development in AISI 316L stainless steel under pulsating water jet treatment. Eng. Sci. Technol. Int. J. **50**, 101630 (2024)
18. Hashish, M.: A modeling study of metal cutting with abrasive waterjets. J. Eng. Mater. Technol. **106**, 88–100 (1984). https://doi.org/10.1115/1.3225682
19. Momber, A.W., Kovacevic, R.: Principles of abrasive water jet machining. Springer Science & Business Media (2012)
20. Finnie, I.: Erosion of surfaces by solid particles. Wear. **3**(2), 87–103 (1960)
21. Arola, D., Ramulu, M.: Material removal in abrasive waterjet machining of metals surface integrity and texture. Wear. **210**, 50–58 (1997)
22. Akkurt, A., Kulekci, M.K., Seker, U., Ercan, F.: Effect of feed rate on surface roughness in abrasive waterjet cutting applications. J. Mater. Process. Technol. **147**, 389–396 (2004)
23. Perec, A., Pude, F., Grigoryev, A., Kaufeld, M., Wegener, K.: A study of wear on focusing tubes exposed to corundum-based abrasives in the waterjet cutting process. Int. J. Adv. Manuf. Technol. **104**, 2415–2427 (2019)
24. Wang, J.: Abrasive waterjet machining of polymer matrix composites–cutting performance, erosive process and predictive models. Int. J. Adv. Manuf. Technol. **15**, 757–768 (1999)
25. Wang, J.: Abrasive waterjet machining of engineering materials. Trans Tech Publications Ltd (2003)
26. Hashish, M.: Characteristics of surfaces machined with abrasive-waterjets. J. Eng. Mater. Technol. **113**, 354–362 (1991)

# Current Approaches in Software Support for Industrial Ergonomics

Darina Dupláková, Matej Cichý(✉), Ján Dupľák, and Viktória Tutokyová

Faculty of Manufacturing Technologies with a Seat in Presov, Technical University of Košice, 080 01 Prešov, Slovakia
matej.cichy@tuke.sk

**Abstract.** Industrial ergonomics plays a fundamental role in improving the safety, health and well-being of workers. At the same time, it ensures the efficiency and sustainability of production processes. Traditional tools and approaches based on manual assessment methods, such as RULA or REBA, have their place, but in a dynamically changing production environment, they prove to be time-consuming, subjective and limited. The advent of digitalisation and the Industry 4.0 concept have stimulated the intensive development of software solutions that use digital human modelling, sensor technologies or computer vision. These tools enable more accurate, faster and often more predictive assessment of working conditions as early as the design stage. Current trends include the integration of digital twins with augmented and virtual reality, support systems for human–robot collaboration, and multi-factor ergonomic assessments that consider not only physical but also cognitive and user-related aspects. Discussions on limitations highlight the need for validation of new technologies in real-world conditions and challenges related to software interoperability. Modern approaches thus shift ergonomics from a supplementary tool to a fundamental component of digital production design, contributing not only to the protection of workers' health but also to the long-term competitiveness and sustainability of the manufacturing industry.

**Keywords:** Industrial ergonomics · Computer vision · Digital human modelling · Human-robot collaboration · Predictive ergonomic analysis

## 1 Introduction

Ergonomics in the manufacturing environment is becoming an increasingly important element in ensuring worker health protection, increasing production efficiency and supporting the long-term sustainability of industrial systems [1, 2]. Traditional approaches, based mainly on manual assessment methods and observation, have played an important role in the past, but in today's highly dynamic and time-efficient manufacturing environment, they are proving to be insufficient. They are often subjective, time-consuming and only partially reflect the complex nature of modern production processes, in which physical, cognitive and organisational aspects of work intersect [3, 4]. With the increasing level of automation and digitisation of production systems, the nature of work is also


A. Hamrol et al. (Eds.): MANUFACTURING 2026, LNME, pp. 259–270, 2026.
https://doi.org/10.1007/978-3-032-21701-1_19

changing, leading to the need for new methods of ergonomic assessment that can respond to changing working conditions in real time. Modern production environments require interconnection between humans, machines and information systems, with ergonomics becoming a key link in this interconnection. At the same time, the understanding of ergonomics is expanding from a narrow focus on physical strain to a broader framework that also includes cognitive, emotional and organisational aspects of work. This shift leads to the need for a multidisciplinary approach, combining knowledge from biomechanics, psychology, data analytics and machine learning, creating space for a more effective and accurate assessment of human interaction with the work environment (Fig. 1).

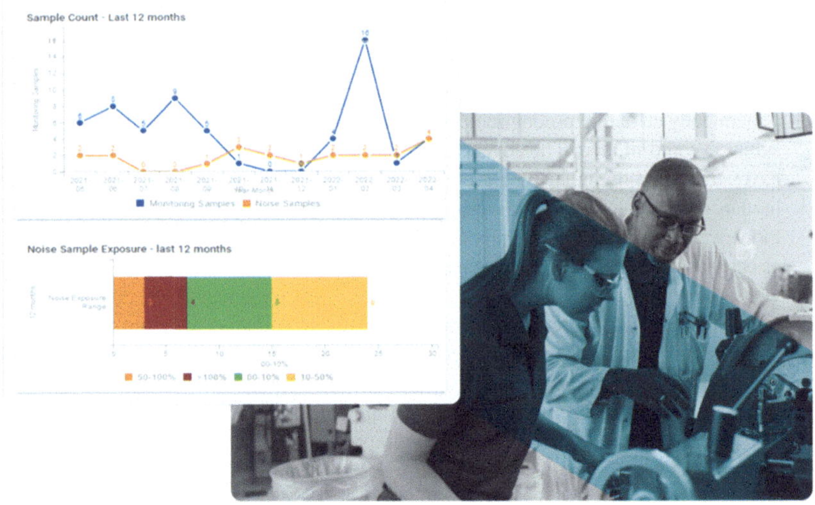


**Fig. 1.** Overview of key elements in industrial ergonomics – integration of workplace design, posture evaluation and digital analysis [5].

The development of digitalisation and the advent of the Industry 4.0 concept have opened up new possibilities in the field of software support for industrial ergonomics. The emergence of new technologies such as digital twins, the Internet of Things (IoT), artificial intelligence (AI) and augmented and virtual reality (AR/VR) has enabled a fundamental change in the way ergonomic analyses are performed, interpreted and applied. Modern approaches not only allow work situations to be visualised and simulated, but also enable the behaviour of workers to be dynamically monitored and evaluated in real time. Digital human modelling, the integration of sensor systems and the use of virtual and augmented reality make it possible to simulate work situations in detail, predict risk factors and propose preventive measures even before production processes are implemented. Standardisation and interoperability of data between different software platforms also play an important role in this process, enabling the integration of ergonomic analyses directly into the digital design of the production system. This makes it possible to identify non-ergonomic elements and optimise workplaces as early as the design phase without the need for physical prototypes, which significantly saves time and costs. Thanks to these technologies, ergonomics is moving from being a supplementary analytical tool to a strategic component of the design and management of production systems [6, 7].

Current approaches to software ergonomics go beyond the assessment of physical strain. Increasing attention is being paid to cognitive factors, human- nd automated technologies, and the integration of digital twins, which enable continuous monitoring and adaptive control of working conditions in real time. This trend clearly reflects the transformation of ergonomics towards a human-centric approach that takes into account not only safety but also the well-being and long-term performance of workers. The increased emphasis on people as the central element of the production system is leading to the formation of so called intelligent ergonomics, which uses data from real operations for feedback learning and optimisation of working conditions. This is also changing the role of ergonomics – from an evaluation and control tool to a proactive system that supports decision-making and innovation in industrial design. The result is a gradual transformation of ergonomics into an active element of digital manufacturing engineering. Modern software solutions now make it possible not only to identify risks, but also to design optimal working conditions that promote health protection, reduce the risk of musculoskeletal disorders and contribute to smoother and more efficient production processes. In this context, it is clear that ergonomics can no longer be seen merely as a supporting tool, but as an integral part of technological innovation and the sustainable development of industrial systems [8, 9].

## 2 Current State of Software Support in Ergonomics

Software tools are used in industrial ergonomics, which can be divided into several main groups. These tools differ in terms of their approach, level of detail, and possibilities for integration into production processes. Traditionally, the most widespread systems are Digital Human Modelling (DHM) systems. Tools such as Tecnomatix Jack, CATIA/DELMIA, and RAMSIS enable the simulation of human movements, the analysis of reach and visibility, the calculation of biomechanical loads, and the application of established assessment methods (RULA, REBA, OWAS, NIOSH) [10]. Their advantage is their integration into CAD/PLM environments, which simplifies their use already in the design phase of workplaces or production lines [11]. They are mainly used in the design of work environments, material handling, or the evaluation of assembly processes. However, they are limited by their focus on predominantly static, common, and predefined situations, which may not fully capture the dynamics of real working conditions, individual differences between workers, or unpredictable situations in the workplace. Recently, research has increasingly focused on automated ergonomic assessment methods that combine computer vision and deep learning models to detect postures and assess risk in real time [6]. For example, one approach involves an algorithm that automatically generates a Rapid Upper Limb Assessment (RULA) score for evaluating the arms and torso from RGB images, achieving high accuracy and speed (e.g., $\approx$29 frames per second) [12]. This procedure can be deployed directly in the workplace without the need for expensive high-performance hardware (Fig. 2).

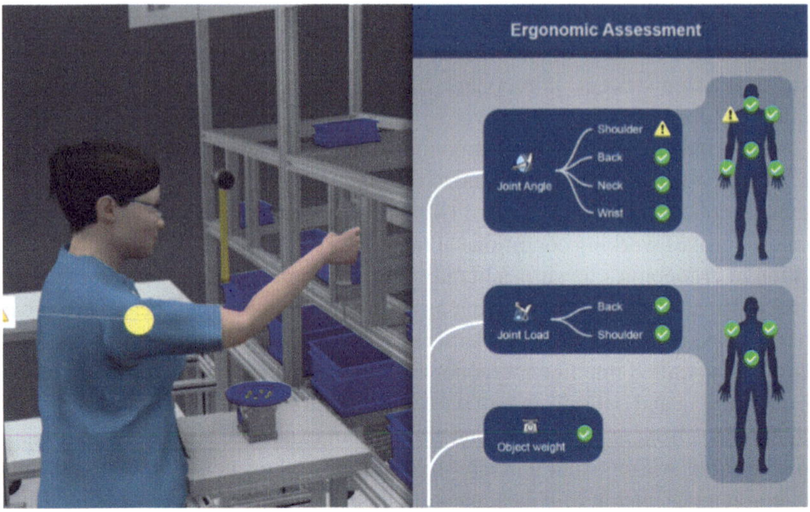


**Fig. 2.** Ergonomic workplace design using DELMIA software [13].

Another category consists of systems based on sensors and computer vision, which are used to collect data directly during the actual production process. These approaches use cameras, depth sensors (e.g. Microsoft Azure Kinect or Intel RealSense) or wearable devices such as IMUs (inertial measurement units) and pressure sensors [8]. The advantage is the ability to record body positions in real time and then automatically assign an ergonomic score according to standard methodologies. The result is a more objective assessment that can serve as immediate feedback for workers or as a basis for production management [9]. The disadvantages remain sensitivity to lighting conditions, coverage of scanned body parts, and the need to process large amounts of data. Further studies are investigating the accuracy and reliability of 2D RGB MoCap (Motion Capture) systems in real production conditions. These systems offer less stress for workers and lower costs compared to traditional marker-based or sensor-based MoCap systems. Several analyses have shown that risk assessments from these systems (e.g., REBA, RULA) are reasonably accurate depending on lighting, workplace layout, and camera configuration [12, 14] (Fig. 3).

In addition to classic digital human modelling and sensor solutions, there is growing interest in hybrid approaches that combine the advantages of both methods. This is an integrated system of digital modelling and real-time computer scanning, which in practice means that the outputs from digital simulations are compared and calibrated against real data from sensors [16]. This integrated system enables not only predictive assessment in the design phase, but also ongoing verification of ergonomics during operation and the introduction of new processes. By combining modelling and real-time measurement, it is possible to achieve greater accuracy in results, tailor assessments to specific groups of workers, and create a more dynamic picture of workplace ergonomics. Analyses focused on the automation of musculoskeletal assessment show that the combination of sensor systems, computer vision and activity recognition algorithms significantly contributes to increased safety and productivity, as it enables real-time decision-making based on

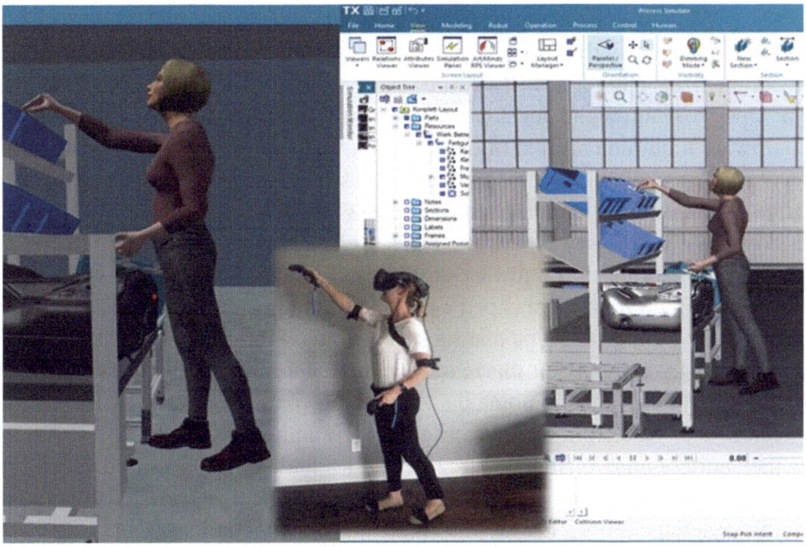

**Fig. 3.** Process simulation and human modelling in Siemens Process Simulate [15].

the data obtained [6, 16]. At the same time, it appears that many existing studies have methodological limitations, such as a small number of participants, inadequate laboratory conditions, or limited possibilities for verification in real-world conditions (Fig. 4).

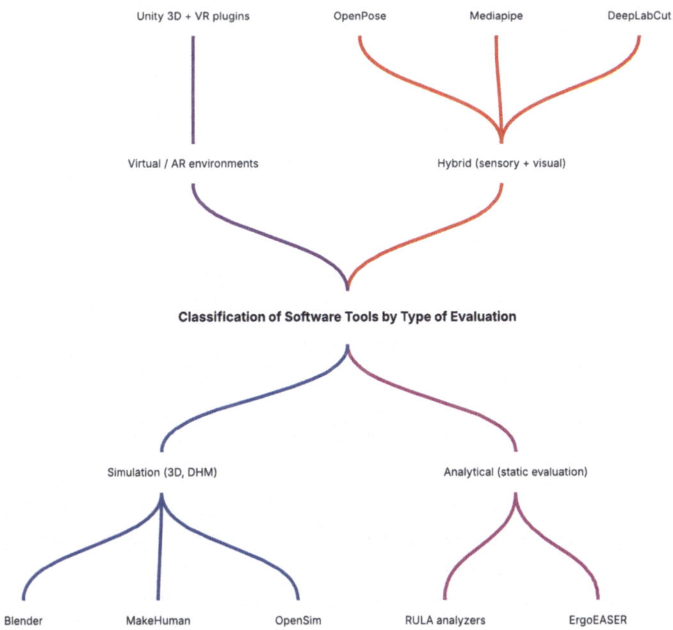


**Fig. 4.** Classification of software tools for industrial ergonomics by type of assessment.

In addition to the above-mentioned solutions, cloud platforms and open-source tools are becoming increasingly popular, significantly reducing the entry barrier for small and medium-sized enterprises. Platforms using libraries such as Google MediaPipe Pose or OpenPose allow the development of custom ergonomic applications with minimal hardware requirements and often only with the help of standard webcams [6, 9]. This widespread availability of technology allows for broader and more flexible use of ergonomic monitoring, although often at the cost of slightly lower accuracy compared to professional systems. A key aspect of all these solutions remains the verification of the accuracy and reliability of the results. Although computer vision-based systems show high agreement with traditional expert assessments, their accuracy can also be affected by other factors such as the type and looseness of work clothing, limited visibility, or atypical working positions. For this reason, modern approaches are increasingly focusing on hybrid systems that combine data from multiple sources. A typical example is the integration of data from inertial sensors (IMU) and video recordings, which eliminates the shortcomings of individual technologies and achieves a more accurate and robust assessment of musculoskeletal risk. This development suggests that the future of ergonomic assessment does not lie in a single universal technology, but in the parallel interconnection of multiple systems that together provide a comprehensive and dynamic picture of the worker's load in real conditions.

These procedures and methodologies also include standardised assessment tools that are integrated directly into the software. In addition to well-known methods (RULA, REBA, NIOSH, OWAS), algorithms based on computer vision and machine learning have also been developed in recent years that can automate these assessments. This significantly reduces the time needed for analysis and minimises the subjectivity that is typical of manual observation methods. The approaches themselves also differ in terms of the extent to which they have been verified in practice. Digital tools such as DHM have been proven over a long period of time and their accuracy has been repeatedly confirmed. However, newer technologies based on computer vision and sensors still require more detailed testing in various production conditions. Nevertheless, they offer significant potential thanks to their high speed, flexibility and the possibility of direct deployment in the production process.

## 3 New Trends in Software Support for Industrial Ergonomics

In recent years, several innovative approaches have emerged that are pushing software support for ergonomics towards greater adaptability, interactivity, and more efficient real-time data evaluation. A significant shift is being brought about by frameworks that combine computer vision, deep neural networks and intent recognition algorithms with dynamic ergonomic assessment, enabling robots or automated devices to respond quickly to changes in a worker's posture or tasks. Such approaches include an intelligent framework for human-robot collaboration (HRC), which combines YOLO models, the Unscented Kalman Filter algorithm and OWAS ergonomic assessment in dynamic mode, achieving high accuracy and fast response times [8, 12]. A specialised area of predictive analytics is worker intent recognition. In the environment of human-robot collaboration, modern algorithms, such as models based on recurrent neural networks combined with an extended Kalman filter, allow not only the analysis of a person's current position and movement, but also the prediction of their future actions. In this way, collaborative robots can automatically adjust their trajectory, speed or grip force in real time, minimising the risk of collision and reducing the physical and mental strain on the operator. This approach is a key step towards creating intuitive and safe human-machine interaction in smart working environments (Fig. 5).

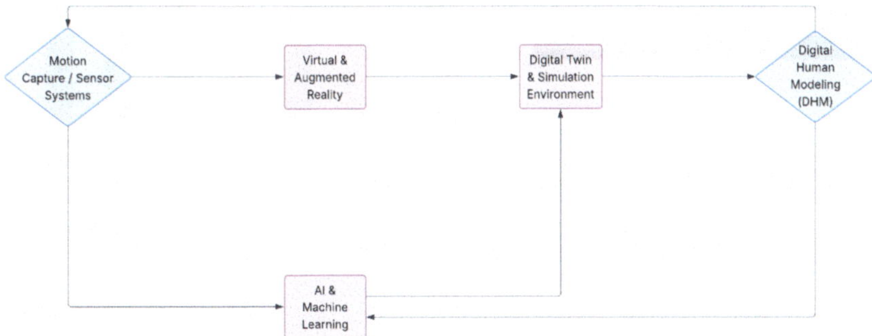


**Fig. 5.** Integration of digital technologies in industrial ergonomics – from sensor systems to AI-driven simulation environments.

Another area of development is visual search and systems capable of automatically generating text descriptions or ergonomic assessments based on image or video inputs. An example is the ErgoChat application, which uses the concept of visual question answering (VQA) to generate comments on workers' postural risks, creating a more interactive and user-friendly way of assessing ergonomics. In parallel with the development of universal solutions, highly specialised systems designed for specific industries are also emerging. In areas such as logistics, assembly or construction, where heavy loads are often handled, solutions using computer vision and posture analysis algorithms are used to detect incorrect lifting techniques. These tools can identify risky movements, such as excessive bending or incorrect torso rotation, in real time and immediately alert the user or supervisor. The result is a preventive system that helps reduce the incidence of musculoskeletal disorders during everyday work activities (Fig. 6).

In addition to traditional visual and sensory approaches, methods based on predictive analytics and machine learning are also being developed that can identify risky patterns of worker behaviour before they manifest themselves in the form of fatigue or musculoskeletal problems. These systems use a combination of historical data, real-time sensing and machine learning algorithms to create predictive risk models, shifting ergonomics from passive assessment to proactive management of working conditions. Particular attention is being paid to the use of digital twins, which in newer solutions are linked to augmented and virtual reality, as well as to industrial standards aimed at harmonising physical and digital systems at the level of machines, people and production flows. These systems offer a real-time digital twin combined with a VR interface for operators and enhanced with cognitive assistance and monitoring via the IoT, with interoperability ensured by prescribed standards such as ISO 23247v [18]. The development of digital twins is increasingly moving towards integration with the field of

**Fig. 6.** n AI-based computer vision system for real-time ergonomic posture assessment in manufacturing environments [17].

cognitive ergonomics. The latest implementations no longer represent only a physical representation of the workplace, but also integrate models of human behaviour such as attention, working memory or decision-making processes. Combined with real-time biometric data – such as heart rate variability or muscle activity – these augmented digital twins enable the cognitive load of workers to be assessed and processes to be optimised to prevent overload and loss of concentration. Such approaches represent a significant step towards intelligent, human-centred production systems. With the growing number of components to be taken into account (physical and cognitive load, interaction with technologies or user behaviour), multi-factor assessment methods are also coming to the fore. An example is the AHP (Analytic Hierarchy Process) method, which supports the technical optimisation of workplaces or VR tools, while also taking into account user preferences and other aspects, including the acceptability of solutions and marketing criteria [11] (Fig. 7).

Another interesting direction is gamification and interactive training platforms that use the principles of virtual reality and artificial intelligence to educate workers in proper work attitudes and safe handling techniques. These tools allow risk scenarios to be simulated in a virtual environment and provide the user with immediate feedback, leading to faster acquisition of correct habits and a long-term reduction in the number of errors or accidents. Current research in the field of software support for ergonomics indicates a shift towards the Industry 5.0 principle, which emphasises human-centric design, sustainability and the resilience of production systems. Software tools are therefore increasingly implementing features aimed at monitoring environmental performance indicators (KPIs), increasing production flexibility and also supporting the physical and mental comfort of workers. The use of artificial intelligence, virtual reality and digital twins is not intended solely to increase efficiency, but above all to improve overall working conditions, reduce fatigue and prevent stress among workers. Current research trends in ergonomic software also point to the growing importance of psychosocial and cognitive factors in the assessment of working conditions. Software solutions are beginning to

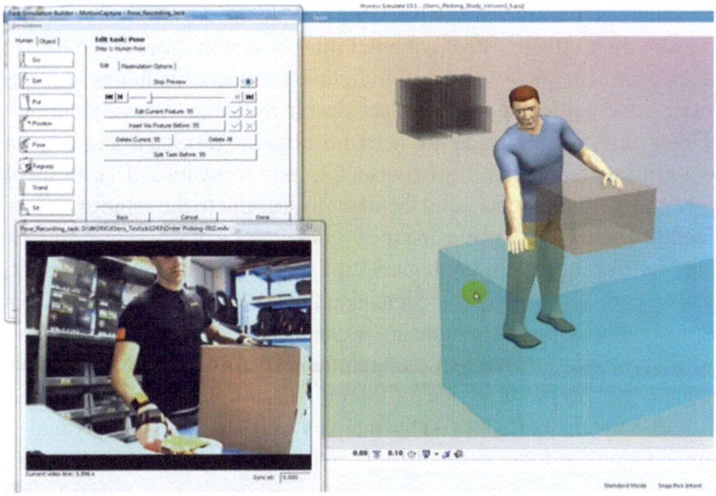

**Fig. 7.** Human-centred design and simulation in industrial environments [19].

include metrics for stress, cognitive load and the emotional state of workers, combining biometric data (e.g. heart rate, galvanic skin response) with digital workplace modelling. This holistic approach provides a more comprehensive assessment of ergonomics, taking into account the physical, mental and cognitive components of work performance equally.

## 4 Conclusion

The field of industrial ergonomics is currently undergoing a fundamental transformation. Traditional assessment approaches based on manual observation and static models are gradually being replaced by intelligent software solutions that combine sensors, artificial intelligence and digital twins. These systems offer greater accuracy, objectivity and the possibility of continuous real-time assessment. The result is a shift from a reactive to a predictive approach, which makes it possible to prevent risks before they arise and optimise the working environment in terms of both physical and cognitive load. Software support in industrial ergonomics has undergone significant development in recent decades – from traditional human digital modelling systems and manual assessment methods to the aforementioned complex solutions using sensors, computer vision and artificial intelligence algorithms [6, 10]. The current state of affairs shows that these technologies enable not only a more efficient and objective assessment of working conditions, but also predictive and adaptive planning of production processes [8, 20]. Integration with digital twins and the use of augmented or virtual reality also represent a significant contribution, opening up new possibilities for interactive employee involvement and workplace optimisation even before their actual implementation. Despite this significant progress, integrating modern software solutions into everyday practice remains a major challenge. The acceptance of technologies by end users – especially employees themselves – is just as important as their technical maturity. Concerns about constant monitoring, changes to established work habits, or mistrust of digital systems can slow down or prevent their effective deployment. Future research should therefore reflect not only technical but also socio-technical and ethical aspects, including transparency of data processing and participatory workplace design that takes into account the views of workers. Other challenges include the need for broader verification in real production conditions, ensuring interoperability between different software and hardware platforms, and taking cognitive and psychosocial factors into account to the same extent as physical strain. Adequate training and increasing the digital literacy of workers are also important prerequisites for the successful implementation of these systems. Software ergonomic solutions should not act as a control mechanism, but as a support tool to improve safety and comfort at work. Open communication channels, clear explanations of objectives and ethical frameworks for data use can significantly influence the level of acceptance of new technologies in the workplace. A promising direction for development is the introduction of Explainable AI (XAI) principles into ergonomic analytical tools. For the recommendations of these systems to be credible and accepted, they must be transparent, interpretable and understandable to both workers and ergonomists themselves. Instead of simple outputs in the form of numerical scores, they should provide specific information about which parts of the body are overloaded and what adjustments to posture or work procedures can reduce the risk. In this way, the technology becomes not only more accurate, but also more human and understandable. Future research and implementation should therefore aim to link digital technologies even more closely with the principles of human-centric design, as well as to reduce the barriers to deploying these solutions in small and medium-sized enterprises by minimising the impact on the financial and human resources of manufacturing companies [9, 16]. Overall, it can be said that modern software approaches to industrial ergonomics represent a significant step towards sustainable, safe and productive manufacturing, where people and technology become equal partners in the creation of effective work systems. Overall, it can be said that developments in the field of software ergonomics are moving away from isolated analytical solutions towards integration into complex cyber-physical systems. These tools no longer serve only to diagnose problems, but are becoming active elements of adaptive and intelligent workplaces that can respond dynamically to human needs. In the context of the transition from Industry 4.0 to Industry 5.0, ergonomics is becoming a bridge between technology and humans – ensuring not only productivity, but also health, satisfaction and the long-term sustainability of work systems.

**Acknowledgement.** This research was supported by the Cultural and Educational Grant Agency of the Ministry of Education, Research, Development and Youth of the Slovak Republic, grant no. KEGA 014TUKE-4/2024 and by the Scientific Grant Agency of the Ministry of Education, Research, Development and Youth of the Slovak Republic and Slovak Academy of Sciences, grant VEGA 1/0302/25.

## References

1. Agostinelli, T., Generosi, A., Ceccacci, S., Mengoni, M.: Validation of computer vision-based ergonomic risk assessment tools for real manufacturing environments. Sci. Rep. **14**, 27785 (2024). https://doi.org/10.1038/s41598-024-79373-4

2. Panda, S., Modak, N.M., Cárdenas-Barrón, L.E.: Does extended warranty depict competitive advantage to a retailer in a retail-e-tail channel supply chain. Comput. Ind. Eng. **149**, 106770 (2020). https://doi.org/10.1016/j.cie.2020.106770
3. Song, D., Kim, E., Bak, H., Shin, G.: Effect of hand loads on upper extremity muscle activity during pushing and pulling motions. Appl. Ergon. **96**, 103504 (2021). https://doi.org/10.1016/j.apergo.2021.103504
4. Fan, C., Mei, Q., Li, X.: 3D pose estimation dataset and deep learning-based ergonomic risk assessment in construction. Autom. Constr. **164**, 105452 (2024). https://doi.org/10.1016/j.autcon.2024.105452
5. Cority: Industrial Ergonomics, https://www.cority.com/wp-content/uploads/2025/08/health-industrial-ergonomics-mosaic.png, last accessed 2025/10/26.
6. Bhattathiri, S.S., et al.: Unlocking human-robot synergy: The power of intent communication in warehouse robotics. Appl. Ergon. **117**, 104248 (2024). https://doi.org/10.1016/j.apergo.2024.104248
7. González-Alonso, J., Martín-Tapia, P., González-Ortega, D., Antón-Rodríguez, M., Díaz-Pernas, F.J., Martínez-Zarzuela, M.: ME-WARD: A multimodal ergonomic analysis tool for musculoskeletal risk assessment from inertial and video data in working places. Expert Syst. Appl. **278**, 127212 (2025). https://doi.org/10.1016/j.eswa.2025.127212
8. Iliuţă, M.E., Moisescu, M.A., Pop, E., Ionita, A.D., Caramihai, S.I., Mitulescu, T.C.: Digital twin—A review of the evolution from concept to technology and its analytical perspectives on applications in various fields. Appl. Sci. (Switz.). **14**, 5454 (2024). https://doi.org/10.3390/app14135454
9. Chaparro-Cárdenas, S.L., et al.: A technological review of digital twins and artificial intelligence for personalised and predictive healthcare. Healthcare. **13**(14), 1763 (2025). https://doi.org/10.3390/healthcare13141763
10. Bitsch, G., Senjic, P., Askin, J.: Integration of legacy systems to cyber-physical production systems using semantic adapters. In: Procedia CIRP, pp. 259–263. Elsevier B.V (2023). https://doi.org/10.1016/j.procir.2023.06.045
11. Iodice, F., De Momi, E., Ajoudani, A.: Intelligent framework for human-robot collaboration: Dynamic ergonomics and adaptive decision-making. J. Intell. Robot. Syst. **112**, 5 (2026)
12. Liu, W., Tian, S., Hu, B., Liang, X., Zheng, M.: A recurrent neural network enhanced unscented Kalman filter for human motion prediction. In: Conference: 2024 International Symposium on Flexible Automation (2024)
13. Dassault Systèmes.: DELMIA Ergonomics–Industrial Engineering Solutions, https://www.3ds.com/products/delmia/industrial-engineering/ergonomics, last accessed 2025/09/26.
14. Plantard, P., Shum, H.P.H., Le Pierres, A.S., Multon, F.: Validation of an ergonomic assessment method using Kinect data in real workplace conditions. Appl. Ergon. **65**, 562–569 (2017). https://doi.org/10.1016/j.apergo.2016.10.015
15. Siemens Digital Industries Software.: Tecnomatix Process Simulate Human Advanced–Ergonomic and Human Simulation, https://plm.sw.siemens.com/en-US/tecnomatix/products/process-simulate-x-human-advanced/, last accessed 2025/09/25.
16. Cardoso, A., Colim, A., Bicho, E., Braga, A.C., Menozzi, M., Arezes, P.: Ergonomics and human factors as a requirement to implement safer collaborative robotic workstations: A literature review. Safety. **7**(4), 71 (2021). https://doi.org/10.3390/safety7040071
17. Retrocausal AI.: RULA Assessment: How to Implement it in Manufacturing, https://i0.wp.com/retrocausal.ai/wp-content/uploads/2024/07/Picture6–3.png?w=1047&ssl=1, last accessed 2025/10/26.
18. Iyer, H., Kim, E., Nam, C.S., Jeong, H.: A scoping review on emerging technologies and automation of musculoskeletal ergonomic assessments. Ergonomics. **22**, 1–26 (2025). https://doi.org/10.1080/00140139.2025.2547286

19. Siemens Digital Industries Software.: Human-Centred Design and Simulation in Industrial Environments, https://plm.sw.siemens.com/en-US/tecnomatix/human-centered-design-simulation/, last accessed 2025/09/25.
20. Park, J.S., Lee, D.G., Jimenez, J.A., Lee, S.J., Kim, J.W.: Human-focused digital twin applications for occupational safety and health in workplaces: A brief survey and research directions. Appl. Sci. **13**(7), 4598 (2023). https://doi.org/10.3390/app13074598

# Research on Multi-Objective Flexible Job Shop Scheduling Problem Based on Improved NSGA-II

Yanfeng Chu(✉) and Lian Xie

Nanjing University of Aeronautics and Astronautics, Nanjiing, China
yanfengc@nuaa.edu.cn

**Abstract.** This study investigates the characteristics of existing flexible job shop scheduling problems and examines relevant algorithms, proposing a scheduling method based on an improved Non-dominated Sorting Genetic Algorithm (NSGA-II) to address these challenges. To mitigate the limitations of NSGA-II, such as low population diversity and slow computational speed, this research introduces an adaptive crossover operator based on crowding distance and incorporates a competitive selection mechanism inspired by bidding principles to enhance solution quality. Experimental simulations are conducted to verify the feasibility and effectiveness of the proposed algorithm.

**Keywords:** Flexible Job Shop Scheduling · Non-dominated Sorting Genetic Algorithm · Adaptive Crossover Operator · Multi-objective optimization

## 1 Introduction

With the advancement of economic globalization, market segmentation has intensified, and competition among enterprises has become increasingly fierce. Evolving customer demands have led to orders characterized by growing personalization and diversification, imposing heightened demands on production methodologies. To survive in this competitive landscape, manufacturing enterprises must optimize their production systems to accommodate multi-variety, small-batch, multi-stage, and multi-task manufacturing processes, thereby securing a sustainable position in future markets. Transitioning from traditional production paradigms and integrating lean production principles to establish a tailored operational model for multi-variety, small-batch, and multi-stage manufacturing presents significant challenges for enterprises. These challenges encompass not only maintaining production continuity but also achieving cost efficiency. Consequently, scientific production management has emerged as a critical focus for manufacturing industries seeking to enhance operational agility and competitiveness.

In the flexible job shop, multiple processing tasks can be executed on various available machines. Unlike traditional job shop scheduling, FJSP allows each operation of a job to select different machines for processing, thereby enhancing production flexibility. The FJSP can be formally defined as follows: A set of workpieces must be processed


A. Hamrol et al. (Eds.): MANUFACTURING 2026, LNME, pp. 271–283, 2026.
https://doi.org/10.1007/978-3-032-21701-1_20

on a set of machines following predefined process routes. Each workpiece comprises at least one operation, and each operation can be assigned to one machine from a group of parallel machines. The processing speed depends on the machine type: parallel machines are classified as homogeneous if they share identical processing speeds or heterogeneous if their speeds differ.

The primary objective of production scheduling is to optimize performance metrics—such as due date compliance, equipment utilization, product quality, and production costs—by generating production orders that allocate resources (machines), determine processing times, and sequence operations while adhering to constraints including process routes, resource availability, and delivery deadlines.

Currently, the most prevalent approach to solving the Flexible Job Shop Scheduling Problem (FJSP) involves intelligent optimization algorithms, such as Particle Swarm Optimization (PSO), Grey Wolf Optimization (GWO), Simulated Annealing (SA), and NSGA-II. For instance, Kong et al. [1] addressed the inefficiency of FJSP solutions by proposing an improved two-layer PSO algorithm incorporating stagnation prevention and concave function decay strategies, validating its efficacy through a model targeting the minimization of makespan. Zhang [2] enhanced the GWO algorithm by introducing novel initialization methods, individual update mechanisms, and local search strategies for optimal individuals to achieve makespan minimization in FJSP. Peng et al. [3] analyzed FJSP characteristics and designed a fast non-dominated sorting algorithm to expedite the elimination of dominated solutions, integrating a cloud strategy model to maintain population diversity and a multi-indicator weighted grey target decision model for selecting Pareto-optimal solutions. Kaplanoglu [4] developed an object-oriented multi-objective FJSP framework using UML class diagrams and a simulated annealing-based optimization method to simultaneously minimize makespan, maximum machine workload, and total machine load. Ziaee et al. [5] formulated a mixed-integer linear programming model for a factoryless FJSP under natural environmental constraints, considering machining selection costs, variable operational costs, transportation costs, and order rejection penalties, and employed heuristic algorithms to derive acceptable solutions. Amiri [6] proposed a meta-model based on multi-objective efficiency simulation optimization, integrating diverse metaheuristics and swarm intelligence algorithms, and utilized artificial neural networks to solve resource allocation and job sequencing challenges in FJSP. Chen et al. [7] enhanced the Brain Storm Optimization (BSO) algorithm by replacing k-means clustering with affinity propagation clustering and introducing a multi-cluster information-based creation operator, achieving superior convergence and success rates in numerical optimization. The Grey Wolf Optimization (GWO) algorithm, introduced by Mirjalili [8] in 2014, mimics the social hierarchy and hunting behavior of grey wolves. Its simplicity, minimal parameterization, and proven superiority over PSO and Differential Evolution (DE) in convergence and exploration-exploitation balance [9] have made it a prominent tool. Mohammed et al. [10] developed a multi-objective chaotic symbiotic organisms search algorithm for query scheduling in cloud computing environments, while Jia et al. [11] proposed an ant colony optimization-based method for energy-aware batch scheduling in multi-objective contexts.

The Non-dominated Sorting Genetic Algorithm (NSGA) is an evolutionary algorithm designed for multi-objective optimization problems. By integrating non-dominated sorting and a crowding distance comparison mechanism, it approximates the Pareto-optimal front while preserving population diversity. For instance, Ouyang et al. [12] addressed inefficiencies in solving multi-objective FJSP by proposing an enhanced NSGA-III. Their improvements included a hybrid allocation strategy to refine initialization, an adaptive elimination strategy based on entropy variation to enhance non-dominated elite preservation, and validation across 11 scheduling algorithms, demonstrating superior efficiency in FJSP resolution. Yang et al. [13] formulated an energy-efficient FJSP model targeting minimization of makespan, total energy consumption, and total machine workload. They employed an NSGA-III variant with enhanced exploration capabilities to solve this multi-objective problem. Wang [14] dynamically adjusted crossover and mutation probabilities in NSGA-II, enabling stage-specific independent operations for sequencing and assigning jobs and machines during FJSP resolution. Zhang Guohui advanced genetic algorithms to address FJSP with dynamic adjustment events, while Du et al. [15] hybridized simulated annealing with whale optimization algorithms for single-objective FJSP (minimizing makespan). For multi-objective FJSP, they refined NSGA-II by modifying elitist retention strategies and incorporating neighborhood search mechanisms to optimize makespan, total machine load, and production costs.

While numerous scholars have investigated the application of NSGA-II to multi-objective flexible job shop scheduling problems, the exploration of bidding mechanisms in this context remains underexplored. To address this gap, this study proposes a novel bidding strategy-based resource allocation framework for flexible job shop scheduling optimization. A dynamic job-machine matching model under multi-constraint conditions is developed to achieve optimal production resource configuration. This approach provides an innovative research pathway for resolving complex FJSP challenges, offering both practical engineering significance for industrial resource management and academic contributions to the advancement of intelligent scheduling methodologies.

## 2 Overview of the Flexible Job Shop Scheduling Problem

FJSP extends the classical job shop scheduling paradigm by permitting multiple workpieces to be processed across multiple machines. Crucially, each operation can be flexibly assigned to alternative machines, significantly expanding the feasible solution space.

FJSP can be described as follows:

There are $n$ workpieces($J_i$, $i \in \{1, 2 \ldots n\}$)processed on m machines ($M_k$, $k \in \{1, 2, \ldots, m\}$). Each workpiece $J_i$ contains one or more operations ($O_{ij}, j \in \{1, 2 \ldots h_i\}$, $h_i$ denotes the total number of operations for workpiece $J_i$). The sequence of operations is predetermined. Each operation can be processed on multiple different machines, and the processing time of an operation varies depending on the selected machine. The scheduling objectives are to select the most suitable machine for each operation, determine the optimal processing sequence and start time for all operations on each machine, and optimize specific performance metrics of the entire system. Therefore, the FJSP consists of two subproblems: determining the processing machine for each workpiece (machine selection subproblem) and determining the processing sequence of operations on each machine (operation sequencing subproblem).

Additionally, the following constraints must be satisfied during processing:

(1) A machine can process only one workpiece at any given time.
(2) The same operation of the same workpiece cannot be processed by multiple machines concurrently.
(3) Once an operation starts, it cannot be interrupted until completion.
(4) All workpieces have identical priority levels.
(5) Operations from different workpieces have no sequential dependencies; Operations within the same workpiece must follow predefined precedence relationships.
(6) All workpieces are available for processing at time zero.

The makespan is the most fundamental metric for evaluating scheduling schemes, primarily reflecting production efficiency in the workshop. The processing time of operations varies across different machines, and the total machine load differs depending on the scheduling scheme. It is essential to minimize the total consumption of all machines while maintaining the same makespan. Just-In-Time production requires consideration of due date constraints, where a workpiece's completion time closer to its delivery deadline indicates better due date performance.

This paper establishes the following optimization objectives: Minimization of Makespan, Minimization of Total Machine Load and Minimization of Maximum Tardiness.

$$f_1 = \min(\max\{T_i | i = 1, 2, \ldots, n\}) \tag{1}$$

$$f_2 = \min\left(\sum_{k=1}^{m}\sum_{j=1}^{n}\sum_{k=1}^{k_i} p_{ijk}x_{ijk}\right) \tag{2}$$

$$f_3 = \min(\max|T_i - d_i|) \tag{3}$$

Constraints:

$$E_{ij} \leq S_{i(j+1)} \tag{4}$$

$$E_{ijk} - S_{ijk} = P_{ijk} \tag{5}$$

$$x_{ijk} = \begin{cases} 1 & \text{Operation } O_{ij} \text{ is processed on } M_k \\ 0 & \text{Operation } O_{ij} \text{ isn't processed on } M_k \end{cases} \tag{6}$$

Notations:

$n$: Total number of workpieces; $m$: Total number of machines; $T_i$: Completion time of workpiece $J_i$; $x_{ijk}$: Binary variable indicating whether operation $O_{ij}$ of workpiece is processed on machine $M_k$; $p_{ijk}$: Processing time of operation $O_{ij}$ of workpiece on machine $M_k$; $d_i$: Due date of workpiece $J_i$; $S_{ij}$: Start time of operation $J$ of workpiece $J_i$; $E_{ij}$: Finish time of operation $j$ of workpiece $J_i$.

Eq. (1) represents the minimization of makespan; Eq. (2) represents the minimization of total machine load; Eq. (3) represents the minimization of maximum tardiness; Eq. (4) enforces precedence constraints between operations; Eq. (5) enforces machine processing constraints (non-interruption once processing begins); Eq. (6) defines the decision variable for whether operation $O_{ij}$ is processed on machine $M_k$.

# 3 Solution of Multi-Objective Flexible Job Shop Scheduling Problems Based on Improved NSGA-II

## 3.1 Scheduling Problem Solving Process

The solution process for multi-objective flexible job shop scheduling based on the improved NSGA-II algorithm is illustrated in Fig. 1. According to the flowchart, the execution primarily comprises two stages: population evolution and solution set refinement. In the population evolution stage, a parent population is initialized with a size of 200 individuals. Genetic operations—including selection, crossover, and mutation—are then performed using an adaptive crossover operator to generate offspring populations. Subsequently, the parent and offspring populations are merged, and individuals are ranked and filtered through fast non-dominated sorting and crowding distance calculation. High-quality individuals are retained to form a new generation. This process iterates until convergence criteria are met, ultimately yielding a Pareto-optimal front solution set composed of mutually non-dominated solutions.

In the solution set refinement stage, this study designs a decision-making method based on a resource competition mechanism. First, a competitive population is selected and established from the Pareto solution set. Next, the machine resource competition intensity metric is quantified by analyzing the processing information of operations in candidate solutions, forming a machine competition intensity distribution matrix. Finally, the optimal solution is filtered through comparative analysis of competition intensities, generating the optimal scheduling scheme. The corresponding scheduling Gantt chart is then produced via a decoding process.

## 3.2 Algorithm Design for Multi-Objective Flexible Job Shop Scheduling Problem Based on Improved NSGA-II

## 3.3 Composite Encoding Design

This study proposes a composite encoding framework that strictly aligns chromosome length with the total number of operations, utilizing an operation sequence encoding segment to define processing priority and a machine association encoding segment to map operations to corresponding processing resources. In intelligent optimization algorithms, the design of the encoding mechanism directly determines the efficiency and quality of problem-solving. Traditional encoding methods in scheduling research are broadly categorized into two types: operation sequence encoding, which employs repeated numerical values to distinguish different operations of the same workpiece (e.g., using the repetition count of numbers in a chromosome to indicate the order of operations for a specific workpiece), and machine association encoding, which adopts multi-dimensional matrix structures to record machine selection information. However, in multi-objective optimization scenarios for flexible job shops, traditional encoding strategies exhibit limitations, failing to effectively coordinate the dual requirements of machine allocation and operation sequencing. The composite encoding design is illustrated in Fig. 2. The sequence of process execution for Operation Sequence Encoding Segment [1–3] is ($O_{31}$, $O_{11}$, $O_{21}$, $O_{22}$, $O_{12}$, $O_{32}$, $O_{23}$, $O_{13}$);The machine-encoded gene string (Machine Gene

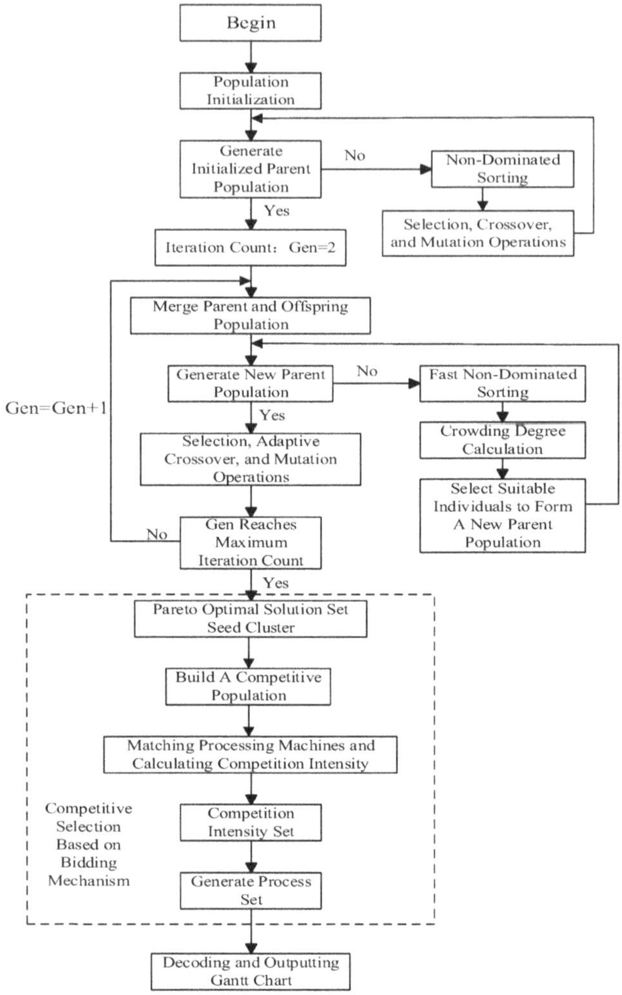


**Fig. 1.** Flowchart for Solving the Multi-Objective Flexible Job Shop Scheduling Problem.

Segment) [1–5] corresponds to the processing machines ()selected for each operation. In the machine gene segment, th $M_2, M_5, M_4, M_3, M_1, M_2, M_3, M_1$ e positional indices of the numerical values strictly adhere to the operation sequencing rules. For example, the value "1" at a specific index corresponds to the selection of machine $M_2$ for the first operation $O_{11}$ of workpiece $J_1$, while the value at index "4" corresponds to the machine $M_5$ selection for the subsequent operation $O_{12}$ of the same workpiece $J_1$, and so on.

### 3.4 Pareto Sortingn

This study employs the fast sorting method to perform Pareto sorting of non-dominated solutions in the population. Within the classical NSGA framework (which lacks the elitist retention mechanism introduced in NSGA-II), non-dominated sorting primarily

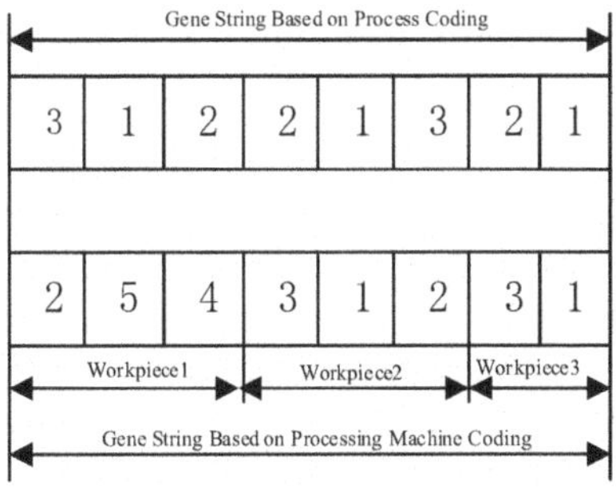


**Fig. 2.** Extended Operation-Based Encoding.

involves two methods: fast sorting and recursive sorting. Research indicates that the fast sorting method exhibits higher computational efficiency in multi-objective optimization scenarios due to its lower time complexity, justifying its adoption in this work. The core of non-dominated sorting lies in establishing a hierarchical structure through dominance relationships, thereby identifying high-quality solution sets within the population. After non-dominated sorting, the population is decomposed into X hierarchically ordered non-dominated frontier sets $\{P_1, P_2, ..., P_X\}$, which satisfy the following mathematical characteristics:

- $U \in \{p_1, p_2, ..., p_X\}p = Pop$, this indicates that the total cardinality of individuals in the sorted non-dominated frontiers $U \in \{p_1, p_2, ..., p_X\}p$ remains consistent with the initial population size *Pop*.
- $\forall h, f \in \{1, 2, ..., X\}$, $h \neq f, p_h \cap p_f = \emptyset$, this signifies that any two distinct frontier sets satisfy mutual exclusivity, i.e., all non-dominated frontiers are mutually independent.

The frontier levels exhibit a strict partial ordering relationship, resulting in X non-dominated frontiers after non-dominated sorting.$p_1 > p_2 > ... > p_X$

Each individual h is associated with two parameters $n_h$ and $S_h$: $S_h$ represents the number of individuals dominated by h; $n_h$ represents the number of individuals that dominate h. So:

$$n_h = |\{z|z > h, h, z \in Pop\}|$$

$$S_h = |\{f|h > f, h, f \in Pop\}|$$

During algorithm execution, individuals satisfying the condition $n_h = 0$ are first assigned to the initial non-dominated frontier set $F_1$ and allocated $rank = 1$. Subsequently, the $n_z$ values of remaining individuals are iteratively updated. Subsequent frontier levels are incrementally constructed by identifying individuals with updated $n_z = 0$, until all individuals are assigned rank values.

### 3.5 Design of Adaptive Crossover Operator Based on NSGA-II Crowding Distance

This study constructs an adaptive crossover operator based on crowding distance by establishing a coupling relationship between recombination probability and individual crowding distance, enabling self-regulating adjustment of recombination intensity. As the algorithm converges toward local minima, the minimum and maximum crowding distance values become increasingly proximate, leading to a corresponding increase in their ratio. While the existing NSGA-II improves convergence efficiency by integrating non-dominated sorting mechanisms and elitist strategies over the original NSGA framework, it suffers from insufficient population diversity. As the core step for generating new individuals, genetic recombination operations directly influence the algorithm's global exploration capability. The parameter settings and execution strategies of these operations play a critical role in maintaining population diversity. Traditional fixed-probability recombination models, which typically use crossover probabilities within the range [0.40–0.99], are prone to local convergence. The design proposed in this study effectively ensures population diversity in the search space and uniform distribution of solution sets by adaptively increasing crossover probabilities, while addressing NSGA-II's tendency toward premature convergence. Crossover operations are performed on chromosomes with controlled probabilities to regulate the algorithm's convergence behavior and solution quality. In conventional methods, crossover probabilities remain fixed within a broad search range. In the proposed algorithm, the crossover probability is an adaptive parameter dependent on the crowding distances of individuals within the population. As crowding degrees dynamically evolve during optimization, the crossover probability adjusts accordingly to preserve diversity around local optima. This mechanism ensures that regions with sparse solutions (larger crowding distances) receive higher crossover probabilities to enhance exploration, while densely populated regions (smaller crowding distances) prioritize exploitation.

Based on the maximum and minimum crowding distances of individuals in the elite population, the adaptive crossover operator is defined as:

$$p_c = i_{d\,\min}/i_{d\,\max} \tag{7}$$

In the equation: $p_c$ is adaptive crossover operator; $i_{d\max}$ is maximum crowding distance; $i_{d\min}$ is minimum crowding distance.

This mechanism dynamically adjusts recombination intensity by monitoring changes in solution set distribution density. In sparse regions of the solution space, recombination intensity is reduced to preserve superior genes, while in densely populated regions, recombination probability is increased to break local convergence. Thereby, it effectively balances global search capabilities and local exploitation.

To address the dual optimization requirements of operation sequencing and machine assignment in FJSP, this study designs a hybrid recombination strategy: For the operation encoding segment, an inheritance-based crossover strategy is adopted, partitioning the workpiece set into complementary subsets to preserve the integrity of parental gene sequences in offspring. For the machine encoding segment, a multi-point recombination strategy is implemented, where the exchange of machine assignment information is controlled through a randomly generated binary mask matrix. Implementation Steps:(1)

Calculate dynamic recombination parameters based on the current generation;(2) Determine gene exchange sites for the operation encoding segment and machine encoding segment, respectively;(3) Generate offspring individuals through parental chromosome information exchange.

Assuming parent chromosomes $P_1$ and $P_2$ undergo crossover operations to generate offspring chromosomes $C_1$ and $C_2$, the operation sequence crossover strategy is as follows: Randomly partition the workpiece set into two subsets $S_1$ and $S_2$; copy the workpieces in $S_1$ from $P_1$ to both $C_1$ and $C_2$, preserving their original gene positions. Copy the workpieces in $S_2$ from to both $C_1$ and $C_2$, preserving their original gene positions. Example as illustrated in Fig. 3.

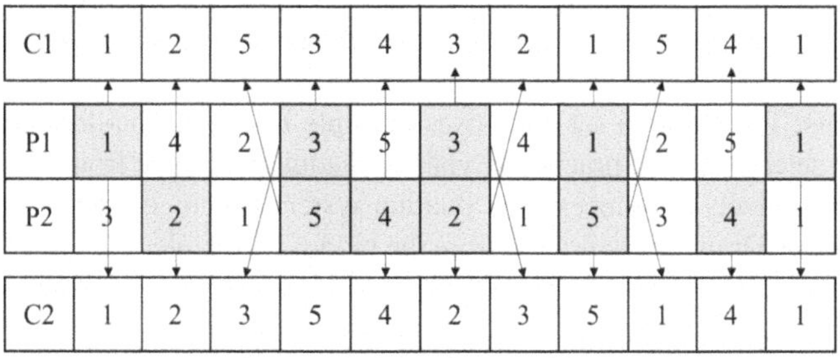


**Fig. 3.** Operation-Based Crossover Operation.

In Fig. 3, assuming the operation gene segment of parent chromosome $P_1$ is: [1–5], the operation gene segment of parent chromosome $P_2$ is [1–5].The workpiece subsets are partitioned as $S_1$= [1, 3], $S_2$= [2, 4, 5].The crossover strategy for the machine encoding segment in chromosomes is as follows: Randomly generate a binary sequence R (composed of 0 and 1) with the same length as the chromosome; Identify the machine codes at positions where R = 1 in parent chromosomes $P_1$ and $P_2$, swap them, and copy the exchanged codes to offspring $C_1$ and $C_2$; For positions where R = 0, directly copy the corresponding machine codes from $P_1$ and $P_2$ to $C_1$ and $C_2$, respectively. This machine-based crossover operation is illustrated in Fig. 4.

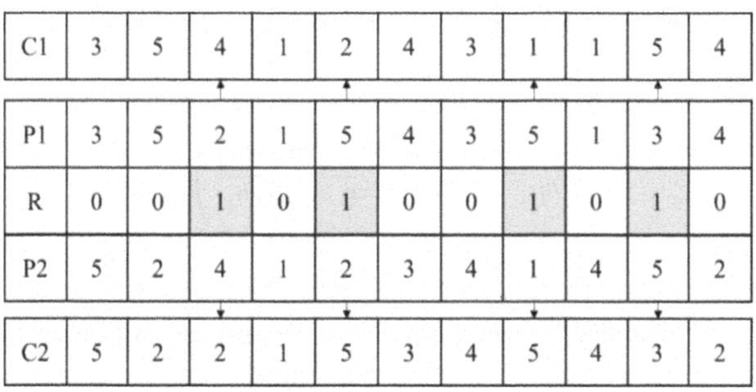


**Fig. 4.** Machine-Based Crossover Operation.

In Fig. 4, assuming the operation gene segment of parent chromosome $P_1$ is: [1–5], the operation gene segment of parent chromosome $P_2$ is [1–5], so R = [0,0,1,0,1,0,0,1,0,1,0].

### 3.6 Competitive Selection Strategy Based on Bidding Mechanism

This study proposes a bidding mechanism-based optimization strategy for selecting the most practical scheduling scheme from the Pareto front solution set obtained upon algorithm termination. Inspired by the competitive, transparent, and fairness-driven nature of bidding mechanisms in economic transactions, this strategy introduces the concept of competition intensity to mitigate local optima caused by information asymmetry between processing machines and pending operations. By aligning machine resource allocation with value principles, competitive mechanisms, and supply-demand dynamics, it achieves near-optimal configuration of processing sequences for both machines and operations. The solution set comprises multiple mutually non-dominated optimal solutions. To select the most practically viable scheduling scheme from these solutions, this study innovatively introduces an evaluation system for processing equipment competition intensity. Drawing inspiration from the bidding principles of market economies, this evaluation model resolves the local optimization issues caused by information asymmetry in traditional scheduling through the establishment of multi-dimensional quantitative metrics, thereby achieving optimal allocation of processing resources. Under flexible manufacturing environments, this strategy provides a scientific basis for selecting the most suitable processing equipment for operations by holistically considering critical factors such as equipment energy consumption, load balancing, and time utilization efficiency.

Specifically, the competition intensity evaluation of processing equipment is grounded in three core criteria:

(1) Criterion $R_1$ is the energy consumption $D_{ijk}$ required to process operation $O_{ij}$ on machine $M_k$.
(2) Criterion $R_2$ is the total processing time $t_{ijk}$ of pending operations on processing machine $M_k$.
(3) Criterion $R_3$ is the idle time $L_{ijk}$ of processing machine $M_k$ before and after operation $O_{ij}$.
(4) Based on $R_1$、$R_2$、$R_3$, the competition intensity from processing machine $M_k$ for operation $O_{ij}$ is:

$$Y_{O_{ij}}^{k} = \sum_{r=1}^{R} \left( \frac{R_{\mathrm{r}}^{\max} - R_r^k(O_{ij})}{R_r^{\max} - R_r^{\min}} \right).$$

And, $Y_{O_{ij}}^{k}$ is the competition intensity of processing machine $M_k$ across all criteria for operation $O_{ij}$; $R_{\mathrm{r}}^{\max}$ and $R_r^{\min}$ represent the maximum and minimum function values, respectively, among all available processing machine $M_k$ for operation $O_{ij}$ under criterion r; $R_r^k(O_{ij})$ is the function value of machine $M_k$ for operation $O_{ij}$ under criterion r.

In summary, the implementation process of the competitive optimization strategy proposed in this study consists of four steps: First, extract solutions from the Pareto solution set to form a candidate population. Second, compute the competition intensity for each processing machine and generate an evaluation matrix. Then, sort solutions in descending order of competition intensity values, where higher values indicate superior solutions. Finally, select the highest-ranked solution as the optimal scheduling plan. Generate a corresponding Gantt chart to visualize the processing sequence of operations on selected machines and the order of operations assigned to the same machine.

# 4 Case Study Simulation and Analysis

Table 1 presents the instance data (non-dimensional) for the FJSP, illustrating the available processing machines and their corresponding processing time for different operations of each workpiece. This study uses this dataset as an example for simulation analysis in Matlab2021b.

**Table 1.** Instance Data for the FJSP.

| Work piece | Operation | Due Date | Processing Time | | | | | | Workpiece | Operation | Due Date | Processing Time | | | | | |
|---|---|---|---|---|---|---|---|---|---|---|---|---|---|---|---|---|---|
| | | | $M_1$ | $M_2$ | $M_3$ | $M_4$ | $M_5$ | $M_6$ | | | | $M_1$ | $M_2$ | $M_3$ | $M_4$ | $M_5$ | $M_6$ |
| $J_1$ | $O_{11}$ | | 15 | — | 14 | — | — | — | $J_4$ | $O_{41}$ | | 11 | 16 | — | — | — | 15 |
| $J_1$ | $O_{12}$ | | — | 11 | — | 15 | 13 | — | $J_4$ | $O_{42}$ | | — | 16 | — | — | — | — |
| $J_1$ | $O_{13}$ | 150 | — | — | — | — | 14 | 12 | $J_4$ | $O_{43}$ | 100 | — | — | 11 | — | — | — |
| $J_1$ | $O_{14}$ | | 11 | 16 | — | — | — | 15 | $J_4$ | $O_{44}$ | | — | 11 | 15 | — | 13 | — |
| $J_1$ | $O_{15}$ | | — | — | 11 | — | — | — | $J_4$ | $O_{45}$ | | — | — | 14 | — | — | 12 |
| $J_1$ | $O_{16}$ | | — | — | 16 | 13 | — | 16 | $J_5$ | $O_{51}$ | | — | 11 | 15 | — | 13 | — |
| $J_2$ | $O_{21}$ | | — | 16 | — | — | — | — | $J_5$ | $O_{52}$ | | 11 | 16 | — | — | — | 15 |
| $J_2$ | $O_{22}$ | | — | — | 11 | — | — | — | $J_5$ | $O_{53}$ | 90 | — | 16 | — | — | — | — |
| $J_2$ | $O_{23}$ | 80 | 12 | — | — | — | — | — | $J_5$ | $O_{54}$ | | 15 | — | 14 | — | — | — |
| $J_2$ | $O_{24}$ | | — | 16 | — | 16 | — | — | $J_5$ | $O_{55}$ | | — | 16 | — | 16 | — | — |
| $J_2$ | $O_{25}$ | | 11 | 16 | — | — | — | 15 | $J_5$ | $O_{56}$ | | — | 16 | 14 | — | — | 16 |
| $J_3$ | $O_{31}$ | | — | 16 | — | — | — | — | $J_6$ | $O_{61}$ | | — | — | 14 | — | — | 12 |
| $J_3$ | $O_{32}$ | | — | — | 14 | — | — | 12 | $J_6$ | $O_{62}$ | | 12 | — | — | — | — | — |
| $J_3$ | $O_{33}$ | 70 | — | 16 | — | 11 | — | 15 | $J_6$ | $O_{63}$ | 70 | — | 16 | — | — | — | 16 |
| $J_3$ | $O_{34}$ | | — | 16 | — | — | 14 | 16 | $J_6$ | $O_{64}$ | | — | — | — | — | 16 | — |
| $J_3$ | $O_{35}$ | | 11 | — | — | — | 15 | — | $J_6$ | $O_{65}$ | | 11 | 16 | — | — | 15 | — |
| | | | | | | | | | $J_6$ | $O_{66}$ | | 13 | — | — | — | 12 | — |

As the iteration count increases, the algorithm demonstrates robust convergence characteristics, with continuous optimization across all scheduling objectives. Specifically, the makespan, tardiness, and machine load exhibit progressive improvement over evolutionary generations. The proposed algorithm ultimately yields a series of Pareto feasible solutions, and the Pareto-optimal solution set is obtained by pruning redundant solutions. Based on the machine strength-based selection strategy proposed in this study, the Pareto-optimal solution set is subjected to competitive selection. The resulting optimal scheduling Gantt chart is illustrated in Fig. 5, with a makespan of 101, zero tardiness, and a total machine load of 419.

In Fig. 5, the horizontal axis represents the processing time on the machines, and the vertical axis denotes different processing machines. Distinct colors indicate different workpieces, with the same color representing the same workpiece. The length of the rectangles corresponds to the processing duration. For example, "2,5 = 11" signifies that Operation 5 of Workpiece 2 is processed on machine M1 with a duration of 11. The original scheduling Gantt chart, shown in Fig. 6, has a makespan of 109, zero tardiness, and a total machine load of 426. Comparing the schemes in Figs. 3 and 4 reveals

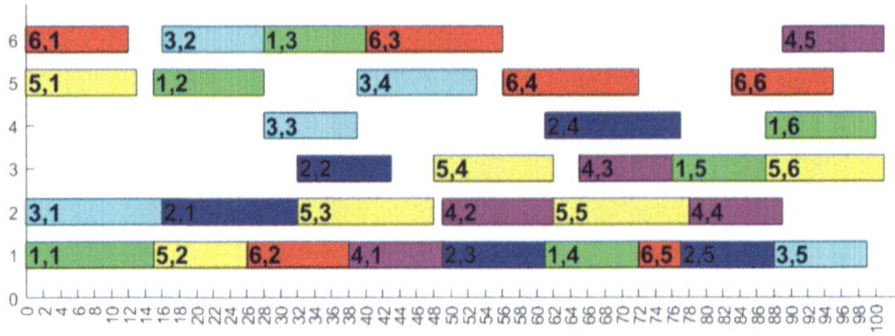


**Fig. 5.** Optimal Scheduling Gantt Chart.

that the makespan is reduced from 109 (pre-improvement) to 101 (post-improvement), optimizing by 8 units, while the total machine load decreases from 426 to 419, optimizing by 7 units. In multi-objective optimization, conflicting objectives often require trade-offs, and this study achieves an overall optimal balance.

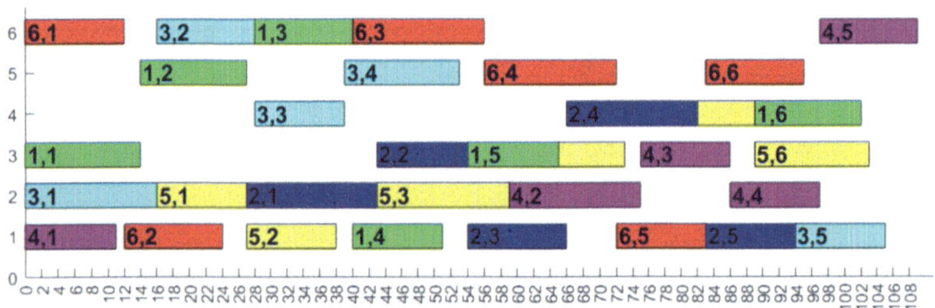


**Fig. 6** Original Scheduling Gantt Chart.

## 5 Conclusion

To achieve intelligent multi-objective scheduling in flexible job shops, this study addresses the issues of NSGA-II's susceptibility to local optima, insufficient population diversity, and low convergence efficiency by proposing an improved algorithm. A crowding degree-based adaptive crossover operator is designed to balance global exploration and local exploitation through dynamically adjusted crossover probabilities: reducing crossover probability in high-crowding regions to avoid solution clustering and increasing it in sparse regions to enhance exploration, thereby maintaining population diversity and uniformity in Pareto front distribution. Additionally, a bidding mechanism is introduced to refine the selection process in non-dominated sorting, prioritizing solutions with higher competition intensity within the same non-dominated front through competitive evaluation metrics, thereby enhancing selection pressure and precision. Experimental simulations validate the feasibility and effectiveness of the proposed algorithm.

**Acknowledgements.** The relevant researches described in this paper are supported by the Fundamental Research Funds for the Central Universities (No. ND2024002)" and Humanities and Social Sciences Research Program of Ministry of Education, China (No. 23YJA630016).

## References


1. Kong, F., Wu, D.H., Ji, Z.C.: Flexible job shop scheduling optimization based on double-layer particle swarm optimization algorithm. J. Comput. Appl. **35**(2), 476–480 (2015)
2. Zhang, Q.W., Wang, C.: An improved grey wolf optimization algorithm for flexible job shop scheduling problem. J. Lanzhou Univ. Technol. **48**(3), 103–109 (2022)
3. Peng, J.G., Liu, M.Z., Zhang, M.X., et al.: Cloud model evolutionary multi-objective flexible job shop scheduling based on improved non-dominated sorting. J. Mech. Eng. **50**(12), 198–205 (2014)
4. Kaplanoglu, V.: An object-oriented approach for multi-objective flexible job-shop scheduling problem. Expert Syst. Appl. **45**, 71–84 (2016)
5. Ziaee, M., Mortazavi, J., Amra, M.: Flexible job shop scheduling problem considering machine and order acceptance, transportation costs, and setup times. Soft. Comput. **25**(1), 1–17 (2021)
6. Amiri, F., Shirazi, B., Tajdin, A.: Multi-objective simulation optimization for uncertain resource assignment and job sequence in automated flexible job shop. Appl. Soft Comput. **75**, 190–202 (2019)
7. Chen, J., Xie, Y., Ni, J.: Brain storm optimization model based on uncertainty information. In: 2014 Tenth International Conference on Computational Intelligence and Security, pp. 99–103 (2014)
8. Mirjalili, S., Lewis, A., et al.: Grey wolf optimizer. Adv. Eng. Softw. **69**, 46–61 (2014)
9. Al-Betar, M., Awadallah, M., Faris, H., et al.: Natural selection methods for grey wolf optimizer. Expert Syst. Appl. **113**, 481–498 (2018)
10. Mohammed, A., Ngadi, M.A., Dishung, S.I., et al.: An efficient symbiotic organisms search algorithm with chaotic optimization strategy for multi-objective task scheduling problems in cloud computing environment. J. Netw. Comput. Appl. **133**, 60–74 (2019)
11. Ouyang, H.C., Zhang, T.R., Wu, D.H.: Multi-objective flexible job shop scheduling based on improved NSGA-III algorithm. Control Eng. China. **30**(1), 105–112 (2023)
12. Yang, C.Y., Deng, Y.B., Sun, M.K.: Research on multi-objective flexible job shop scheduling problem based on NSGA-III algorithm. Inf. Technol. Inform. **12**, 121–123 (2021)
13. Wang, G., Gao, S., Fang, S.J.: Adaptive NSGA-II-based multi-objective flexible job shop scheduling. Mach. Tool Hydraul. **50**(18), 129–135 (2022)
14. Du, X.L., Zhang, N., Meng, F.Y., et al.: Improved NSGA2 algorithm for flexible job shop scheduling problem. Mod. Mach. Tool Autom. Manuf. Tech. **5**, 182–186 (2022)
15. Jia, Z.H., Wang, Y., Wu, C., et al.: Multi-objective energy-aware batch scheduling using ant colony optimization algorithm. Comput. Ind. Eng. **131**, 41–56 (2019)

# The Influence of Additional Heat Supplied to an FDM-Manufactured Product on its Dimensional Accuracy

Bohdan Wojciechowski(✉), Adam Hamrol, and Wiesław Kuczko

Poznan University of Technology, Faculty of Mechanical Engineering, Poznan 60-965, Poland
wojciechowskibohdan@gmail.com

**Abstract.** This article presents research focused on improving dimensional accuracy and reducing thermal deformations in Fused Deposition Modeling (FDM) 3D printing by locally supplying additional heat in the form of a directed stream of hot air to the product shaping zone. The experiments were conducted using two materials: PET-G and ABS. Dimensional accuracy was evaluated by comparing the actual dimensions of the printed models with the dimensions of the CAD model, with the differences interpreted as height deviations. The deviations obtained during printing without additional airflow were compared to those obtained when using post-heating with air at a temperature of 95 °C. The results showed that the height deviations (H1 and H2) were reduced when using the heated air stream, especially for PET-G and ABS. These promising results indicate the potential for further development of this method. Future work will focus on investigating various parameters of heat delivery to the printed surface, including temperature, flow rate, application speed, and the distance from the nozzle to the heat source. The long-term goal is to develop guidelines for improving the geometric accuracy of prints produced on 3D printers that do not have a heated build chamber.

**Keywords:** FDM · PETG · ABS · Heat Shrink

## 1 Introduction

Fused Deposition Modeling (FDM) technology is widely applied in rapid prototyping as well as in the fabrication of finished products [1–4] made from thermoplastic materials. Under the influence of temperature and pressure, these materials exhibit the behavior of a viscous fluid [5], and upon cooling, they solidify while retaining the imposed geometry. Depending on the intended application and the required mechanical performance, both visual and functional prototypes can be produced using this method [6].

The most commonly used thermoplastics in FDM technology include PLA (polylactide), PET-G (polyethylene terephthalate glycol), ABS (acrylonitrile–butadiene–styrene), and PA (polyamide) [7]. Among them, PLA is particularly popular due to its ease of processing (low shrinkage, suitability for open-chamber printers) and favorable environmental characteristics — it is derived from starch fermentation, biodegradable,


A. Hamrol et al. (Eds.): MANUFACTURING 2026, LNME, pp. 284–295, 2026.
https://doi.org/10.1007/978-3-032-21701-1_21

and compostable. However, PLA is not suitable for manufacturing functional prototypes. For such applications, materials such as ABS, PET-G, and PA are preferred. These polymers exhibit significantly higher glass transition temperatures and superior impact resistance, which considerably expand their potential applications compared to PLA [8, 9].

The main drawback of these materials is their substantial thermal shrinkage, which can cause deformation, cracking, or spontaneous detachment of the printed part from the build platform during the manufacturing process [9]. To mitigate these effects, enclosed and heated build chambers are commonly employed. While effective, this approach increases the complexity and cost of the printers, as well as the energy consumption of the production process [10].

Deformation phenomena in FDM-manufactured components and the related issues of heat transfer during the process have been extensively studied. For instance, reference [8] investigated the effect of build-plate temperature on the geometry of a cube-shaped specimen made from PLA material.

It was demonstrated that the higher the build-plate temperature — and consequently, the greater the amount of heat supplied to the printed part — the smaller the geometric deviations of the finished product (Fig. 1).

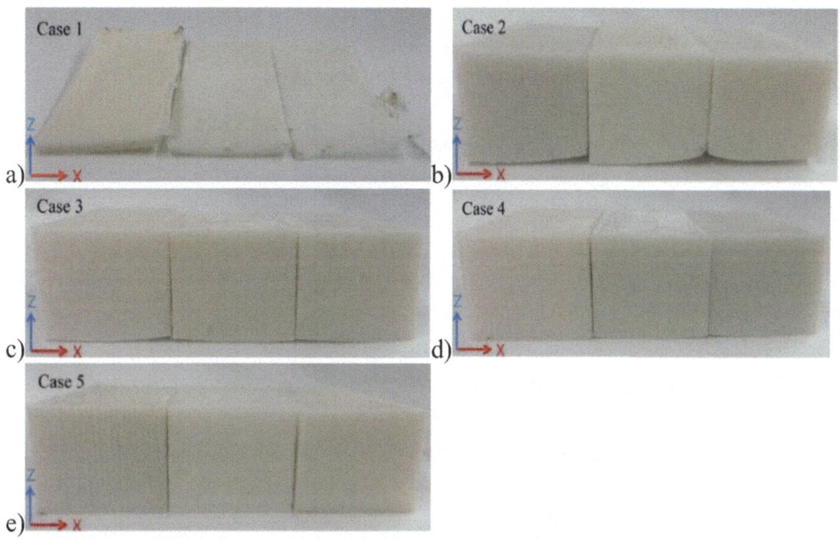


**Fig. 1.** View of the specimens manufactured with different build-plate temperatures: a) 40 °C, b) 50 °C, c) 70 °C, d) 90 °C, e) 110 °C [11].

In study [11], the authors focused on the influence of printing parameters on residual stresses in Fused Deposition Modeling (FDM) using numerical simulation. The analysis considered key process parameters, including printing temperature, build-plate temperature, chamber temperature, and printing speed (Fig. 2).

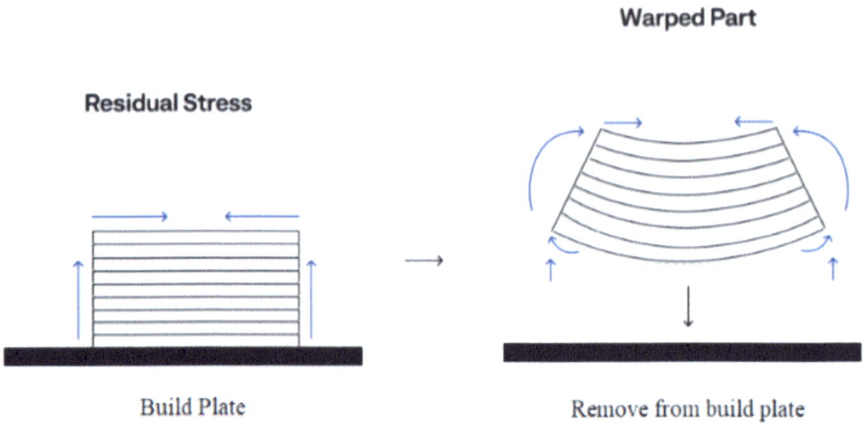


**Fig. 2.** Internal generation of residual stresses due to thermal gradients in 3D printing [11].

According to the results, it was determined that the printing temperature has the greatest influence on minimizing residual stresses (59.3%), followed by the chamber temperature (20.5%) and the printing speed (19.4%) (Fig. 3).

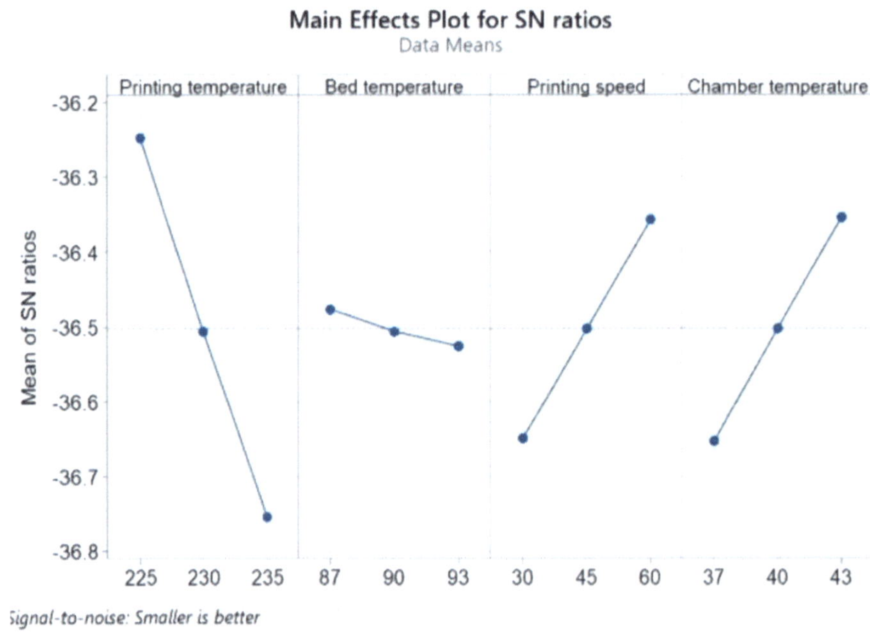


**Fig. 3.** Signal-to-noise ratio chart for residual stresses [11].

Based on the conducted study, the authors concluded that during the printing process, the phase change of the material is a significant cause of part deformation, resulting from uneven temperature distribution on the build plate and heat conduction within the part at the initial stages of the process.

In study [12], heat propagation and product deformation were analyzed using a model developed with the Finite Element Method (FEM) (Fig. 4). In the highest point of the specimen (Fig. 5), a dimensional deviation of approximately 7% was observed.

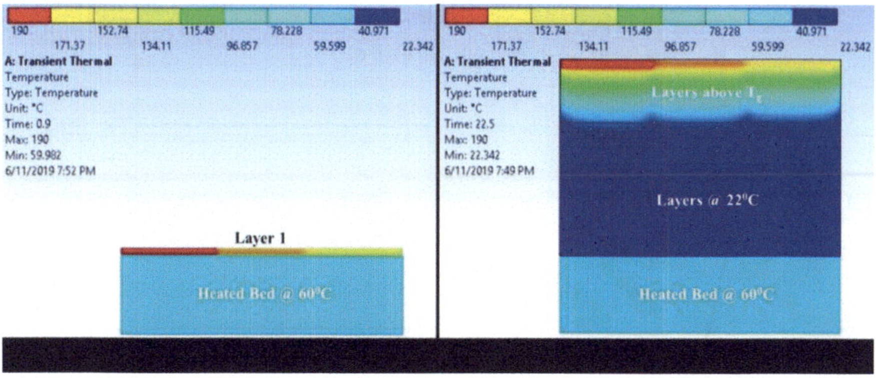


**Fig. 4.** Theoretical model of heat propagation in the specimen [12].

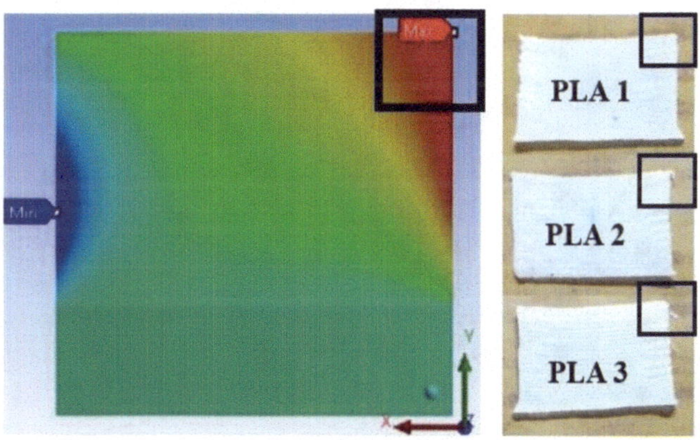


**Fig. 5.** Deviation map showing the difference between the theoretical and actual dimensions in the upper right corner of the specimen [12].

In study [13], the authors was developed experimental method using a ~ 40 W halogen bulb as an auxiliary heat source to compare the configurations of no heating, preheating, and postheating at print speeds ranging from 3 to 50 mm/s, in order to fabricate single-wall tensile specimens with a wall thickness of 0.8 mm and a layer height of 0.3 mm.

Comparing the results (Fig. 6), preheating demonstrated better strength performance at higher print speeds, whereas postheating was more effective at lower speeds. At higher

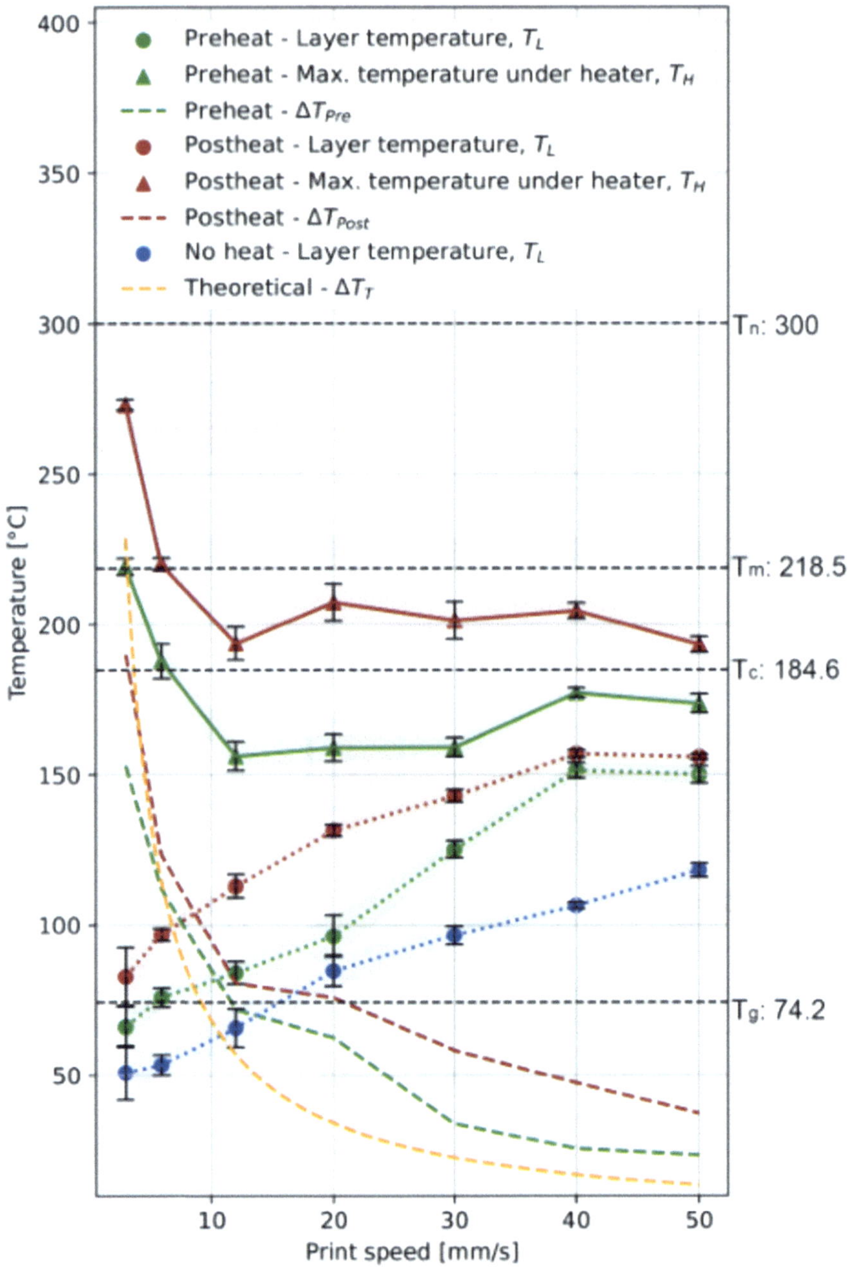


**Fig. 6.** Thermal measurements at different print speeds [13].

speeds, however, the thermal energy does not have sufficient time to penetrate through the top layer in the postheating scenario, while the cooling time after preheating decreases, resulting in stronger specimens in the preheating case.

Based on the studies presented above, as well as others not discussed here [11, 12, 13], it can be concluded that during the printing process, the phase change of the material is a significant cause of part deformation, resulting from uneven build-plate temperature distribution and heat conduction within the part. This indicates that deformations in FDM-manufactured components constitute an important, yet still relatively under-researched issue, and a research gap exists regarding the lack of externally applied heat in studies, which substantially affects the practical application of this technology. The author will focus on delivering directed, externally applied, and locally concentrated heat to specific areas of the printed model.

## 2 Aim and Research Methodology

One method to reduce deformation in parts manufactured using FDM technology is to maintain consistent thermal conditions within the printer's build chamber. This is typically achieved using enclosed, properly insulated, heated chambers with forced air circulation. Such an approach requires supplying energy to heat the chamber volume and maintain a constant temperature throughout the entire manufacturing process. A less energy-intensive alternative is localized heating of the part in areas particularly prone to deformation caused by thermal shrinkage.

The aim of the study presented in this work is to investigate the extent to which applying an additional, appropriately directed heat flux to an FDM-manufactured part can reduce thermal deformation of the print. The specimen selected for the study is shown in Figs. 7 and 8. It is a customized mounting component for securing a lighting fixture in the Keyence vision system.

Due to its geometry, the use of printing supports was necessary; these supports simultaneously conducted heat away from the heated build plate and acted as a radiator, dissipating heat from the model.

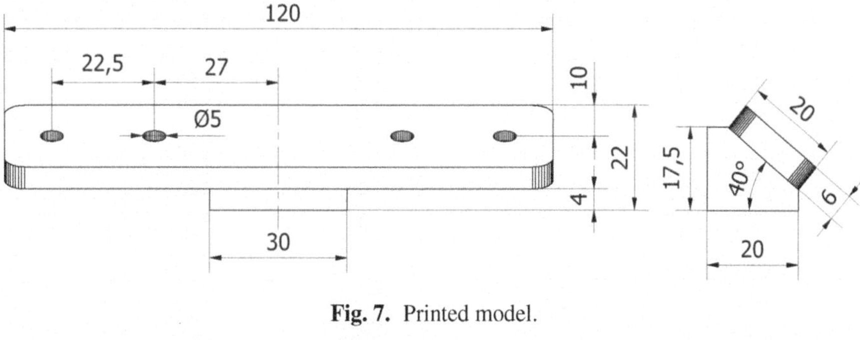


**Fig. 7.** Printed model.

The method of supplying heat to the specimen is shown in Fig. 9. The heat source was a hot-air blower with electronically adjustable temperature in the range of 40–600 °C

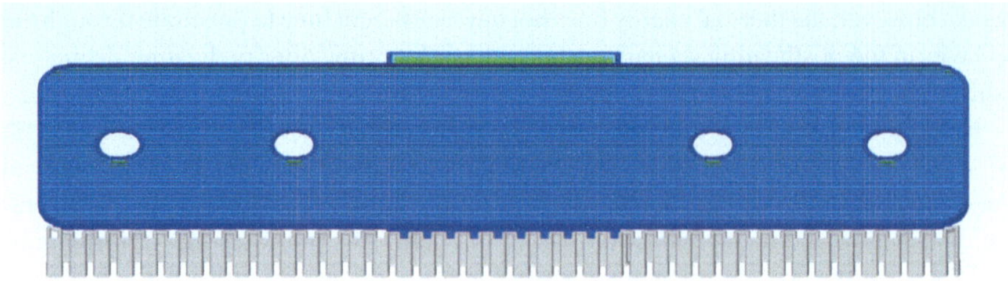

**Fig. 8.** Printed model with support structures (gray color).

and an air flow rate of 400 L/min. Three types of nozzles were used in the experiments to deliver heat to the model. Type 1 was a non-directional circular nozzle, resulting in uniform heating of the entire model. Type 2 directed the heated air; however, thermal imaging analysis demonstrated that type 3 most effectively redirected heat toward the right side of the model.

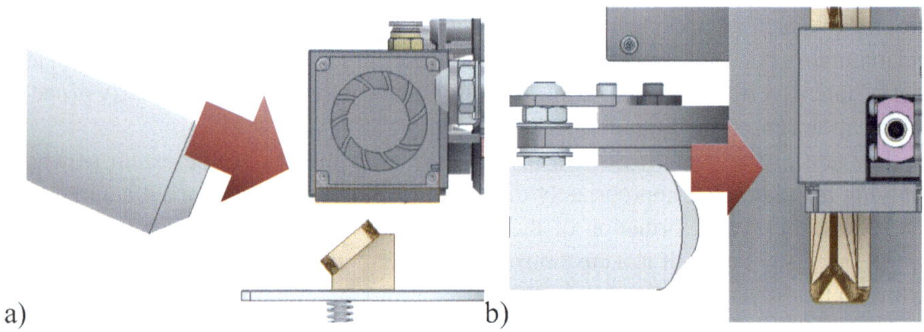


**Fig. 9.** Method of directing the air flow onto the specimen: a) side view b) top view.

Nozzles with varied geometries were used to direct the air flow (Fig. 10). The specimens were fabricated using a Creality Ender 3 printer with a standard 0.4 mm brass nozzle. G-code files were prepared using Simplify3D software (version 4.1.2, Simplify3D, LLC, Cincinnati, OH, USA).

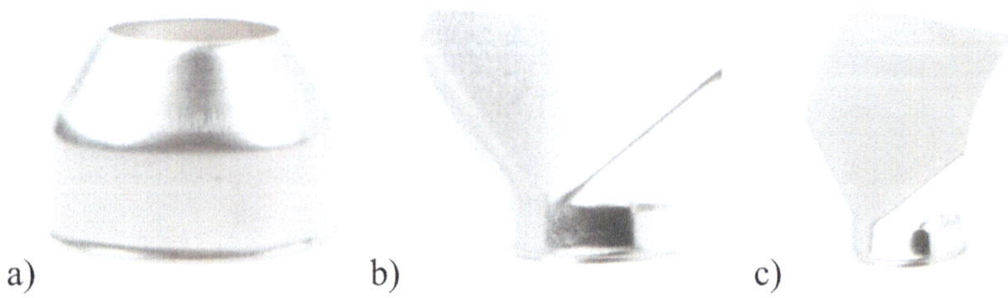


**Fig. 10.** Nozzles for directing hot air: a) type 1, b) type 2, c) type 3.

Two material variants of the model were tested: Fiberlogy PET-G and Fiberlogy ABS (Fiberlab S.A., Brzezie, Poland). The materials used in the study exhibit significant thermal shrinkage. During the initial trials conducted without additional heat input, it was not possible to complete the printing of the model using ABS due to deformation. PETG demonstrates lower thermal shrinkage; however, following the investigation of ABS, the author identified it as the next material worthy of examination. The nozzle and build-plate temperatures were set as follows: for PET-G, 245 °C and 70 °C, respectively; for ABS, 240 °C and 100 °C. The following printing parameters were applied: layer deposition speed 50 mm/s, layer height 0.2 mm, number of wall contours 4, number of top and bottom layers 4, and 20% infill with a rectilinear pattern.

The temperature of the air stream was measured using a FLIR T530 thermal camera with an 18 mm lens (Fig. 11). The measurement accuracy was ±2 °C. Five air-flow variants were tested:

I. Nozzle type 1 – air temperature 45 °C.
II. Nozzle type 2 – air temperature 55 °C.
III. Nozzle type 3 – air temperature 75 °C.
IV. Nozzle type 3 – air temperature 85 °C.
V. Nozzle type 3 – air temperature 95 °C.

a)
b)

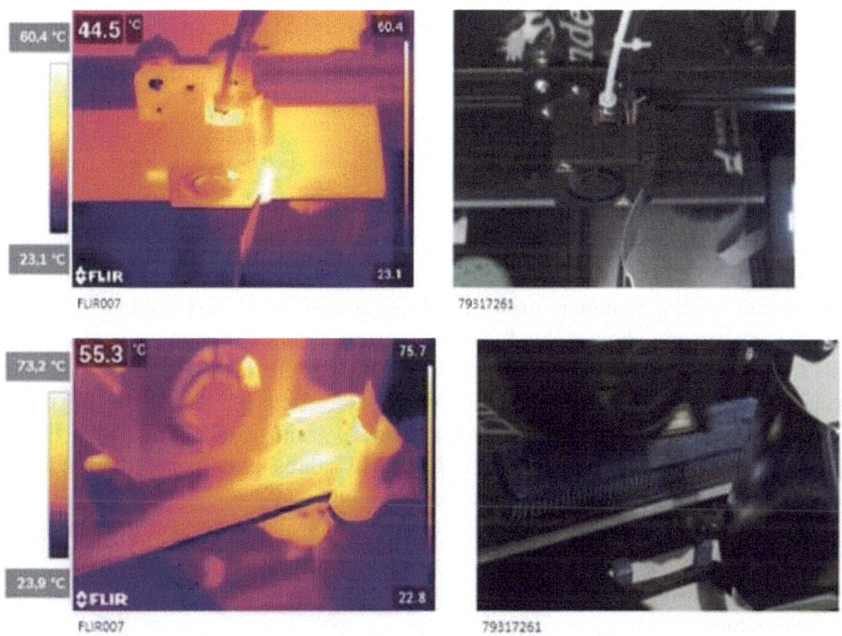


**Fig. 11.** Hot air flow view: a) 100 °C, nozzle type 2; b) 120 °C, nozzle type 3.

The manufactured model was scanned using a dual-source Phoenix v|tome|x m X-ray micro-computed tomography (µCT) system (Waygate Technologies, Wunstorf, Germany). From the scan, the edge heights were determined using GOM Inspect software (version V8, ZEISS Industrial Metrology, Braunschweig, Germany), measured from the base plane of the specimen in the cross-sections shown in Fig. 1. The obtained dimensions were compared with the CAD model dimensions, and their differences were treated as height deviations. Height deviations of the edges H1 and H2 of the model were determined (Fig. 12). The magnitude of deviations was compared for prints produced without additional air flow and with air flow applied.

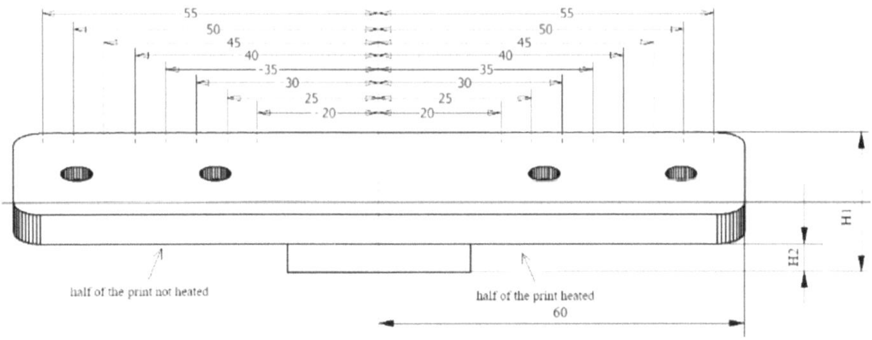


**Fig. 12.** Studied specimen indicating edge heights H1 and H2, measurement locations, and areas heated with an additional heat source.

## 3 Results

The results demonstrate that localized thermal assistance applied during the FDM process can significantly reduce thermally induced distortions of printed parts. By supplying heat directly to regions prone to shrinkage, the temperature gradients within the printed model are reduced, leading to a more uniform cooling process and lower residual stresses.

The experiments showed that for PET-G material, surface heating according to variants I, II, and III did not produce positive effects—the deformations of the model during printing without additional air flow and with air flow were comparable. Positive results were obtained only for air-flow variants IV and V, with variant V yielding the most favorable outcome. This indicates the existence of a threshold thermal input required to effectively mitigate shrinkage-induced deformation.

For ABS printing, the process without additional heat or with air flow in variants I to IV was unstable. The specimens deformed to such an extent that completing the model print was impossible. Only with hot air flow according to variant V were successful prints obtained.

The graphs in Fig. 13 refer to printing with ABS and PET-G when the surface of the printed model was heated according to variant V (i.e., air flow at 95 °C delivered through nozzle type 3) and to printing without surface heating. They present the magnitudes of height deviations H1 and H2 obtained for the respective materials.

The graphs indicate that supplying a suitably directed heat flux to a part produced by FDM can improve its geometric accuracy compared to fabrication without additional

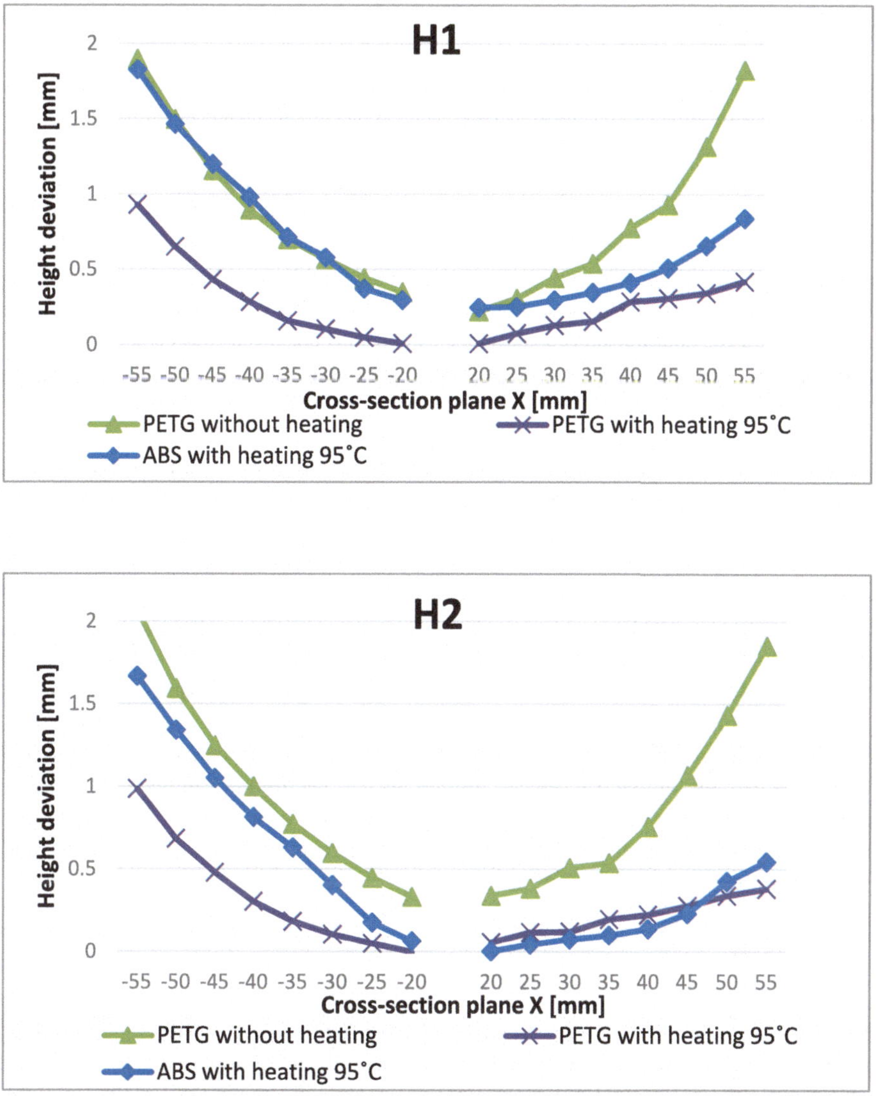


**Fig. 13.** Graphs of height deviations H1 and H2.

heat input. When printing without supplementary heating, the deviation in height H1 at the end of the model (cross-sect. 55 mm) was nearly 2 mm, whereas with additional heat supply it was less than 0.5 mm. A similar improvement was observed for height H2. It should be noted that, for both heights, deviations in the unheated section also decreased (from nearly 2 mm to 1 mm). For ABS material, in the heated variant, height deviations H1 and H2 were similar to those observed for PETG. However, due to the aforementioned process instability, these results cannot be directly compared with the non-heated variant.

## 4 Discussion

The effectiveness of the proposed approach strongly depends on both the temperature and the directionality of the applied heat flux. Among the tested configurations, the type 3 nozzle provided the most efficient heat transfer to the critical region of the model, which was confirmed by thermographic analysis and geometric deviation measurements. The geometry and directionality of the applied heat flux play a decisive role in the effectiveness of the proposed method. The type 1 nozzle, providing uniform and diffuse heating, was insufficient to compensate for localized heat losses caused by the presence of support structures. The type 2 nozzle improved flow concentration; however, only the type 3 nozzle enabled precise delivery of thermal energy to the critical region of the model, as confirmed by thermographic analysis and quantitative geometric deviation measurements. The application of localized heating influenced not only the directly heated region but also the global deformation of the part. Height deviations in non-heated regions were also reduced, suggesting that the additional heat input contributed to a more homogeneous thermal field within the entire component.

The presence of support structures significantly enhanced heat dissipation from the printed model to the build platform, acting as a thermal sink. The localized hot-air flow effectively compensated for this heat loss, indicating that the proposed method is particularly suitable for geometrically complex parts requiring extensive support material.

Compared to fully heated build chambers, the investigated approach offers a potentially energy-efficient alternative, as thermal energy is supplied only where it is required. This suggests that localized heating could be implemented in open-frame or low-cost FDM systems without major hardware modifications. The findings indicate that further improvements may be achieved through adaptive control of the heat source, including real-time adjustment of air temperature and flow direction based on part geometry and printing progress.

## 5 Conclusions

The obtained results are therefore encouraging for further research. In subsequent studies, the authors intend to focus on investigating variants of heat delivery to the surface of the printed part, differing in heat flux temperature and intensity, heating rate, distance of the application point from the nozzle, and other parameters. The ultimate goal of this work will be to develop guidelines enabling the improvement of geometric accuracy in prints produced on printers not equipped with a heated build chamber.

## References

1. Singh, S., Singh, G., Prakash, C., Ramakrishna, S.: Current status and future directions of fused filament fabrication. J. Manuf. Process. **55**, 288–306 (2020)
2. Bakar, N.S.A., Alkahari, M.R., Boejang, H.: Analysis on fused deposition modelling performance. J. Zhejiang Univ. Sci. A. **11**, 972–977 (2010)

3. Prince, J.D.: 3D Printing: An industrial revolution. J. Electron. Resour. Med. Libr. **11**(1), 39–45 (2014). https://doi.org/10.1080/15424065.2014.877247
4. Chua, C.K., Leong, K.F.: 3D printing and additive manufacturing: Principles and applications, 5th edn. World Scientific Publishing Company, Location (2017)
5. Novakova-Marcincinova, L., Novak-Marcincin, J., Barna, J., Torok, J.: Special materials used in FDM rapid prototyping technology application. In: 2012 IEEE 16th International Conference on Intelligent Engineering Systems (INES), pp. 73–76 (2012)
6. Dey, A., Roan Eagle, I.N., Yodo, N.A.: Review on filament materials for fused filament fabrication. J. Manuf. Mater. Proc. **5**(3), 69 (2021)
7. Bellini, A., Guceri, S.: Mechanical characterization of parts fabricated using fused deposition modeling. Rapid Prototyp. J. **9**(4), 252–264 (2003)
8. Cano-Vicent, A., et al.: Fused deposition modelling: Current status, methodology, applications and future prospects. Addit. Manuf. **47**, 102378 (2021)
9. Dey, A., Roan Eagle, I.N., Yodo, N.A.: Review on filament materials for fused filament fabrication. J. Manuf. Mater. Process. **5**, 69 (2021)
10. Dey, A., Yodo, N.A.: Systematic survey of FDM process parameter optimization and their influence on part characteristics. J. Manuf. Mater. Proc. **3**(3), 64 (2019)
11. Hussein, A., Peter, F., Lajos, B.: Thermal analysis of FDM process parameters and their effects on residual stress. In: VIII. Gépészeti Szakmakultúra Konferencia (2023)
12. Rodríguez-Panes, A., Claver, J., Camacho, A.M.: The influence of manufacturing parameters on the mechanical behaviour of PLA and ABS pieces manufactured by FDM: A comparative analysis. Materials. **11**, 1333 (2018)
13. Nesheim, O.S., Eikevåg, S.W., Steinert, M., Elverum, C.W.: Identifying thermal effects during 3D printing by comparing in-layer infrared pre- and postheating of carbon fiber reinforced polyamide. Addit. Manuf. **101**, 104705 (2025)
14. Spoerk, M., Gonzalez-Gutierrez, J., Sapkota, J., Schuschnigg, S., Holzer, C.: Effect of the printing bed temperature on the adhesion of parts produced by fused filament fabrication. Plast. Rubber Compos. **47**(1), 17–24 (2018). https://doi.org/10.1080/14658011.2017.1399531
15. Hopkins, N., Jiang, L., Brooks, H.: Energy consumption of common desktop additive manufacturing technologies. Clean. Eng. Technol. **2**, 100068 (2021). https://doi.org/10.1016/j.clet.2021.100068
16. Choi, Y.-H., Kim, C.-M., Jeong, H.-S., Youn, J.-H.: Influence of bed temperature on heat shrinkage shape error in FDM additive manufacturing of the ABS-engineering plastic. World J. Eng. Technol. **4**, 186–192 (2016)
17. Manikandan, N., Vignesh, T., Prasath, C., Ismail, M.: Thermo-Mechanical analysis of fused filament fabrication process. IOP Conf. Series Mater. Sci. Eng. **764**, 012034 (2020)
18. He, K., Zhang, Q., Hong, Y.: Profile monitoring based quality control method for fused deposition modeling process. J. Intell. Manuf. **30**, 947–958 (2019)
19. Vanaei, H.R., Deligant, M., Shirinbayan, M., et al.: A comparative in-process monitoring of temperature profile in fused filament fabrication. Polym. Eng. Sci. **61**(1), 68–76 (2020). https://doi.org/10.1002/pen.25555
20. Kuznetsov, V., Solonin, A., Tanitov, A., Urzhemtsev, O.: Increasing of strength of FFF 3D printed parts by influencing on temperature-related parameters of the process. Rapid Prototyp. J. **25**(1), 144–155 (2019)
21. Cicala, G., Giordano, D., Tosto, C., Filippone, G., Recca, I.: Polylactide (PLA) filaments a biobased solution for additive manufacturing: Correlating rheology and thermomechanical properties with printing quality. Materials. **11**, 1191 (2018)

# Determinants of Energy Transition Efficiency in the Era of Industry 4.0

Alla Polyanska[1,2](✉), Dariusz Sala[1], and Vladyslav Psyuk[2]

[1] AGH University of Krakow, av. Adama Mickiewicza 30, 30-059 Krakow, Poland
polyanska@agh.edu.pl

[2] Ivano-Frankivsk National Technical University of Oil and Gas, Karpatska, 15, 76019 Ivano-Frankivsk, Ukraine

**Abstract.** The article examines the energy transition through factors' analysis defining its outcome. The set of factors is identified based on review of academic literature and expert reports in the field of energy transition. the shift from Industry 4.0 to Industry 5.0 strengthens the focus on assessing both the efficiency of the energy transition and its driving forces has become increasingly important. In this context, energy productivity is employed as an analytical outcome that captures the combined economic and energy-related effects of the transition. Using factor analysis, including elasticity estimates, standardised beta coefficients, correlation matrices, and forecasting models, the study identifies critical bottlenecks in the energy transition, highlights the most influential determinants of energy transformation, and summarises the key risks that constrain progress. The results indicate that energy-system readiness, ICT use in the workplace, and gross domestic expenditure on R&D constitute innovation–digital core of energy productivity, providing the strongest positive contributions to its improvement. At the same time, significant risks are associated with high carbon intensity of power generation, energy import dependence, and price pressures, which systematically affect performance at both national and enterprise levels. These findings provide an empirical basis for designing targeted energy transition policies and supporting measures.

**Keywords:** Energy Transition · RES · Factor Modelling Analysis · Energy Productivity · Risks · Industry 4.0

## 1 An Introduction

The transition from Industry 4.0 to Industry 5.0 intensifies the focus on sustainable development and the energy transition, necessitating the identification of both the drivers of progress and the constraints that limit transformative change. The energy transition is a process of global transformation of energy systems, involving a gradual shift away from fossil fuels toward renewable energy sources and improved energy efficiency. In this context, the study of the energy-transition environment becomes particularly relevant [1]. For effective transition planning, it is essential to assess the environment of renewable energy production and consumption, the development of infrastructure, the scale of innovative solutions, and the degree of readiness for systemic change.


A. Hamrol et al. (Eds.): MANUFACTURING 2026, LNME, pp. 296–312, 2026.
https://doi.org/10.1007/978-3-032-21701-1_22

Ambitious energy transition goals include rapid $CO_2$ emissions reduction and a movement toward a net-zero trajectory, the large-scale deployment of renewables, increased energy efficiency, the electrification of sectors using clean electricity, the modernization of grids to integrate renewables, as well as universal access, affordability, and reliability of the energy supply [2].

Understanding the growth dynamics of green energy sources enables the modeling of how realistic it is to achieve a targeted level of decarbonization by 2030 or 2050. Studying the factors that determine the outcomes of the energy transition enables the assessment of the effectiveness of government programs that promote renewable energy, tax incentives, tariff regulations, and support mechanisms for clean technologies.

The energy transition reshapes labor markets, influences energy prices, and affects energy security. Reports regularly published by international organizations characterize the transition process and highlight the key factors shaping it [3]. Using these data is valuable for validating research that relies on a broad spectrum of determinants and for building models that can serve as an analytical basis for examining the factors underlying the transition [4].

This perspective provides a foundation for studying country-specific cases to understand the features, differences, barriers, and innovative practices of the energy transition. Therefore, the topic of the article is timely and highly relevant.

Therefore, the aim of the article is to examine the determinants that shape the progress of the energy transition and to assess their influence on the outcome represented by energy productivity. To achieve this aim, the study pursues the following objectives: to identify and analyze the factors affecting the implementation of energy-transition goals; to determine the most influential determinants; and to highlight the constraints and risks that require additional support and deeper awareness on the part of decision-makers.To avoid potential misunderstanding, the energy productivity in this article is employed as an indicator of system-level energy efficiency, not as a measure of energy production, generation, or supply.

To achieve the stated objectives, the article is structured as follows. The introduction outlines the relevance of the research and its prospective contribution. The literature review examines quantitative models used in previous studies to assess the factors influencing energy transition processes in the context of Industry 4.0, as well as scholarly works that identify key determinants shaping the energy transition. The methodology section presents the selected analytical approach, which systematically evaluates the impact of the chosen factors on energy transition outcomes, examines the elasticity and potential risks associated with these effects, and identifies the relative priority of the determinants considered. The results section presents and interprets the empirical findings to identify the most influential factors. The discussion section addresses the main discussion points and outlines the study's limitations. Finally, the conclusions summarize the key results and highlight their implications.

## 2 Literature Review

An analysis reporting on progress toward the Sustainable Development Goals (particularly Goal 7) shows positive dynamics, including improvements in productivity, reductions in energy consumption, and measurable progress toward achieving the energy-transition target of increasing the share of renewables to 42.5% by 2030 [5]. However, despite global advances, the pace of the energy transition remains insufficient to meet the objectives of the Paris Agreement. According to the Global Energy Transition Outlook (2025), the trajectory toward net-zero continues to shift toward the end of the century, while the risk of warming near 2.2 °C persists. The report concludes that countries must urgently strengthen energy resilience, scale domestic renewable energy alongside grid infrastructure, storage, and demand flexibility, and ensure prudent fiscal policies to absorb shocks and accelerate decarbonization [6].

In studying the impact of Industry 4.0 on the energy transition and the associated transformational changes, scholars have applied various quantitative analysis models. In particular, the article [7] explores a model of digital maturity of energy enterprises, which is described through a system of interrelated factors covering development strategy, human capital, organizational culture, technology, organizational structure, and marketing. Empirical studies [8] provide mixed evidence on the environmental effects of Industry 4.0. Using panel data for 15 countries and 8 manufacturing sectors over the period 2012–2020, it was found that different dimensions of digitalisation have heterogeneous impacts on manufacturing energy intensity. In particular, robot density is associated with a reduction in energy intensity, while higher digital capital intensity is linked to an increase in energy intensity; however, the share of ICT specialists shows no statistically significant effect. These results suggest that the influence of Industry 4.0 on energy efficiency is not uniform and depends on the specific type of digital technology applied. Studies [9] have highlighted the role of institutional and structural factors in shaping the relationship between innovation and energy efficiency within the context of Industry 4.0. Using data for MENA countries over the period 1990–2016, the authors find that technological innovation has a positive impact on energy efficiency, while the expansion of the shadow economy has a negative impact, and structural economic transformation contributes positively to energy efficiency outcomes.

Empirical research on 41 countries from 2003 to 2021 confirms that geopolitical risks hinder the energy transition by amplifying price volatility, disrupting supply chains, and diverting political priorities. These effects are moderated in countries with strong renewable energy capacity, robust fiscal mechanisms, and flexible labor markets, whereas resource-dependent and militarized economies tend to lag behind. Policy recommendations emphasize strengthening resilience, expanding domestic renewable energy, and maintaining sound fiscal management [10].

A broader assessment of the transition environment has enabled the development of models that rank countries by the effectiveness of their efforts and identify best practices relevant for shaping national energy-transition policies [11]. Study [12] emphasizes the importance of mobilizing investment and strengthening the financial prerequisites for low-carbon development, drawing on global experience in clean energy finance. Modern technological solutions in the energy sector are widely recognized as a pivotal accelerator of the transition to sustainable development [13], while energy efficiency remains a key driver of progress [14]. Research [15, 16] also emphasizes the importance of incorporating energy-pricing mechanisms into policy design, highlighting the role of comprehensive financial architectures that encompass public and private finance, market mechanisms, innovative tools, risk-mitigation instruments, and institutional capacity building. Information technologies are another critical enabler, allowing the processing of large-scale data required for managing the complexity of the energy transition [17]. A separate stream of research indicates that the shift to renewable energy reshapes workforce dynamics and organizational culture, and that Green HRM can accelerate renewable energy integration by embedding sustainable practices and aligning human capital development with environmental goals [18].

Further evidence indicates that the transition is shaped by a wide set of determinants, including capital-market conditions, regulatory frameworks, energy efficiency, intermittency and cost of renewables, infrastructure structure, electricity-pricing mechanisms, ownership models, cross-border electricity trade, environmental implications, competition dynamics, and macro-shocks such as extreme weather, pandemics, and financial crises. Consumer behaviour and international trade channels exert additional influence [19]. Another study proposes a multifactor model of renewable-energy development, emphasizing the need for country-specific analysis due to significant local heterogeneity (in energy efficiency, import dependence, and geography) and cross-sectoral differences across electricity, heating, cooling, and transportation. Thus, single-country case studies are methodologically justified for evaluating transition outcomes [20]. Scholars also point to the shortage of single-country and sector-level analyses, which limits the validity of causal inferences in the energy-growth nexus. A deeper sectoral approach is therefore required [21]. Readiness-to-change assessments must also be incorporated, as they provide the basis for effective decision–making under environmental instability and help identify interdependencies with stakeholders [22].

Overall, progress in the energy transition is evident, yet the process remains slow and constrained. Major gaps include insufficient country-level and sector-specific research that incorporates dynamic lags, primarily descriptive assessments of factor impact without quantitative modeling, an underestimation of the role of human capital and digitalization, and a limited understanding of the relationship between energy and productivity. Bridging these gaps requires combining panel and desk-based research approaches, testing standardized influence metrics, and identifying bottlenecks more precisely - steps necessary for accelerating progress toward achieving the 42.5% renewable energy target by 2030.

## 3 Methods and Data

To achieve the stated aim, the study applies a метод OLS (Ordinary Least Squares) - factor analysis that we apply for the energy transition investigation, reducing a large set of indicators to a smaller number of core drivers that shape system behaviour:

$$Y_t = \beta_0 + \sum \beta_i \bullet X_{i,t} + \varepsilon_t \quad (1)$$

where $\varepsilon_t$ - random error.

The research design includes the following key stages. First, data was collected to characterise the factors influencing the outcomes of the energy transition. Data collection was based on a review of prior empirical studies relevant to the problem area. For this study, variables $X_1$ through $X_{10}$ were compiled (Table 1).

For validation purposes, $CO_2$ emissions ($Y_{val}$) were selected as an additional indicator, a common approach for assessing the proportionality between economic growth and carbon-emission intensity.

After identifying the factors, the next step involved analyzing their influence on the outcome variable, defined here as energy productivity (Y). The factor analysis was conducted using the following analytical components:

- construction of a correlation matrix, which provides a quantitative assessment of the linear relationship between all pairs of variables in the dataset. Each element $R_{ij}$ of the correlation matrix R represents the Pearson correlation coefficient between variable $X_i$ and variable $X_j$. The element $R_{ij}$, reflecting the correlation between variables i and j, is calculated as:

$$R_{ij} = \frac{\sum_{k=1}^{n} (X_{i,k} - \overline{X}_i)(X_{j,k} - \overline{X}_j)}{\sqrt{\sum_{k=1}^{n} (X_{i,k} - \overline{X}_i)^2}\sqrt{\sum_{k=1}^{n} (X_{j,k} - \overline{X}_j)^2}} \tag{2}$$

$R_{ij}$ - Pearson correlation coefficient between the variable $X_i$ and $X_j$;
n - number of observations;
$X_{i,k}$ and $X_{j,k}$ - k-th observation for variables $X_i$ and $X_j$ relatively;
$\overline{X}_i$ and $\overline{X}_j$- arithmetic mean values of variables $X_i$ and $X_j$.

- applying the elasticity coefficient ($E_i$) shows the percentage change in Y in response to a 1% change $X_i$:

$$E_i = \beta_i \left( \frac{\overline{X}_i}{\overline{\overline{Y}}} \right) \tag{3}$$

$\beta_i$ - unstandardized coefficient for $X_i$;
$\overline{X}_i$ - average value Xi;
$Y^{-}$ - average value Y.

- the use of the standardized Beta coefficient ($\beta_i$), which makes it possible to assess the influence of variable factors on the resulting indicator:

$$\beta_i = B_i \left( \frac{\sigma X_i}{\sigma Y} \right) \tag{4}$$

$B_i$ - unstandardized coefficient for $X_i$;
$\sigma X_i$ - standard deviation $X_i$;
$\sigma Y$ - standard deviation Y.

- the calculation of a risk index that shows how much stronger the influence of negative factors is compared to positive ones, as well as the moment when the balance of their

**Table 1.** Initial data for factor analysis. Source: data collected from Eurostat.

| Years | Composite index of energy system readiness/efficiency for the transition (reliability, accessibility, sustainability) | Share of net imports in the country's total available energy, % | Share of energy from renewable sources in gross final energy consumption, % | Capital investment in generation, grid infrastructure, and renewable energy projects, million eur | Average electricity price for households or businesses (with/without taxes), euro per kWh | $CO_2$ emissions per unit of electricity generated, g $CO_2e$/kWh | Energy efficiency (energy consumption relative to output), million tons of oil equivalent | Share of employees using computers/the internet for work purposes, by NACE Rev.2 sectors (enterprises ≥10 employees), % | R&D personnel by performing sector (business, government, higher education, non-profit), headcount or full-time equivalent | Gross domestic expenditure on R&D by performing sector and source of funding, million euro | Energy productivity, euro per kilogram of oil equivalent | Greenhouse gas emission intensity from energy consumption |
|---|---|---|---|---|---|---|---|---|---|---|---|---|
| | $X_1$ | $X_2$ | $X_3$ | $X_4$ | $X_5$ | $X_6$ | $X_7$ | $X_8$ | $X_9$ | $X_{10}$ | $У$ | $У_{вал}$ |
| 2013 | 49 | 29,42 | 11,43 | 2237,1 | 0,14 | 9 | 89,49 | 27,19 | 0,56 | 3864,02 | 4,248 | 92,6 |
| 2014 | 49 | 29,42 | 11,61 | 3113,21 | 0,14 | 9,2 | 89,49 | 27,19 | 0,62 | 3864,02 | 4,248 | 92,2 |
| 2015 | 49 | 29,85 | 11,88 | 3724,37 | 0,14 | 9,2 | 90,05 | 27,19 | 0,65 | 4316,51 | 4,386 | 92,3 |
| 2016 | 48,8 | 30,76 | 11,40 | 2234,05 | 0,13 | 9,3 | 94,83 | 28,34 | 0,67 | 4112,34 | 4,308 | 90,9 |
| 2017 | 48,3 | 38,27 | 11,06 | 2291,27 | 0,15 | 9,6 | 99,08 | 28,65 | 0,86 | 4834,04 | 4,327 | 90,3 |
| 2018 | 51 | 43,51 | 14,94 | 2936,86 | 0,14 | 9,7 | 104,06 | 30,33 | 0,98 | 6018,49 | 4,399 | 86 |
| 2019 | 51 | 45,24 | 15,38 | 3891,76 | 0,13 | 9,6 | 100,19 | 32,23 | 0,97 | 7046,92 | 4,769 | 83,6 |
| 2020 | 52,9 | 42,77 | 16,10 | 3178,15 | 0,15 | 9,3 | 96,84 | 37,6 | 1,03 | 7292,84 | 4,812 | 82,2 |
| 2021 | 58 | 40,46 | 15,62 | 3047,86 | 0,16 | 10,1 | 103,94 | 40,08 | 1,05 | 8252,84 | 4,832 | 82,2 |
| 2022 | 59 | 46,03 | 16,85 | 3303,29 | 0,15 | 9,4 | 98,4 | 41,44 | 1,01 | 9539,36 | 5,329 | 82,2 |
| 2023 | 59,7 | 48,02 | 16,49 | 2899,62 | 0,18 | 8,6 | 93,27 | 41,96 | 1,02 | 11694,4 | 5,686 | 82,2 |
| 2024 | 61,3 | 48,02 | 16,48 | 2899,62 | 0,21 | 8,6 | 93,27 | 44,57 | 1,02 | 11694,4 | 5,686 | 82,2 |

impact on the final model shifts. The Z-score indicates how much each indicator in each year deviates from its mean value:

$$z(X_J(\mathrm{t})) = \frac{\mathrm{X_J(t)} - \overline{X_j}}{\sigma_{X_j}} \tag{5}$$

constructing a factor-analysis model and forecasting changes in the outcome variable depending on variations in the most influential factors (Table 2).

**Table 2.** Forecasting models.

| Model | Formula |
|---|---|
| Quadratic Regression Model | $Y_t = \beta_0 + \beta_1 t + \beta_2 t^2$ |
| Logistic Growth Model | $Y_t = \frac{K}{1+Ae^{-Bt}}$ |
| Damped Holt's Trend Method | $\hat{Y}_{t+h} = L_t + \left(\varphi^h\right)T_t$ |
| Gompertz Growth Model | $Y_t = K \cdot \exp\left(-Ae^{-Bt}\right)$ |

To conduct the study, the variable factors presented in Table 2 were selected.

Thus, the proposed methodology enables a comprehensive examination of the influence of factors on energy productivity, specifically the volume of energy produced. Given the wide range of indicators studied by researchers and experts in the field of energy transition, this approach enables the selection of those factors that exert the strongest influence, have the capacity to alter energy productivity, and ultimately help identify bottlenecks that hinder progress.

The proposed methodology represents a logical continuation of the conducted research, within which groups of countries were classified according to the achieved efficiency of the energy transition based on a multifactor Data Envelopment Analysis (DEA) [11].

## 4 Results

Investigation of the energy transition leads us to select and analyze the determinants that we worked out based on the method described above. Based on the collected data, a correlation matrix was constructed to identify the relationships among the factors and to examine the influence of the selected determinants on the outcome variable, specifically energy productivity. Table 3 presents the correlation matrix calculated using formula (2).

The strongest associations with energy productivity are observed for innovation- and digitalisation-related factors. Specifically, $X_{10}$ (gross domestic expenditure on R&D) and $X_8$ (ICT use at work) exhibit very high positive correlations (0.96–1.00). They are followed by $X_1$ (energy system readiness) and $X_7$ (energy efficiency indicator), which also show strong positive relationships ($r = 0.82$–$0.94$). $X_3$ (share of renewables in

**Table 3.** Correlation matrix.

| | $X_1$ | $X_2$ | $X_3$ | $X_4$ | $X_5$ | $X_6$ | $X_7$ | $X_8$ | $X_9$ | $X_{10}$ | Y | $Y_{val}$ |
|---|---|---|---|---|---|---|---|---|---|---|---|---|
| $X_1$ | 1.00 | 0.765 | 0.85 | 0.18 | 0.77 | −0.28 | 0.19 | 0.98 | 0.74 | 0.96 | 0.94 | −0.83 |
| $X_2$ | 0.77 | 1.00 | 0.90 | 0.29 | 0.54 | −0.07 | 0.54 | 0.82 | 0.94 | 0.88 | 0.82 | −0.93 |
| $X_3$ | 0.85 | 0.90 | 1.00 | 0.45 | 0.51 | −0.07 | 0.44 | 0.90 | 0.90 | 0.89 | 0.84 | −0.97 |
| $X_4$ | 0.18 | 0.29 | 0.45 | 1.00 | −0.04 | 0.15 | 0.12 | 0.19 | 0.31 | 0.23 | 0.23 | −0.36 |
| $X_5$ | 0.77 | 0.54 | 0.51 | −0.04 | 1.00 | −0.62 | −0.17 | 0.74 | 0.44 | 0.78 | 0.79 | −0.49 |
| $X_6$ | −0.28 | −0.07 | −0.07 | 0.15 | −0.62 | 1.00 | 0.77 | −0.24 | 0.18 | −0.35 | −0.47 | −0.03 |
| $X_7$ | 0.19 | 0.54 | 0.44 | 0.12 | −0.17 | 0.76 | 1.00 | 0.26 | 0.70 | 0.22 | 0.07 | −0.55 |
| $X_8$ | 0.98 | 0.82 | 0.90 | 0.19 | 0.74 | −0.24 | 0.26 | 1.00 | 0.82 | 0.96 | 0.94 | −0.91 |
| $X_9$ | 0.74 | 0.94 | 0.90 | 0.31 | 0.44 | 0.18 | 0.70 | 0.82 | 1.00 | 0.81 | 0.71 | −0.95 |
| $X_{10}$ | 0.96 | 0.88 | 0.89 | 0.23 | 0.78 | −0.35 | 0.22 | 0.96 | 0.81 | 1.00 | 0.98 | −0.88 |
| Y | 0.94 | 0.82 | 0.84 | 0.23 | 0.79 | −0.47 | 0.07 | 0.94 | 0.714 | 0.98 | 1.00 | −0.82 |
| $Y_{val}$ | −0.83 | −0.93 | −0.97 | −0.36 | −0.49 | −0.03 | −0.55 | −0.91 | −0.95 | −0.88 | −0.82 | 1.00 |

final energy consumption) exhibits a positive correlation with productivity (0.84–0.90), although its effect warrants further investigation.

The strongest negative influence is associated with $X_6$ ($CO_2$ per kWh), which demonstrates a notable negative correlation with productivity (down to −0.47), reflecting the fact that higher carbon intensity suppresses improvements in efficiency. $X_5$ (electricity price) exhibits weak or mixed correlations with the outcome variable, indicating that it primarily functions as a demand-side stress factor rather than a productivity driver. The $X_5$-$X_2$ relationship represents a risk-moderating mechanism.

The factors themselves form a dense development cluster, comprising $X_1$, $X_2$, $X_3$, $X_8$, $X_9$, and $X_{10}$, which are highly correlated with one another (0.82–0.98). This is a clear indication of multicollinearity, necessitating a VIF assessment. Computed VIF values for most variables far exceed the critical threshold of 10 (Table 4), confirming that these indicators reflect closely related transition processes and largely duplicate one another. Under such conditions, the interpretation of individual regression coefficients becomes unstable, and analytical emphasis should instead be placed on elasticity estimates of key factor groups (innovation capacity, digitalisation, carbon intensity, and energy efficiency) as well as the decomposition of changes in ln(Y) into factor contributions over 2013–2024.

**Table 4.** VIF coefficients table.

| VIF | 181.37 | 51.14 | 52.83 | 10.04 | 10.76 | 83.83 | 76.92 | 118.31 | 102.88 | 249.07 |
|---|---|---|---|---|---|---|---|---|---|---|
| Variable | $X_1$ | $X_2$ | $X_3$ | $X_4$ | $X_5$ | $X_6$ | $X_7$ | $X_8$ | $X_9$ | $X_{10}$ |

Thus, the core cluster for analysing productivity consists of $X_{10}$, $X_8$, and $X_1$. The main negative influences are associated with factors $X_6$ and $X_3$, whose effects should be evaluated with time lags. Due to the strong cross-correlations among $X_1$, $X_3$, $X_8$, $X_9$, and $X_{10}$, these variables can be used for regularized regression techniques with partial $R^2$. Accounting for these dependencies ensures stable conclusions and valid triggers for the expert system.

To determine the intensity of changes in the outcome variable under the influence of the factors selected within the research framework of the energy transition, we applied formula (3) to compute elasticities. The calculation results are presented in Table 5.

**Table 5.** Calculated elasticity values.

| Factor | Elasticity | cilo | cihi | pvalue |
|---|---|---|---|---|
| X1 | 0.1607 | 0.0074 | 0.3139 | 0.0477 |
| X2 | 0.0958 | 0.0413 | 0.1503 | 0.0285 |
| X3 | −0.3433 | −0.3942 | −0.2924 | 0.0074 |
| X4 | 0.1414 | 0.1116 | 0.1713 | 0.0106 |
| X5 | −0.1670 | −0.2003 | −0.1337 | 0.0100 |
| X6 | −1.1045 | −1.3427 | −0.8664 | 0.0108 |
| X7 | 0.6561 | 0.3938 | 0.9184 | 0.0200 |
| X8 | 0.3869 | 0.2945 | 0.4793 | 0.0120 |
| X9 | −0.1650 | −0.2260 | −0.1039 | 0.0185 |
| X10 | 0.1739 | 0.1201 | 0.2277 | 0.0155 |

The calculated elasticity values enable the determination of the strength and direction of the influence of each factor $X_i$, while the ci values indicate the reliability and precision of the estimate. The p-value reflects formal statistical significance. Taken together, these columns enable the prioritization of factors and help assess how confidently each can be used for decision-making. The magnitude of the impact of the analysed factors is illustrated graphically in Fig. 1.

Elasticity-based assessment revealed two groups of leading factors and a clear polarity of effects:

Strong factors ($|\beta| \geq 0.5$): $X_6$ (electricity carbon intensity) exhibits the largest negative effect, $\beta = -1.10$, indicating that reducing $CO_2$/kWh yields the strongest gains in energy productivity; $X_7$ (energy efficiency) is the main positive driver, $\beta = +0.66$.

Medium factors ($0.2 \leq |\beta| < 0.5$): $X_8$ (ICT use at work) supports productivity, $\beta = +0.39$, whereas $X_3$ (renewables share) is formally negative, $\beta = -0.34$; this "minus" is likely due to multicollinearity with $X_1$, $X_2$, $X_8$, and $X_{10}$, and should be interpreted with caution.

Moderate factors ($|\beta| < 0.2$): positive effects for $X_{10}$ (R&D investment) $\beta = +0.17$, $X_1$ (composite readiness) $\beta = +0.16$, $X_4$ (capital expenditure) $\beta = +0.14$, $X_2$ (net imports) $\beta = +0.10$; negative effects for $X_5$ (electricity price) $\beta = -0.17$ and $X_9$ (R&D personnel) $\beta = -0.17$ (likely overlap with $X_{10}$).

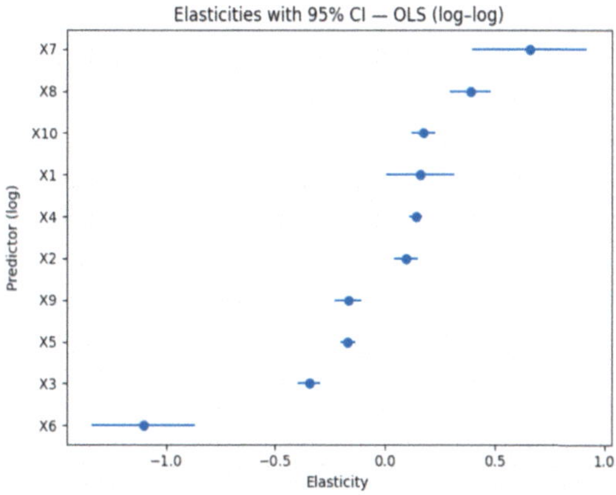


**Fig. 1.** The elasticity of factors influencing energy productivity.

All coefficients are statistically significant ($p < 0.05$), and the signs in the figure are consistent with Table 5. Priority should be given to decarbonising generation ($X_6$) and scaling system-wide energy efficiency ($X_7$). Digitalisation ($X_8$) is an important additional level. R&D investment ($X_{10}$) and capital expenditure ($X_4$) provide supplementary but modest gains, whereas higher tariffs ($X_5$) worsen outcomes. According to the proposed methodology and Formula (3), we examine the effects of the factors on the outcome indicator in Table 6.

**Table 6.** Factor effects.

| Factors | Contribution to ΔlnY | Share of total % |
|---|---|---|
| X3 | −0.1258 | −43.1327 |
| X9 | −0.0989 | −33.9296 |
| X5 | −0.0644 | −22.0977 |
| X7 | 0.0271 | 9.3102 |
| X1 | 0.0360 | 12.3420 |
| X4 | 0.0367 | 12.5841 |
| X2 | 0.0470 | 16.1032 |
| X6 | 0.0502 | 17.2227 |
| X8 | 0.1912 | 65.5762 |
| X10 | 0.1926 | 66.0549 |

*Notes: The logarithmic specification is used because a log-log model allows the estimated coefficients to be interpreted as elasticities, and changes in ln(X) and ln(Y) to be treated as approximations to percentage changes in the variables.*

Table 6 reports the contribution of each factor to the total change in the logarithm of the outcome over 2013–2024. A positive value means the factor pushed energy productivity up, a negative value means it pushed it down. The sum of all contributions equals the observed change $\Delta \ln Y$=0.2916. Table 6 also shows each contribution's share as a percentage of the total $\Delta \ln Y$. Positive percentages indicate the portion of growth delivered by the factor; negative percentages indicate how much it offset growth. Because of the opposing signs, the percentages add up to 100%. For example, $\ln_{X10} = 0.1926$ and 66.05% means that the growth in R&D expenditure contributed +0.1926 log points, i.e., approximately two-thirds of the total increase in $\Delta \ln Y$. By contrast, $\ln X3 = -0.1258$ and − 43.13% means that this factor reduced the net increase, absorbing approximately 43% of what was achieved by the other factors. Figure 2 presents a graphical distribution of the examined factor contributions.

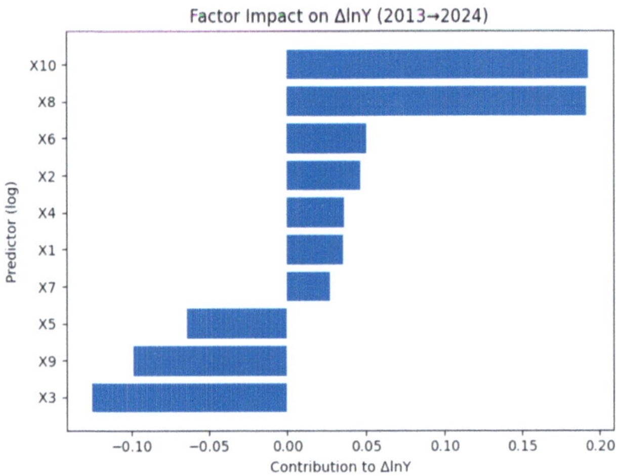


**Fig. 2.** Factor contributions to the outcome indicator.

Figure 2 illustrates the contribution of individual factors to changes in the indicator from 2013 to 2024. The largest gains were achieved through expenditure on research and development, as well as the use of information technologies at work. Additional positive contributions came from lower carbon intensity of power generation, higher net energy imports, capital investment, overall system readiness of the energy sector, and improvements in energy efficiency. Negative contributions to this specification were attributed to the share of renewable energy, the headcount of R&D personnel, and higher electricity prices. Overall, progress was driven by investment in research and by digitalisation, while it was constrained by price pressures, the structural features of the renewable energy mix, and misalignment in the deployment of scientific human capital.

In the article, we also used the risk score. Figure 3 presents the dynamics of risks. The R score is a weighted sum of the annual z-score changes of negative factors-carbon intensity of generation ($X_6$), import dependence ($X_2$), and price pressure ($X_5$) - net of the mitigating effect of renewables ($X_3$). When $R > 0$, negative drivers dominate and risk increases; when $R < 0$, mitigating factors prevail and risk declines. The dashed line marks the zero balance. Year by year, 2013–2017 constitutes a zone of elevated risk (0.47 to 0.68, with a peak in 2017); 2018 is near neutral (0.09); 2019–2020 show a transition to a more stable zone (down to −0.69); 2021 returns to risk (0.12); 2022–2023 exhibit the best balance in the series (−0.83 and − 1.01); and 2024 shows a partial rebound of risk that remains below zero (−0.50). The risk analysis, based on the computed results, enables the identification of risk regimes that require specific actions.

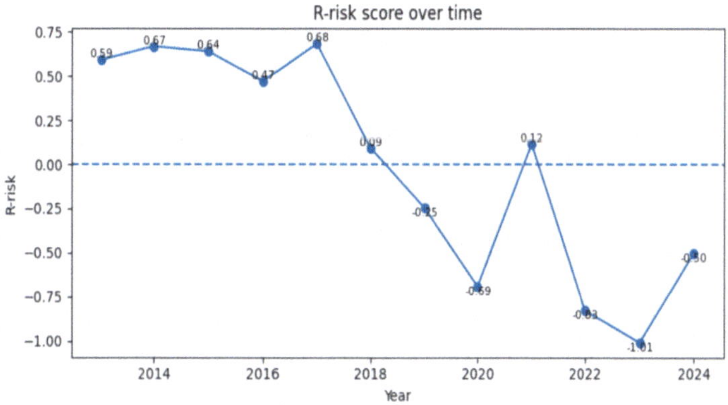


**Fig. 3.** Dynamics of the R score in 2023–2024.

When $R > 0$, crisis-response measures should be activated; when $R < 0$, programs for renewables and efficiency should be scaled up to maintain the favorable trend.

For validation, we use $Y_{val}$, the $CO_2$ emissions intensity from energy consumption. This indicator is appropriate because, as productivity rises, the dynamics of $CO_2$ emissions must also be considered: growth in energy productivity under an energy transition should substantially outpace any increase in $CO_2$ emissions. The additional indicator $Y_{val}$ shows a downward trend (−11% over 2013–2024) and has a strong inverse relationship with the key indicators of the energy transition and innovation development (renewable share, ICT use, R&D investment, R&D personnel, composite readiness), as well as with the integrated indicator $Y$ ($r \approx -0.82$). By contrast, $Y_{val}$ is strongly positively correlated with the integrated risk in $R_{risk}$ ($r \approx 0.90$).

Given the results of the factor analysis, it is appropriate to build a model that can forecast changes in energy productivity while accounting for changes in the determining factors. Based on the set of forecasting models (Table 1), Table 7 reports the $R^2$ values used to assess the adequacy of the energy-productivity forecasts.

Figure 4 presents the dynamics of the forecasting models for energy productivity.

The polynomial model provides the most accurate estimates ($R^2 = 0.9597$). Thus, according to the results of the multifactor analysis, the primary systemic constraint on energy efficiency growth is the carbon intensity of electricity generation ($X_6$). Slow decarbonisation of the fuel mix and limited grid flexibility reduce the effectiveness of other policy measures; therefore, without accelerated RES deployment and infrastructure modernisation, productivity remains structurally constrained. The variables $X_{10}$, $X_8$, and $X_1$ are well-justified as forming the productivity core, as they combine the innovation,

**Table 7.** Value $R^2$ by model.

| Factor | $R^2$ |
|---|---|
| Quadratic Regression Model | 0.9597 |
| Logistic Growth Model | 0.8705 |
| Damped Holt's Trend Method | 0.8691 |
| Gompertz Growth Model | 0.8686 |
| Holt_linear | 0.8908 |

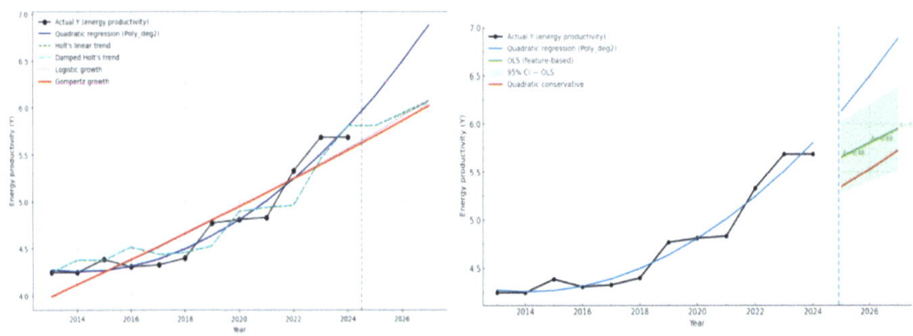


a) b)

**Fig. 4.** Forecasting energy productivity for 3 years (2025–2028): a) using the representation of different types of regressions; b) using the representation of the OLS methodology and the regression with the highest coefficient of determination.

digital, and institutional dimensions, provide the largest positive contribution to ΔlnY, and create a robust set of triggers for the expert model.

Energy efficiency ($X_7$) is the strongest individual driver; however, its impact is limited by the slow modernisation of fixed assets and restricted access to financing and ESCO mechanisms. Labour digitalisation ($X_8$) exhibits a stable and moderately strong positive effect; however, its impact depends on the digital maturity of enterprises and access to high-quality ICT infrastructure. The apparent negative effect of renewables ($X_3$) in the partial model is driven by multicollinearity with $X_1$, $X_2$, $X_8$, and $X_{10}$, as well as by network-integration lags. The static linear specification does not capture the time, dispatching requirements, and grid preparation needed for renewable benefits to materialize.

Within the R&D block, an asymmetry of effects is observed: R&D investment ($X_{10}$) demonstrates a clear positive influence, whereas changes in the number of R&D personnel ($X_9$) do not produce a systematic effect due to weak conversion from laboratory to market, insufficient integration with industry, and delayed implementation of research outcomes. This explains the divergent directions of their effects without implying any contradiction between the roles of $X_{10}$ and $X_9$.

The Fig. 4 compares two models for forecasting energy productivity Y: a factor-based ordinary least squares (OLS) specification and a quadratic time trend presented with two trajectories (baseline and conservative). The OLS model, estimated on the most influential variables - such as system readiness and the lagged share of renewables-projects a moderate increase in Y from approximately 5.66 in 2025 to approximately 5.95 in 2027, and displays uncertainty through a 95% confidence band, which makes it suitable for scenario management in which changes in explanatory factors translate into changes in outcomes. The quadratic trend implies a faster dynamic: the baseline trajectory rises from approximately 6.14 in 2025 to approximately 6.89 in 2027, while the conservative trajectory increases from approximately 5.35 to approximately 5.76; in an extended variant, a fast trajectory would range from roughly 6.93 to 8.05 over the same horizon. The vertical dashed line indicates the start of the forecast period, which begins after 2024. In practical terms, OLS should serve as the primary decision-support model for policy and investment analysis. The quadratic baseline and conservative paths, on the other hand, provide a sanity check and outer bounds for plausible trajectories. If desired, both approaches can be combined into a single blended estimate.

## 5 Disscussions

There is very high multicollinearity across the institutional-innovation block, as system readiness, renewable energy share, digitalization, and research and development move together (the variance inflation factor indicator peaks at $X_{10} \approx 203$, $X_1 \approx 144$, and $X_8 \approx 106$). Under these conditions, coefficient estimates become unstable, so the negative partial effect for the renewable share ($X_3$) is likely an artefact of joint variation rather than a true drag on energy productivity.

For robust interpretation, the specification should be simplified or regularised. A practical core model includes the carbon intensity of generation ($X_6$) as the main constraining factor, energy efficiency ($X_7$) as the main positive driver, one representative from the innovation-digital block ($X_8$ or $X_{10}$), plus two meaningful moderators: electricity price ($X_5$) and capital expenditure ($X_4$). Other highly correlated indicators ($X_1$, $X_2$, $X_3$, $X_9$) should be dropped or aggregated into a single index based on standardised scores.

Capital expenditure, research and development, and renewable share require accounting for delayed effects; a static linear model underestimates them, so partial coefficients of determination and lag structures should be tested. The price factor acts as a moderator of sensitivity to volatility and demand flexibility incentives, warranting separate regime-shift testing. Despite the small annual sample, the largest potential gain in energy productivity comes from rapid decarbonisation (lower $X_6$) and scaling up energy-efficiency programmes (stronger $X_7$); the next tier of impact comes from digitalisation and research and development ($X_8$ and $X_{10}$), while tariff design and capital expenditure operate as supportive levers with time lags.

It seems debatable to consider a single country for studying the impact of this factor on the energy transition. However, we note that the study is a logical continuation of our previous study of the effectiveness of the energy transition in the EU. Accordingly, groups of countries were identified based on their effectiveness in the energy transition, which requires a more detailed consideration of the experience of a particular country to gain a deeper understanding of the structural factors driving energy changes within the country. Factor analysis across heterogeneous national energy system risks conflates structurally different transition pathways, thereby reducing the interpretability of latent factors.

One limitation of the OLS-based forecasting model is the possible presence of endogeneity. Some explanatory variables, such as system readiness or the share of renewable energy, may both influence and be influenced by energy productivity. For example, higher energy productivity can create better conditions for investment in renewables and system modernization, which makes the direction of causality ambiguous. In addition, unobserved factors, such as institutional quality or technological changes related to Industry 4.0, may affect both energy productivity and the explanatory variables. As a result, the estimated coefficients should be interpreted as associative rather than strictly causal effects. Nevertheless, the OLS model remains useful for scenario analysis, as it transparently shows how changes in key factors are linked to changes in energy productivity, while alternative time-trend trajectories provide a complementary benchmark for assessing the plausibility of the forecasts.

## 6 Conclusion

Based on the results, Poland's energy-transition policy priorities should proceed as follows: rapid decarbonisation of carbon-intensity ($X_6$) by phasing out coal and heavy fuel oil; acceleration of renewables together with storage and demand-side flexibility, supported by increased grid investment; scaling up energy-efficiency programmes in buildings and industry, standards, and ESCO models; speeding up the digitalisation of production and energy-management systems; closing the gap between R&D and market deployment; reforming tariffs to reduce the damaging effects of price volatility and to incentivise storage; and shifting analytical practice to regularised and dynamic models with lags for renewables and capital expenditure, using robust errors to verify sign stability. This configuration reduces risk pressure, strengthens transition momentum, and ensures sustained growth in energy productivity and progress toward energy-transition targets.

The results obtained in this study complement existing analyses of energy transition efficiency in EU countries and identify the factors that have enabled the current level of energy transition to be achieved [11]. They also define scenario-based perspectives in line with a country's physical capacities and the policies being implemented. The proposed methodology for assessing the factors shaping the energy transition in the context of Industry 4.0 can be applied to analyze the situation in other EU countries.

Future research should explicitly incorporate the time-lagged effects of investments, digitalization, and policy measures into the modeling framework to better capture delayed industrial responses and translate the projected post-2025 increase in energy productivity into concrete, sector-specific technological, organizational, and policy interventions that can effectively moderate or redirect this growth trajectory.

## References


1. Wang, Q., Wang, X., Li, R.: Energy transition and environmental sustainability: The interplay with natural resource rents and trade openness. Humanit. Soc. Sci. Commun. **12**, 1152 (2025). https://doi.org/10.1057/s41599-025-05521-4
2. International Renewable Energy Agency (IRENA): World energy transitions outlook 2024. IRENA Report. https://www.irena.org/-/media/Files/IRENA/Agency/Publication/2024/Nov/IRENA_World_energy_transitions_outlook_2024.pdf, last accessed 2025/10/11.
3. World Bank, ESMAP: Tracking SDG 7: The Energy Progress Report 2025. https://trackingsdg7.esmap.org/sites/default/files/download-documents/chapter6_the_ooutlook_for_sdg7.pdf, last accessed 2025/10/5.
4. Janikowska, O., Generowicz-Caba, N., Kulczycka, J.: Energy poverty alleviation in the era of energy transition – Case study of Poland and Sweden. Energies. **17**(21), 5481 (2024). https://doi.org/10.3390/en17215481
5. Eurostat: Sustainable development in the European Union. https://ec.europa.eu/eurostat/documents/15234730/19397895/KS-05-24-071-EN-N.pdf, last accessed 2025/10/3.
6. DNV: Global energy transition outlook 2025. Report. https://www.dnv.com/energy-transition-outlook/download/, last accessed 2025/10/15.
7. Polyanska, A., Savchuk, S., Zapukhliak, I., Zaiachuk, Y., Stankovska, I.: Digital maturity of the enterprise as an assessment of its ability to function in Industry 4.0. In: Trojanowska, J., Kujawińska, A., Machado, J., Pavlenko, I. (eds.) Advances in Manufacturing III. Manufacturing 2022 Lecture Notes in Mechanical Engineering, pp. 209–227. Springer, Cham (2022). https://doi.org/10.1007/978-3-030-99310-8_17
8. Matthess, M., Kunkel, S., Dachrodt, M.F., Beier, G.: The impact of digitalization on energy intensity in manufacturing sectors – A panel data analysis for Europe. J. Clean. Prod. **397**, 136598 (2023). https://doi.org/10.1016/j.jclepro.2023.136598
9. Chen, M., Sinha, A., Hu, K., Shah, M.I.: Impact of technological innovation on energy efficiency in Industry 4.0 era: Moderation of shadow economy in sustainable development. Technol. Forecast. Soc. Chang. **164**, 120521 (2021). https://doi.org/10.1016/j.techfore.2020.120521
10. Zhu, Z., Hunjra, A.I., Alharbi, S.S., Zhao, S.: Global energy transition under geopolitical risks: An empirical investigation. Energy Econ. **145**, 108495 (2025). https://doi.org/10.1016/j.eneco.2025.108495
11. Polyanska, A., Sala, D., Psyuk, V., Pazynich, Y.: A multicriteria approach to the study of the energy transition results for EU countries. Energies. **18**(20), 5406 (2025). https://doi.org/10.3390/en18205406
12. Polyanska, A., Pazynich, Y., Mykhailyshyn, K., Buketov, V.: Energy transition: The future of energy based on smart specialization. Sci. Bull. Natl. Min. Univ. **4**, 89–95 (2023). https://doi.org/10.33271/nvngu/2023-4/089. (in Ukrainian)
13. Lukashevych, Y., Evdokimov, V., Polukhin, A., Maksymova, I., Tsvilii, D.: Innovation in the energy sector: The transition to renewable sources as a strategic step towards sustainable development. Afr. J. Appl. Res. **10**(1), 43–56 (2024). https://doi.org/10.26437/ajar.30.06.2024.03
14. Polyanska, A., Pazynich, Y., Mykhailyshyn, K., Babets, D., Toś, P.: Aspects of energy efficiency management for rational energy resource utilization. Min.-Geol.-Pet. Bull. **39**(3), 13–26 (2024). https://doi.org/10.17794/rgn.2024.3.2
15. Liu, T., Hamori, S.: Spillovers to renewable energy stocks in the US and Europe: Are they different? Energies. **13**(12), 3162 (2020). https://doi.org/10.3390/en13123162
16. Long, P., Tram, N., Ngoc, P.: Financial mechanisms for energy transitions: A review article. Fulbr. Rev. Econ. Policy. **4**(2), 126–153 (2024). https://doi.org/10.1108/FREP-07-2024-0039

17. Psyuk, V., Polyanska, A.: The usage of artificial intelligence in the activities of mining enterprises. E3S Web Conf. **526**, 01016 (2024). https://doi.org/10.1051/e3sconf/202452601016
18. Chauhan, S.: Empowering the green transition: Human resource strategies for workforce transformation in renewable energy integration. Power Syst. Technol. **49**(3), 350–360 (2025)
19. Genc, T.S., Kosempel, S.: Energy transition and the economy: A review article. Energies. **16**(7), 2965 (2023). https://doi.org/10.3390/en16072965
20. Firlej, K.A., Stanuch, M.: Economics and environment. Econ. Environ. **3**(86), 89–113 (2023). https://doi.org/10.34659/eis.2023.86.3.583. (in Polish)
21. AlKhars, M., Miah, F., Qudrat-Ullah, H., Kayal, A.: A systematic review of the relationship between energy consumption and economic growth in GCC countries. Sustainability. **12**, 3845 (2020). https://doi.org/10.3390/su12093845
22. Zapukhliak, I., Zaiachuk, Y., Polyanska, A., Kinash, I.: Applying fuzzy logic to assessment of enterprise readiness for changes. Manag. Sci. Lett. **9**, 2277–2290 (2019). https://doi.org/10.5267/j.msl.2019.7.026

# The Impact of Digital Transformation on Quality Management Systems in Manufacturing Organizations

Katarzyna Piotrowska[1(✉)], Jakub Pizoń[2], and Robert Waszkowski[3]

[1] Lublin University of Technology, Faculty Mechanical Engineering, Lublin, Poland
k.piotrowska@pollub.pl

[2] Lublin University of Technology, Faculty of Management, Lublin, Poland

[3] Military University of Technology, Cybernetics Faculty, Warsaw, Poland

**Abstract.** The paper explores how digital transformation reshapes quality management systems (QMS) in manufacturing, highlighting the role of Industry 4.0 technologies in advancing modern practices of operational excellence.

It contains the contribution of tools such as the Internet of Things (IoT), artificial intelligence (AI), machine learning (ML), big data analytics, and integrated ERP/MES platforms to the ongoing shift in quality management. The discussion also includes organizational determinants of effective digital adoption, particularly employee capabilities, institutional readiness, and collaboration across departments. Drawing on contemporary research and practical examples, the paper demonstrates that well-executed digital transformation enhances alignment with ISO 9001:2015, increases process reliability, minimizes deviations, and strengthens continuous improvement efforts. Nevertheless, companies must address obstacles including system-integration difficulties, cybersecurity concerns, investment demands, and workforce skill gaps. The analysis emphasizes that achieving the full potential of Quality 4.0 requires a deliberate, staged implementation strategy that connects technological change with organizational growth. Key recommendations include developing digital skills, establishing robust data-management practices, employing predictive-analytics tools, and building cooperative relationships with technology partners to create intelligent, adaptable, and sustainable quality-management environments.

**Keywords:** Digital transformation · Quality Management System · Predictive Analytics · Process Optimization

## 1 Introduction

In the era of dynamic development of digital technologies, such as the concept of Industry 4.0, automation, the Internet of Things (IoT) and artificial intelligence (AI), manufacturing organizations are faced with the need to transform their processes towards more integrated, flexible and efficient operating models [1]. In this context, the key question is how these new technologies affect quality management systems (QMS) – both in terms of improving production processes and ensuring compliance, continuous improvement and customer satisfaction.


A. Hamrol et al. (Eds.): MANUFACTURING 2026, LNME, pp. 313–325, 2026.
https://doi.org/10.1007/978-3-032-21701-1_23

Digital transformation enables real-time data collection and analysis, automatic detection of quality deviations, and prediction of potential production problems – which contributes to improving the quality of products and processes [2]. For example, IoT implementations in manufacturing are conducive to monitoring process parameters and identifying areas that need to be corrected. As a result, it is possible to reduce the number of product defects, reduce the costs associated with corrections, and improve the competitiveness of the company.

In addition, quality management systems in manufacturing organizations must be adapted to their role in the digital environment: integration with ERP/MES systems, documentation automation, predictive analysis, tracking of the full product lifecycle – these are just some of the challenges that arise during transformation. As shown, digitalization contributes to better quality management through the integration of QMS into the production ecosystem [3].

Finally, due to global trends – growing customer demands, pressure to reduce costs and the need to react quickly to market changes – the topic of digital transformation in the context of quality management is of particular importance [4]. Therefore, this article addresses a research gap that will allow to supplement knowledge in the area of theoretical foundations of these changes, as well as practical implications for quality management systems in manufacturing companies.

## 2 The Essence and Importance of Quality Management in Production Organizations

### 2.1 The Evolution of Quality Management: From Quality Control (QC) to Comprehensive Quality Management (TQM)

Quality management in manufacturing companies has undergone a significant evolution. In the literature, there are four stages of the development of the so-called "eras" of quality: Quality Inspection (QI), Quality Control (QC), Quality Assurance (QA) and Total Quality Management (TQM). Initially, the main focus was on inspecting products after the manufacturing process – whether they meet specifications. Next, the emphasis was shifted to production processes, their statistical control and the reduction of deviations (QC) [5, 6].

The next step was quality assurance and a systems approach, which ultimately led to TQM's philosophy: quality as a company strategy, involving all employees and management in a culture of continuous improvement. In the TQM era, quality has been treated not as a function of control, but as the foundation of business management – encompassing strategic goals, management engagement, organizational culture and internal communication. Today, quality management in production is actively related to digital technologies, data analytics, Lean and Six Sigma approaches – which is the next stage in the development of quality in organizations [7–9].

### 2.2 Main Standards and Models

In manufacturing organizations that strive for a high level of quality, there are international standards and quality management models, such as [10–13]:

ISO 9001 – An international standard that sets requirements for a quality management system that emphasizes, m.in example, process approach, data-driven decision-making, continuous improvement and management involvement.

EFQM Excellence Model – A model of organizational excellence that integrates aspects of leadership, strategy, resources, processes and results, and is widely used as a framework for improving the quality and effectiveness of an organization.

Kaizen – Japanese philosophy of continuous improvement; In practice, it means the participation of employees in improving processes, eliminating waste and gradually improving quality.

Lean Management – A concept focusing on eliminating waste, increasing customer value and stabilizing production processes; Combined with quality, it is a key part of modern quality systems. The combination of standards such as ISO 9001 with models and practical methods (Kaizen, Lean) allows manufacturing organizations to manage quality more effectively not only in terms of the product, but of the entire production process and value chain.

### 2.3 The Importance of Organizational Culture and Employee Involvement in the Quality Management System

A key element of an effective quality management system is organizational culture and employee engagement. The success of such systems depends largely on whether quality becomes part of the company's day-to-day operations rather than just a set of procedures [14].

Standards and literature indicate that:

Management must demonstrate strong leadership and commitment to building a quality culture – managers must initiate, support and engage employees in quality processes.

Employees at all levels should be involved in quality improvement activities, analyzing the causes of non-compliance, proposing improvements and continuous improvement. Such participation increases the sense of responsibility, motivation and effectiveness [15].

### 2.4 The Role of Quality in Ensuring Process Stability, Cost Reduction and Increasing Customer Satisfaction

An effective quality management system in production organizations brings numerous, measurable benefits. These undoubtedly include the stabilization of processes – through standardization, monitoring and continuous improvement, companies achieve predictability of production processes, which translates into smaller quality deviations and better efficiency [16]. Another benefit is cost reduction – thanks to, among other things, reducing the number of defective products, reducing revisions and complaints, optimizing the use of resources [17]. For example, a standard quality system can reduce process errors by around 30%. Another important element is the increase in customer satisfaction – companies that provide products of high, consistent quality, build customer trust, increase their loyalty and positively influence their market image [18]. In light of these mechanisms, quality management becomes a strategic tool – not just an operational one – that allows a company to stand out in a competitive market, increase efficiency and build a lasting relationship with the customer.

## 3 Digital Transformation in Industry – The Essence and Scope

### 3.1 Definition and Significance for Manufacturing

Digital transformation means integrating digital technologies into all aspects of business operations, fundamentally changing the way organizations operate and deliver value. In manufacturing, this includes Industry 4.0 concepts: smart factories where cyber-physical systems monitor physical processes, create virtual replicas, and make decentralized decisions using IoT, AI, and cloud computing [19].

### 3.2 Key Technologies in Industry 4.0

Industry 4.0 technologies have already entered production practice for good. They allow for the implementation of solutions that have a direct impact on the level but also the efficiency of production. What is important, the Industry 4.0 solution is always composed adequately to the needs. This means that it is an interdisciplinary solution and its implementation requires analysis [20]. In addition, in the field of maintenance, Industry 4.0 solutions provide a way and techniques that can automate, using artificial intelligence, improve the complex process of activities in this area [21] (Fig. 1).

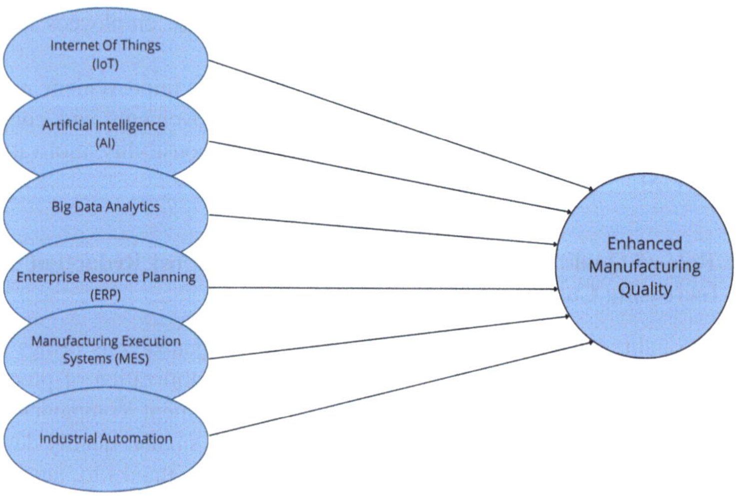


**Fig. 1.** Industry 4.0 to support quality.

The following are the key technologies from the perspective of Industry 4.0:

**Internet of Things (IoT)**
Connected sensors and devices collect real-time data from production equipment, enabling continuous monitoring of temperature, pressure, humidity, vibration, and other quality-critical parameters [22]. The industrial IoT market is showing strong growth, with 46% of manufacturers deploying IoT solutions at the facility or network level [23].

**Artificial Intelligence and Machine Learning**
AI algorithms analyze huge data sets to identify patterns, predict quality issues, and automate inspection processes [24]. Applications include computer vision for defect detection (achieving 99.86% accuracy), predictive quality analysis, and root cause analysis. AI-based quality control can run 1000 times faster than human inspection while maintaining consistency [25].

**Big Data Analytics**
Manufacturing generates massive amounts of data from machines, sensors, quality systems, and supply chains. Advanced analytics extract actionable insights for quality improvement, process optimization, and predictive maintenance. Organizations see a 36% increase as a result of performance correlation between Big Data plants [26].

**Enterprise Resource Planning (ERP)**
Integrated ERP systems provide centralized quality data repositories, real-time synchronization between departments, and automated quality flows. A modern cloud-based ERP enables scalability and remote access while reducing IT infrastructure costs [27].

**Manufacturing Execution Systems (MES)**
FEM combines enterprise planning with production operations. It provides real-time production control, quality management, tracking and compliance documentation, with integration capabilities that combine ERP systems, laboratory systems and quality platforms [28].

**Industrial Automation & Robotics**
Automated inspection systems with machine vision, 3D scanning, and AI-assisted defect detection enable 100% inspection compared to statistical sampling. Collaborative robots (cobots) work safely with human operators while maintaining precise quality measurements [29].

### 3.3 Barriers and Success Factors

The research points to significant implementation challenges: organizational barriers (cultural resistance, rigid hierarchical structures), technical challenges (complexity of integration of legacy systems, insufficient IT/OT infrastructure, data quality issues, cybersecurity risks), financial constraints (high upfront costs, unclear payback periods) and human capital gaps (48% of manufacturers have difficulties in filling production and operational roles, 35% with challenges in the improving the qualifications of employees) (Fig. 2) [30–32].

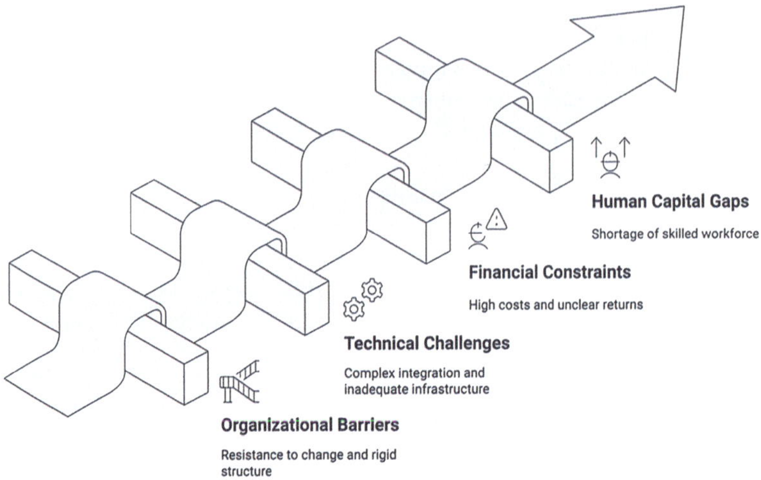


**Fig. 2.** Barriers and Success Factors.

Key success factors include management engagement, a clear strategic vision related to business value, appropriate investments, phased implementation starting with high-impact use cases, robust change management, employee training, and ecosystem partnerships using external knowledge [33].

# 4 Impact of Digital Transformation on Quality Management Systems

## 4.1 Comprehensive Technology-ISO 9001:2015 Clause Mapping

This chapter presents a detailed mapping of digital technologies to specific ISO 9001:2015 clauses along with implementation mechanisms and documented benefits (Table 1) [34–36].

**Table 1.** Digital Technologies Supporting ISO 9001:2015 Clauses.

| ISO Clause | Primary Technologies | Applications | Documented Benefits |
|---|---|---|---|
| **Clause 4: Context of Organization** | Big Data Analytics, Market Intelligence Tools | PESTLE/SWOT analysis, external/internal context monitoring, stakeholder analysis | 38% less compliance drift, 22% faster issue remediation |
| **Clause 5: Leadership** | Digital Dashboards, KPI Visualization, Real-time Reporting | Management review data, quality policy communication, resource allocation transparency | 70% higher audit pass rates, 2x fewer recurring incidents |

(*continued*)

**Table 1.** (*continued*)

| ISO Clause | Primary Technologies | Applications | Documented Benefits |
|---|---|---|---|
| **Clause 6: Planning** | Predictive Analytics, AI Risk Assessment, Risk Management Software | Risk/opportunity identification, quality objectives tracking, SMART goal management | 12% higher target achievement, proactive quality planning |
| **Clause 7: Support** | Document Management Systems, Knowledge Management Systems, Digital Training Platforms | Document control, knowledge capture/sharing, competence management, resource monitoring | 21% productivity gain, prevention of knowledge loss |
| **Clause 8: Operation** | MES, ERP Integration, IoT Sensors, Automated Inspection | Real-time operations control, automated quality checks, traceability, process enforcement | 10% reduction in returns, 44–66% production rate increases |
| **Clause 9: Performance Evaluation** | Real-time Monitoring, Big Data Analytics, Digital Audit Tools | KPI tracking, customer satisfaction measurement, internal audits, performance analytics | Faster decision-making, real-time deviation detection |
| **Clause 10: Improvement** | AI Root Cause Analysis, CAPA Systems, Improvement Analytics | Automated RCA, nonconformance tracking, trend analysis, continuous improvement platforms | 50% faster corrective action completion, prevention of recurrence |

## 4.2 Detailed Analysis by ISO 9001:2015 Clause

Digital technologies transform each clause of ISO 9001:2015 through specific implementation mechanisms. Clause 4 (Context) uses Big Data analytics to comprehensively understand context with 38% less compliance drift. Clause 5 (Leadership) uses digital dashboards, enabling a 70% higher audit throughput rate. Clause 6 (Planning) uses predictive analytics to achieve 12% higher goal achievement. Clause 7 (Support) implements document and knowledge management systems, bringing a 21% increase in productivity. Clause 8 (Operation) integrates MES, ERP, IoT sensors, and automated inspection, resulting in a 10% reduction in returns and a 44–66% increase in production rate. Clause 9 (Performance Evaluation) uses real-time monitoring and big data analytics for faster decision-making. Clause 10 (Improvement) applies AI-based root cause analysis, CAPA systems, achieving 50% faster execution of corrective actions [37, 38].

### 4.3 From Manual to Real-Time Data-Driven Quality Management

The cumulative effect of digital technologies in the clauses of ISO 9001 means a paradigm shift from manual, periodic, reactive quality management to automated, continuous, predictive quality management. Traditional QMS mechanics have relied on manual data logging, periodic inspections, reactive problem solving, segregated information, paper documentation, statistical sampling, and delayed decision-making [39] (Fig. 3).

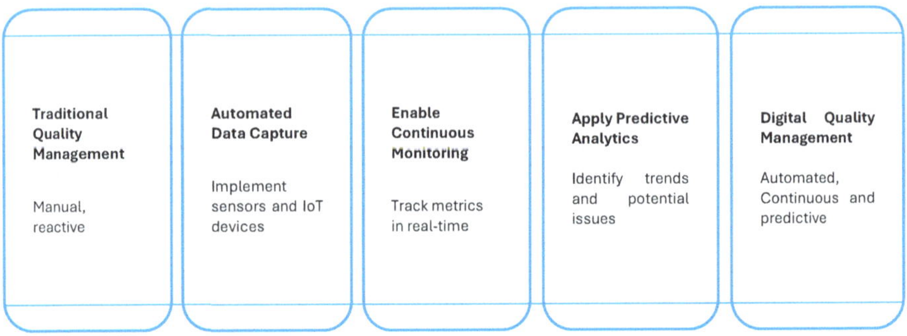


**Fig. 3.** Transforming quality management with digitalization.

Digital QMS enables automatic data capture, continuous monitoring, predictive analytics, integrated IT systems, instantly accessible electronic documentation, 100% inspection capabilities, and real-time decision-making. As a result, organizations can significantly enhance process transparency, traceability, and compliance with regulatory and normative requirements. Moreover, the integration of digital quality tools supports proactive risk management and fosters a culture of continuous improvement aligned with the principles of Quality 4.0 [40].

## 5 Conclusions and Implications for the Management Systems

Digital transformation is transforming responsibility for the quality of work. Quality professionals move from manual inspection to data analysis, training AI models, and generating insights. This role becomes more strategic and technically advanced, requiring collaboration with IT for system integration and real-time process control operations. Quality departments are shifting their focus from reactive problem solving to predictive modeling, risk assessment, and preventive actions [41].

This research comprehensively analyzed the impact of digital transformation on quality management systems in manufacturing organizations, with particular emphasis on the integration of Industry 4.0 technologies with the requirements of ISO 9001:2015. Key findings include systematic mapping of technology and ISO clauses (each ISO 9001:2015 clause uses specific digital technologies), measurable efficiency improvements (10–20% of production, reduction of defects by 22–32%, reduction of quality costs by 25–45%, inspection accuracy up to 99.86%, ROI of 300–500%), paradigm shift in quality management (transition from reactive to proactive and then predictive), implementation challenges (resistance of organizational culture, competence gaps, high costs, complexity of system integration, cybersecurity risks) and key success factors (management involvement, appropriate investments, phased implementation, robust change management, comprehensive training, ecosystem partnerships).

Based on the results of the research, we recommend a three-step implementation approach:

**Phase 1:** Evaluation and foundation

Conduct a digital maturity assessment, align current QMS with ISO 9001 requirements, define a Quality 4.0 vision aligned with the business strategy, secure management support, develop business cases for priority applications with clear ROI projections, and assess the technology landscape and integration requirements.

**Phase 2:** Pilot deployment

Start with high-impact and high-governance pilot projects (IoT sensors on critical equipment, AI-powered visual inspections, digital document management system), select technology partners, build a data infrastructure, implement a pilot project with clear success criteria, conduct comprehensive training, and document lessons learned.

**Phase 3:** Scaling and optimization

Scaling effective pilots on additional hardware, products or facilities, integrating systems (ERP-MES-QMS-LIMS) for comprehensive digital quality management, implementing advanced capabilities (predictive analytics, AI root cause analysis, digital twins), extending training to a wider workforce, implementing continuous improvement processes, and regularly measuring and communicating benefits.

Recommendations for technology adoption include the gradual introduction of solutions with varying levels of maturity. It is worth starting the process with the use of Internet of Things sensors, which are a well-developed technology characterized by a proven return on investment and the possibility of implementation in less than two years. The next step should be to implement documentation management systems, as they meet the basic requirements of clause 7 of the quality standards and immediately improve productivity. Another element is the integration of ERP/MES systems with the quality management system (QMS), which is a key condition for the effective implementation of the requirements of clause 8 and enables the flow of quality data in real time. Next, it is recommended to use artificial intelligence and machine learning algorithms in areas with high value-added potential, such as control of key components, predictive approach to quality or root cause analysis. For complex and multi-stakeholder supply chains, it is also worth considering the use of blockchain technology, especially when full traceability of processes is critical (Fig. 4).

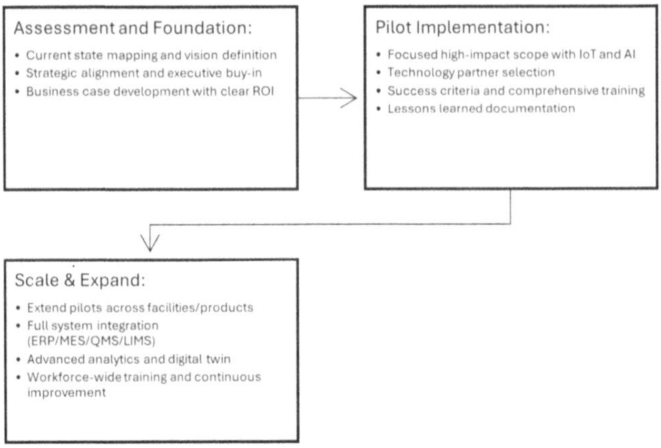


**Fig. 4.** Three-phase implementation approach.

From an organizational perspective, investment in human capital is of particular importance, with the recommendation to allocate 30–40% of the digital transformation budget to training and change management activities. Another recommendation is to eliminate silos by creating interdisciplinary teams combining qualitative, IT and operational competencies. It is also crucial to develop a data-driven strategy, including data quality assurance, standardisation and clear governance rules. It is also recommended to establish strategic partnerships with external entities in order to acquire specialized knowledge while developing internal competences. The final element is the implementation of a system for monitoring and communicating results, including regular monitoring of key performance indicators, calculation of return on investment and systematic communication of results to stakeholders.

## 5.1 Limitations and Future Research Directions

The digital transformation of quality management systems is one of the most important developments in production management over the past decade. As technology matures and is more widely adopted, organizations implementing a Quality 4.0 approach will increasingly stand out from the competition with higher quality, operational efficiency, and customer satisfaction.

The study is mainly based on a literature review and published case studies, which may reflect a tendency towards selection towards successful implementations. Future research directions include: artificial intelligence in quality decision-making (study of governance frameworks, ethical issues, validation requirements), cybersecurity in quality management systems (study of specific threats, effectiveness of security controls, resilience strategies), applications in the service sector (extending Quality 4.0 research beyond manufacturing), SME-specific approaches (development of a framework tailored to SMEs with limited resources), sustainability integration (studying how Quality 4.0 technologies support environmental sustainability goals) and long-term impact on the organization, conducting the studies examining the impact on organizational structure, culture, and workforce composition over a period of 5–10 years.

## References


1. Vial, G.: Understanding digital transformation: A review and a research agenda. J. Strateg. Inf. Syst. **28**(2), 118–144 (2019)
2. Khan, N., et al.: Big data: Survey, technologies, opportunities, and challenges. Sci. World J. **2014**, 712826 (2014). https://doi.org/10.1155/2014/712826
3. Ponsignon, F., Kleinhans, S., Bressolles, G.: The contribution of quality management to an organisation's digital transformation: A qualitative study. Total Qual. Manag. Bus. Excell. **30**(sup1), S17–S34 (2019). https://doi.org/10.1080/14783363.2019.1665770
4. Schilirò, D.: Digital transformation and its impact on organizations. Int. J. Bus. Manag. **19**, 71 (2024). https://doi.org/10.5539/ijbm.v19n6p71
5. Goetsch, D.L., Davis, S.B.: Quality Management: Introduction to Total Quality Management for Production, Processing, and Services. Prentice-Hall, New Jersey (2003)
6. Conti, T.: Quality management: Current issues and future trends. In: Conti, T., Kondo, Y., Watson, G.H. (eds.) Quality into the 21st Century: Perspectives on Quality and Competitiveness for Sustained Performance, pp. 237–244. ASQ Quality Press, Wisconsin (2003)
7. Yong, J., Wilkinson, A.: The long and winding road: The evolution of quality management. Total Qual. Manag. **13**(1), 101–121 (2002). https://doi.org/10.1080/09544120120098591
8. Majstorovic, V., Weckenmann, A., Akkasoglu, G., Werner, T.: Quality management – History and trends. TQM J. **27**(3), 281–293 (2015). https://doi.org/10.1108/TQM-11-2013-0125
9. Maguad, B.A.: The modern quality movement: Origins, development and trends. Total Qual. Manag. Bus. Excell. **17**(2), 179–203 (2006). https://doi.org/10.1080/14783360500450608
10. Abu, F., Fahmi, A., Abu Al-Rub, F., Shibhab, P.: Quality Management Systems (ISO 9001:2015). Gavin eBooks, Lisle (2020). https://doi.org/10.29011/978-1-951814-01-4-002
11. Chen, C.K., Lee, J.D., Dahlgaard, J.J.: A stepwise ISO-based TQM implementation approach using ISO 9001:2015. Manag. Prod. Eng. Rev. **4**, 65–75 (2016)
12. Knapčíková, L., Dupláková, D., Behúnová, A., Antosz, K.: Fundamentals of LEAN support in the era of sustainable manufacturing. In: Antosz, K., Trojanowska, J., Machado, J., Stadnicka, D. (eds.) Advances in Lean Manufacturing, Volume 1: LEAN in Action: Digital Transformation and Sustainable Practices. Springer, Cham (2025). https://doi.org/10.1007/978-3-032-09806-1_6
13. Wolniak, R., Sułkowski, M.: The reasons for the implementation of quality management systems in organizations. Sci. Pap. Siles. Univ. Technol. Organ. Manag. Ser. **92**, 443–455 (2016) (in Polish)
14. Greenhalgh, T., Russell, J., Swinglehurst, D.: Narrative methods in quality improvement research. BMJ Qual. Saf. **14**, 443–449 (2005)
15. Pacana, A., Lew, G., Kulpa, W.: Rating the quality of implementation of environmental management systems. J. Bus. Retail Manag. Res. **11**(2), 165–169 (2017)
16. Hamrol, A., Grabowska, M., Starzyńska, B.: Consumers' willingness to adopt pro-environmental attitudes. Sustainability. **17**, 5948 (2025). https://doi.org/10.3390/su17135948
17. Watson, G.H.: Six-sigma: An evolving stage in the maturity of quality. In: Conti, T., Kondo, Y., Watson, G.H. (eds.) Quality into the 21st Century: Perspectives on Quality and Competitiveness for Sustained Performance, pp. 219–236. ASQ Quality Press, Wisconsin (2003)
18. Olkiewicz, M., Wolniak, R., Grebski, E.M., Olkiewicz, A.: Comparative analysis of the impact of the business incubator center on the economic sustainable development of regions in USA and Poland. Sustainability. **11**, 1–22 (2019)
19. Misztal, A.: The importance of business ethics in quality management. Sci. Pap. Univ. Szczecin Finance Financ. Mark. Insur. **64**, 365–372 (2013) (in Polish)

20. Roy, A.: The fourth industrial revolution. Currency Books, New York (2017). J. Int. Consum. Mark. **32**, 1–3 (2020). https://doi.org/10.1080/08961530.2020.1727164
21. Sumaedi, S., Yarmen, M.: The effectiveness of ISO 9001 implementation in food manufacturing companies: A proposed measurement instrument. Procedia Food Sci. **3**, 436–444 (2015). https://doi.org/10.1016/j.profoo.2015.01.048
22. Chowdhury, P., Pagidoju, R.T., Lingamgunta, R.K.K.: Generative AI for MES optimization: LLM-driven digital manufacturing configuration recommendation. Int. J. Appl. Math. **38**, 875–890 (2025)
23. Pizoń, J., Gola, A.: The meaning and directions of development of personalized production in the era of Industry 4.0 and Industry 5.0. Lecture Notes. Mech. Eng., 1–13 (2023). https://doi.org/10.1007/978-3-031-09360-9_1
24. Criveanu, M.M.: Investigating digital intensity and e-commerce as drivers for sustainability and economic growth in the EU countries. Electronics. **12**, 2318 (2023). https://doi.org/10.3390/electronics12102318
25. Russell, S., Norvig, P.: Artificial Intelligence: A Modern Approach, 3rd edn. Pearson Education, Upper Saddle River (2010)
26. Xu, Q., Yang, J., Cao, W., Hu, Z., He, A., Liu, K.: AI-based digital maintenance technology for aircraft engines. In: Data-Driven Intelligence Empowers Next-Generation Aircraft Engine Innovation. Springer, Singapore (2025). https://doi.org/10.1007/978-981-95-0942-3_8
27. Gbadegeshin, S.A.: The effect of digitalization on the commercialization process of high-technology companies in the life sciences industry. Technol. Innov. Manag. Rev. **9**, 49–63 (2019)
28. Rallabandi, M.: Risk and compliance considerations in ERP implementations. Int. J. Sci. Res. Arch. **16**, 151–160 (2025). https://doi.org/10.30574/ijsra.2025.16.1.1964
29. Kiedrowicz, M., Nowicki, T., Waszkowski, R., Wesołowski, Z., Worwa, K.: Software simulator for property investigation of document management system with RFID tags. MATEC Web Conf. **76**, 04012 (2016). https://doi.org/10.1051/matecconf/20167604012
30. Waszkowski, R., Kiedrowicz, M., Nowicki, T., Wesołowski, Z., Worwa, K.: Business processes in the RFID-equipped restricted access administrative office. MATEC Web Conf. **76**, 04003 (2016). https://doi.org/10.1051/matecconf/20167604003
31. Safitra, M.F., Lubis, M., Fakhrurroja, H.: Counterattacking cyber threats: A framework for the future of cybersecurity. Sustainability. **15**(18), 13369 (2023). https://doi.org/10.3390/su151813369
32. Strzałkowski, P., et al.: Management commitment to compliance with occupational safety and health laws and regulations in Polish rock mining companies. Sustainability. **17**(20), 9168 (2025). https://doi.org/10.3390/su17209168
33. Quattrini, R., Berrocal Menárguez, A.B., Zamorano Martín, C.: Heritage and railways: Sustainable tourism opportunities boosted by digital transformation. Sustainability. **15**(21), 15585 (2023). https://doi.org/10.3390/su152115585
34. Schiavone, F., Pietronudo, M.C., Sabetta, A., Ferretti, M.: Total quality service in digital era. TQM J. **35**(5), 1170–1193 (2023). https://doi.org/10.1108/TQM-12-2021-0377
35. Salazar, F., Tigre, F., Tubón-Núñez, E., Carrillo, S., Buele, J.: Implementation of the quality management system (ISO 9001:2015) in the bodywork industry. J. Inf. Syst. Eng. Manag. **4**, 1–10 (2019). https://doi.org/10.29333/jisem/5890
36. Banaszak, Z., Kłos, S., Mleczko, J.: Integrated Management Systems. PWE, Warsaw (2016) (in Polish)
37. Sharma, S.: Tools and techniques in quality management and quality management models. In: Quality Demystified: A Comprehensive Guide to Quality Improvement, Patient Safety, & Clinical Audit in Healthcare. Springer, Singapore (2025). https://doi.org/10.1007/978-981-95-0339-1_11

38. Meller, A., Piechowski, M., Gola, A.: Indicators of hierarchical structure model of supporting the production management process – A framework. In: International Conference on Intelligent Systems in Production Engineering and Maintenance, pp. 356–367. Springer Nature, Switzerland, Cham (2023). https://doi.org/10.1007/978-3-031-45021-1_27
39. Waszkowski, R., Nowicki, T., Saniuk, A.: Human-computer interaction in sanitary inspection simulation exercises. Adv. Ergon. Model. Usability Spec. Popul. **486**, 245–254 (2017). https://doi.org/10.1007/978-3-319-41685-4_22
40. Rojek, I., Jasiulewicz-Kaczmarek, M., Piszcz, A., Galas, K., Mikołajewski, D.: Review of the 6G-based supply chain management within Industry 4.0/5.0 paradigm. Electronics. **13**(13), 2624 (2024). https://doi.org/10.3390/electronics13132624
41. Hosseini Baghanam, A., Nourani, V., Norouzi, E., Vakili, A.T., Gökçekuş, H.: Application of wavelet transform for bias correction and predictor screening of climate data. Sustainability. **15**(21), 15209 (2023). https://doi.org/10.3390/su152115209

# Author Index





A. Hamrol et al. (Eds.): MANUFACTURING 2026, LNME, pp. 327–328, 2026.
https://doi.org/10.1007/978-3-032-21701-1